Structural Health Monitoring: Advanced Sensing, Diagnostics and Prognostics

Structural Health Monitoring: Advanced Sensing, Diagnostics and Prognostics

Guest Editors

Bing Li
Yongbo Li
Khandaker Noman

MDPI

Basel • Beijing • Wuhan • Barcelona • Belgrade • Novi Sad • Cluj • Manchester

Guest Editors
Bing Li
School of Aeronautics
Northwestern Polytechnical University
Xi'an
China

Yongbo Li
School of Aeronautics
Northwestern Polytechnical University
Xi'an
China

Khandaker Noman
School of Civil Aviation
Northwestern Polytechnical University
Xi'an
China

Editorial Office
MDPI AG
Grosspeteranlage 5
4052 Basel, Switzerland

This is a reprint of the Special Issue, published open access by the journal *Sensors* (ISSN 1424-8220), freely accessible at: https://www.mdpi.com/journal/sensors/special_issues/2PHS4C9KEF.

For citation purposes, cite each article independently as indicated on the article page online and as indicated below:

Lastname, A.A.; Lastname, B.B. Article Title. *Journal Name* **Year**, *Volume Number*, Page Range.

ISBN 978-3-7258-3705-2 (Hbk)
ISBN 978-3-7258-3706-9 (PDF)
https://doi.org/10.3390/books978-3-7258-3706-9

Contents

About the Editors

Bing Li

Dr. Bing Li is a professor in the School of Aeronautics at Northwestern Polytechnical University. His current research interests include structural health monitoring, dynamics of elastic/mechanical metamaterials, wave mechanics. He has published >100 scientific papers in top-tier peer-reviewed journals and renowned international conferences.

Yongbo Li

Dr. Yongbo Li received a Ph.D. degree in General Mechanics from Harbin Institute of Technology (HIT), Harbin, China in 2017. He is currently a professor in the School of Aeronautics, Northwestern Polytechnical University, China. Prior to joining Northwestern Polytechnical University in 2017, he was a visiting student with the University of Alberta, Edmonton, AB, Canada, and the University of Huddersfield, England. He has published more than 90 peer-reviewed papers on international journals. His published papers have been cited more than 4000 time in total. He also managed more than 10 national projects from prominent funding institutions including the National Natural Science Foundation of China, Military Commission Science and Technology Commission, Ministry of Science and Technology China, and Shaanxi Provincial R&D projects. He has also delivered 10 invited reports in international/domestic academic conferences and won several key awards for his research. Currently, he is a Senior Member of IEEE and is working as the Associate Editor of the internally recognized journal Measurement (Elsevier) and the IEEE journals *IEEE Sensors Journal*, *IEEE Transaction on Instrumentation and Measurement*, and *IEEE Transactions on Industrial Informatics*.

Khandaker Noman

Dr. Khandaker Noman completed his Ph.D. degree from Shanghai Jiao Tong University, Shanghai, China in 2021. He is currently working as an Associate Professor at School of Civil Aviation of Northwestern Polytechnical University, Xi'an, China, since May 2023. Before starting to work as an Associate Professor, he worked as a post-doctoral research fellow in the School of Aeronautics of Northwestern Polytechnical University, Xi'an, China. He has authored and co-authored more than 40 peer-reviewed papers in international journals. He has also managed more than five scientific projects from key funding institutions such as the National Natural Science Foundation of China, Ministry of Science and Technology, China (Taicang Science and Technology Plan project). In his professional service, he is serving as the Associate Editor of the *Journal of Intelligent Manufacturing* (Springer) and is an Advisory Board Member of the journal *Measurement* (Elsevier). He has also worked as the Guest Editor of several journals including *Measurement*, *Sensors*, and *Machines*.

Preface

The subject of Structural Health Monitoring (SHM) has become increasingly crucial in an era where the complexity of infrastructure and machinery is constantly on the rise. Our aim with this reprint is to present the latest advancements in SHM, spanning from cutting-edge sensing technologies to innovative diagnostic and prognostic methods. The motivation behind this scientific work stems from the urgent need to enhance the safety, reliability, and longevity of structures across various industries. As structures age and operate in more demanding environments, traditional monitoring methods fall short. The articles in this reprint offer solutions by exploring new techniques and technologies. For instance, the integration of advanced sensors with intelligent data-driven algorithms can provide real-time insights into structural conditions, enabling the early detection of potential failures. This reprint is addressed to a diverse audience. It is a valuable resource for researchers in the field of engineering, who can gain inspiration from the novel approaches presented. Practicing engineers can also benefit, as the research findings can be applied to improve their current monitoring and maintenance strategies. Furthermore, students interested in structural engineering and related fields will find this collection a great source of knowledge to understand the latest trends. The contributions in this reprint are the result of the hard work of numerous authors. Each author has brought their unique expertise and perspective, whether it is in developing new theoretical models, conducting experimental studies, or applying SHM in real-world scenarios. We would like to express our sincere gratitude to all the authors for their outstanding contributions, whose research has made this reprint a rich repository of knowledge. We also want to acknowledge the invaluable assistance from the reviewers. Their detailed feedback and constructive criticism have significantly enhanced the quality of the articles. Without their efforts, this reprint would not have reached its current level of excellence. Finally, we are grateful for the support of the Editorial Board of *Sensors*, which has facilitated the smooth publication process, ensuring that these important studies reach a wide readership.

Bing Li, Yongbo Li, and Khandaker Noman
Guest Editors

sensors

Editorial

Structural Health Monitoring: Advanced Sensing, Diagnostics and Prognostics

Bing Li [1], Yongbo Li [1,2,*] and Khandaker Noman [3,*]

1 School of Aeronautics, Northwestern Polytechnical University, Xi'an 710072, China; bingli@nwpu.edu.cn
2 Aircraft Strength Research Institute of China, Xi'an 710065, China
3 School of Civil Aviation, Northwestern Polytechnical University, Xi'an 710072, China
* Correspondence: yongbo@nwpu.edu.cn (Y.L.); khandakernoman93@nwpu.edu.cn (K.N.)

Structural Heath Monitoring (SHM) can be considered one of the most prominent emerging components of modern engineering applications. Ensuring the reliability and safety of machinery and infrastructure has become more challenging due to their increasing complexity. Over the course of time, SHM has evolved significantly beyond its conventional foundations in aeronautical, civil, and mechanical engineering, with the discovery of applications in domains such as nuclear energy, maritime constructions, and wind turbine technology. The ability to detect failures earlier and forecast the remaining usable life (RUL) of structures has become crucial over the years. Early identification not only prevents catastrophic failures but also enhances maintenance strategies, saving both money and time. In recent years, different advanced sensing technologies, intelligent data-driven strategies, and innovative diagnostic and prognostic methodologies have witnessed revolutionary advancement. The integration of different analytical methods like non-destructive testing (NDT), artificial intelligence-based detection, vibration, and wave analysis has improved the precision and efficiency of conditioning monitoring. Advancements in these areas provide substantial insights into structural behavior, improving dependability, optimizing performance, and reducing maintenance costs. This Special Issue of *Sensors* aims to gather recent research findings and present the latest advancements in Structural Health Monitoring (SHM) in relation to advanced sensing, diagnostics, and prognostics. Overall, 13 different research contributions are featured in this collection, presenting groundbreaking applications of advanced sensing, diagnostics, and prognostics in the field of Structural Health Monitoring (SHM).

Received: 8 February 2025
Accepted: 18 February 2025
Published: 21 February 2025

Citation: Li, B.; Li, Y.; Noman, K. Structural Health Monitoring: Advanced Sensing, Diagnostics and Prognostics. *Sensors* **2025**, *25*, 1313. https://doi.org/10.3390/s25051313

To enhance the finite element model updating (FEMU) process for aging highway viaducts, Hekič et al. used both acceleration- and strain-based assessments on a multi-span concrete viaduct over 50 years old (Contribution 1). In this research, the authors used mid-span strain readings from large trucks for strain-based FEMU and frequencies/mode shapes for acceleration-based FEMU. Optimization methods, such as residual reduction and error-domain model falsification (EDMF), were used to enhance structural parameters. The findings demonstrate that the integration of strain data improves the accuracy of FEMU, showing an estimated 20% increase in the viaduct's design stiffness and a 25–50% overestimation of internal girder stiffness. The advantages of EDMF in producing physically significant updates in bridge model calibration are highlighted.

Deriving coupled ordinary differential equations for the first two modes in in-plane, out-of-plane, and torsional directions, Cui et al. investigated the spatial galloping behavior of iced conductors under multimodal coupling (Contribution 2). This study analyzed critical conditions within the wind speed–sag parameter space and classified galloping patterns into five distinct regions. The results obtained indicate that single-mode galloping exhibits elliptical motion, whereas coupled-mode galloping follows an "8"-shaped trajectory. The theoretical understanding of multimodal galloping in transmission lines, which provides a basis for designing anti-galloping measures, was enhanced by this study.

Hong et al. proposed a tensor optimization-based robust interval prediction method in order to forecast intermittent demand for spare parts (Contribution 3). The authors performed the integration of tensor decomposition with a stacked autoencoder to smooth abnormal demand variations while preserving intrinsic evolutionary trends. To enhance prediction reliability, an adaptive prediction interval algorithm was designed using LightGBM estimators and a dynamic update mechanism. Improved forecasting for small-sample intermittent time series and provides a reliable elastic prediction interval. This study offers a robust solution for intelligent inventory management.

Janeliukstis et al. proposed a wavelet-based output-only damage detection method for composite structures, using continuous wavelet transform (CWT) methods to extract modal features such as resonant frequencies and damping ratios (Contribution 4). The extracted features were used to construct a statistical damage detection scheme based on kernel density estimation (KDE), where deviations in modal features were identified via the Euclidean distance between KDE centroids. Experimental validation on glass-fiber-reinforced polymer cylindrical specimens demonstrated that the proposed method achieved comparable accuracy to the Mahalanobis distance metric while providing a simpler and more interpretable damage indicator.

Lu et al. simulated the random traffic flow of heavy vehicles via the incorporation of the R-vine Copula model and an improved Latin hypercube sampling (LHS) method (Contribution 5). Weigh-in-motion data were used in this study to examine correlations in vehicle weight, establishing an ideal R-vine Copula model for describing the relationships among vehicle weight characteristics. An improved LHS method was introduced for enhancing sampling accuracy to ensure a more authentic distribution of traffic flow characteristics. Finally, as visible from the load effect analysis, the consideration of vehicle weight correlations yields more conservative and realistic structural safety and assessments compared to the traditional Monte Carlo methods.

Estimating the torsional stiffness using an adaptive extended Kalman filter (AEKF) with a forgetting factor update, Park et al. monitored crack development in rotating shafts. To implement the AEKF, a dynamic system model was developed that allowed the real-time detection of torsional stiffness reduction due to cracks (Contribution 6). Results from the simulation and experiment demonstrated that the method successfully tracked the stiffness changes. Also, the method quantitatively evaluated the fatigue crack growth. This approach relies on the cost-effective rotational speed sensors, which makes it a viable solution for the structural health monitoring of rotating machinery.

Single-sensor engine multi-type fault detection, conducted via the integration of a variational mode decomposition (VMD) method with a Random Forest (RF) classifier, was investigated by Tang et al. (Contribution 7). Through decomposition under multiple operating conditions, the spectral energy distribution of engine signals was obtained. They also optimized the mode number and penalty term to enhance the mode separation and decomposition efficiency. The construction of a future set, including unit bandwidth energy, center frequency, and a maximum singular value, was completed and input into RF for classification. The results from comparative experiments showed that the proposed IVMD-RF method outperformed all the other deep learning approaches in both accuracy and training efficiency, demonstrating effectiveness in cross-speed fault diagnosis, with minimal training data and low hardware requirements.

To detect the axially loaded beams subjected to seasonal thermal variations via principal component analysis (PCA), Berardengo et al. developed a short-training damage detection method (Contribution 8). In this particular approach, PCA is applied to vibration-based damage features to filter out temperature effects and improve detection reliability, even given the constraints of a limited training set. Both numerical simulations and experimental studies on a tie-rod structure, demonstrating superior robustness compared to the conventional Mahalanobis squared distance (MSD)-based approach, validated this method. The results obtained indicate that the proposed PCA-based strategy effectively isolates the damage-sensitive components while suppressing environmental variations, making it a suitable solution for structural health monitoring under varying thermal conditions.

Zhuang et al. used an IBA-ISMO-based method integrating variational mode decomposition (VMD) and sample entropy and diagnosed the rolling bearing faults (Contribution 9). VMD algorithms were applied to decompose vibration signals into intrinsic mode components (IMFs). The sample entropy was extracted as a feature for fault identification. During the optimization of the sequence minimization optimization (ISMO) classifier's parameters, enhanced classification accuracy was ensured via an improved bat algorithm (IBA). Experimental verification using the CWRA dataset showed that the proposed IBA-ISMO model outperformed conventional methods in fault recognition, demonstrating robustness in detecting different bearing fault types under variable working conditions.

Park et al. extracted degradation features for the prognostics of an extruder screw using multi-source monitoring data from a real micro-extrusion system (Contribution 10). The operational data were utilized in this work to develop a prognostic method for predicting screw wear, addressing the challenge of obtaining real-world run-to-failure data. Based on physical and mechanical properties, integrating motor load, head pressure, and puller speed to estimate screw deterioration, degradation features were derived. It is visible from experimental validation that the extracted feature exhibited monotonic degradation behavior, enabling the accurate prediction of remaining useful life. A practical solution for monitoring extrusion systems health in industrial applications is also provided by the proposed method.

In Jia et al.'s work (Contribution 11), high-precision features were extracted for aircraft attitude sensor fault diagnosis using a RepVGG-based convolutional neural network with an SENet attention mechanism. In this particular method, the transformation of time domain sensor signals was performed to yield time–frequency representations, and SENet was used to allocate weights for both signal domains. Following that, the weighted features were fed into RepVGG for deep feature extraction and classification. Experimental observation validates the achievement of the proposed model as an optimal balance between diagnostic accuracy and computational efficiency, making it suitable for real-time aircraft fault diagnosis.

A TCP acceleration algorithm was developed by Liu et al. (Contribution 12) for aerospace–ground service networks. This method leverages historical transmission characteristics and congestion control optimizations. This study introduced BoostTCP, a learning-based algorithm that dynamically adjusts transmission rates based on end-to-end delay variations, feedback packet intervals, and random packet loss factors, and addressed the inefficiencies of standard TCP in high-bandwidth, long-delay networks. The comparative evaluations with conventional TCP congestion control algorithms demonstrated that BoostTCP significantly improved throughput, fairness, and bandwidth utilization in both simulated and real-world aerospace networks, and the results suggest that BoostTCP provides a promising solution for high-speed satellite data transmission.

Yang et al. (Contribution 13) reviewed the status and applications of hydraulic pump fault diagnosis, summarizing existing methodologies in fault detection, prediction, and health management. The classification of hydraulic pump fault diagnosis methods into signal processing-based approaches, artificial intelligence-driven techniques, and mechanism analysis-based methods was performed in this paper. This work also addressed several key challenges such as sensor selection, model construction, and multi-source data fusion, and the growing role of AI in improving fault recognition accuracy was highlighted by the comparative analysis of reviewed methods. Insights into future trends, emphasizing the need for hybrid techniques that integrate physical modeling with data-driven strategies for enhanced fault diagnostics and prognostics, were also provided.

The editors express their sincere gratitude to all the contributing authors for the outstanding research contributions. Special thanks are also due to the reviewers for their valuable feedback, which significantly helped to enhance the overall quality of this Special Issue. Lastly, we also appreciate the support of the editorial board of *Sensors* for facilitating the dissemination of these important studies.

Conflicts of Interest: The authors declare no conflict of interest.

List of Contributions:

1. Hekič, D.; Ribeiro, D.; Anžlin, A.; Žnidarič, A.; Češarek, P. Improved Finite Element Model Updating of a Highway Viaduct Using Acceleration and Strain Data. *Sensors* **2024**, *24*, 2788. https://doi.org/10.3390/s24092788.
2. Cui, F.; Zheng, K.; Liu, P.; Wang, H. Spatial Galloping Behavior of Iced Conductors under Multimodal Coupling. *Sensors* **2024**, *24*, 784. https://doi.org/10.3390/s24030784.
3. Hong, K.; Ren, Y.; Li, F.; Mao, W.; Gao, X. Robust Interval Prediction of Intermittent Demand for Spare Parts Based on Tensor Optimization. *Sensors* **2023**, *23*, 7182. https://doi.org/10.3390/s23167182.
4. Janeliukstis, R.; Mironovs, D. Wavelet-Based Output-Only Damage Detection of Composite Structures. *Sensors* **2023**, *23*, 6121. https://doi.org/10.3390/s23136121.
5. Lu, H.; Sun, D.; Hao, J. Random Traffic Flow Simulation of Heavy Vehicles Based on R-Vine Copula Model and Improved Latin Hypercube Sampling Method. *Sensors* **2023**, *23*, 2795. https://doi.org/10.3390/s23052795.
6. Park, Y.H.; Lee, H.B.; Kim, G.W. Crack Monitoring in Rotating Shaft Using Rotational Speed Sensor-Based Torsional Stiffness Estimation with Adaptive Extended Kalman Filters. *Sensors* **2023**, *23*, 2437. https://doi.org/10.3390/s23052437.
7. Tang, D.; Bi, F.; Cheng, J.; Yang, X.; Shen, P.; Bi, X. Single-Sensor Engine Multi-Type Fault Detection. *Sensors* **2023**, *23*, 1642. https://doi.org/10.3390/s23031642.
8. Berardengo, M.; Lucà, F.; Vanali, M.; Annesi, G. Short-Training Damage Detection Method for Axially Loaded Beams Subject to Seasonal Thermal Variations. *Sensors* **2023**, *23*, 1154. https://doi.org/10.3390/s23031154.
9. Zhuang, D.; Liu, H.; Zheng, H.; Xu, L.; Gu, Z.; Cheng, G.; Qiu, J. The IBA-ISMO Method for Rolling Bearing Fault Diagnosis Based on VMD-Sample Entropy. *Sensors* **2023**, *23*, 991. https://doi.org/10.3390/s23020991.
10. Park, J.K.; Lee, H.; Kim, W.; Kim, G.M.; An, D. Degradation Feature Extraction Method for Prognostics of an Extruder Screw Using Multi-Source Monitoring Data. *Sensors* **2023**, *23*, 637. https://doi.org/10.3390/s23020637.
11. Jia, Z.; Wang, K.; Li, Y.; Liu, Z.; Qin, J.; Yang, Q. High Precision Feature Fast Extraction Strategy for Aircraft Attitude Sensor Fault Based on RepVGG and SENet Attention Mechanism. *Sensors* **2022**, *22*, 9662. https://doi.org/10.3390/s22249662.
12. Liu, C.; Zhao, J.; Mao, F.; Chen, S.; Fu, N.; Wang, X.; Cao, Y. A TCP Acceleration Algorithm for Aerospace-Ground Service Networks. *Sensors* **2022**, *22*, 9187. https://doi.org/10.3390/s22239187.
13. Yang, Y.; Ding, L.; Xiao, J.; Fang, G.; Li, J. Current Status and Applications for Hydraulic Pump Fault Diagnosis: A Review. *Sensors* **2022**, *22*, 9714. https://doi.org/10.3390/s22249714.

sensors

MDPI

Article

Improved Finite Element Model Updating of a Highway Viaduct Using Acceleration and Strain Data

Doron Hekič [1,2,*], Diogo Ribeiro [3], Andrej Anžlin [2], Aleš Žnidarič [2] and Peter Češarek [1]

1 Faculty of Civil and Geodetic Engineering, University of Ljubljana, Jamova cesta 2, 1000 Ljubljana, Slovenia; peter.cesarek@fgg.uni-lj.si
2 Department of Structures, Slovenian National Building and Civil Engineering Institute, Dimičeva ulica 12, 1000 Ljubljana, Slovenia; andrej.anzlin@zag.si (A.A.); ales.znidaric@zag.si (A.Ž.)
3 CONSTRUCT-LESE, School of Engineering, Polytechnic of Porto, 4249-015 Porto, Portugal; drr@isep.ipp.pt
* Correspondence: doron.hekic@fgg.uni-lj.si

Abstract: Most finite element model updating (FEMU) studies on bridges are acceleration-based due to their lower cost and ease of use compared to strain- or displacement-based methods, which entail costly experiments and traffic disruptions. This leads to a scarcity of comprehensive studies incorporating strain measurements. This study employed the strain- and acceleration-based FEMU analyses performed on a more than 50-year-old multi-span concrete highway viaduct. Mid-span strains under heavy vehicles were considered for the strain-based FEMU, and frequencies and mode shapes for the acceleration-based FEMU. The analyses were performed separately for up to three variables, representing Young's modulus adjustment factors for different groups of structural elements. FEMU studies considered residual minimisation and the error-domain model falsification (EDMF) methodology. The residual minimisation utilised four different single-objective optimisations focusing on strains, frequencies, and mode shapes. Strain- and frequency-based FEMU analyses resulted in an approximately 20% increase in the overall superstructure's design stiffness. This study shows the benefits of the intuitive EDMF over residual minimisation for FEMU, where information gained from the strain data, in addition to the acceleration data, manifests more sensible updated variables. EDMF finally resulted in a 25–50% overestimated design stiffness of internal main girders.

Keywords: finite element model updating (FEMU); optimisation; calibration; monitoring; concrete highway viaduct; structural health monitoring (SHM); error-domain model falsification (EDMF)

Citation: Hekič, D.; Ribeiro, D.; Anžlin, A.; Žnidarič, A.; Češarek, P. Improved Finite Element Model Updating of a Highway Viaduct Using Acceleration and Strain Data. *Sensors* **2024**, *24*, 2788. https://doi.org/10.3390/s24092788

Academic Editor: Mohammad Noori

Received: 23 February 2024
Revised: 17 April 2024
Accepted: 25 April 2024
Published: 27 April 2024

1. Introduction

The bridge management sector is facing many challenges strongly linked to climate change, which, in recent years, has accelerated the rate of material and structural degradation. For example, increased temperatures strengthen the corrosion rates [1] and amplify other risks [2], posing a significant threat to bridges' safety and durability. Despite the uncertainties associated with the magnitude of the changes [3], it is accepted that they negatively affect infrastructure [4], which is subjected to longer and warmer dry spells and more frequent and severe flooding events, leading to economic losses [5].

The increased traffic capacity demands add to the challenges. ITF Transport Outlook states that tonne-kilometres of freight traffic worldwide will nearly double between 2019 and 2050 [6]. Furthermore, under the current ambition scenario, the share of road modes will increase from 22% to 27% in 2050. Traffic count data near the case study viaduct, designated in the following as the Ravbarkomanda viaduct, show that 3.6 million vehicles over 3.5 tonnes crossed the viaduct in 2022 [7,8], nearly a three-time increase since 2002 when 1.3 million vehicles had been recorded.

At times of increasing loads, the infrastructure is ageing. The average age of European and other developed countries' bridges exceeds 50 years, as indicated in [9], affecting their condition. Many bridges before 1970 were designed for a service life of 50 years and are thus approaching the end of their design life [10]. Moreover, once considered long-lasting reinforced concrete structures have not met these expectations, particularly those built in the 1970s [11–13]. A 2019 review [14] reported that 12% of highway bridges in Germany were in a very poor, insufficient, or inadequate condition, a figure that a 2022 report [15] has updated to nearly 13%.

Joint Research Centre (JRC) Science for Policy report [16] states that Europe's ageing transport infrastructure needs effective and proactive maintenance to ensure its safe operation throughout its entire life cycle and ensure sufficient serviceability and safety. This can be achieved with adequate investments in inspections and structural health monitoring (SHM) systems and by prioritising interventions for critical structures with sustainable retrofitting solutions. Applying such an approach requires further research, particularly in benchmarking different SHM concepts. This is vital for standardisation and making informed decisions about the most suitable solutions for various applications, as initiated in the recent EU project IM-SAFE [17]. In the wake of significant events like the Morandi bridge collapse in Genoa, Italy [18], it has embarked on one of the most extensive SHM campaigns to date [19]. Such projects facilitate real-time monitoring that supplies critical data to assure safety and structural integrity.

Ageing infrastructure and increasing loads underline the necessity for preventive maintenance and inspection, visually and through SHM. Within SHM, various finite element model updating (FEMU) strategies are employed based on static and/or dynamic responses. Ereiz et al. [20] provide general guidelines about using SHM data to perform FEMU accurately. The process of FEMU is described step by step, namely (i) the selection of updating parameters (design variables); (ii) the definition of the model updating problem; and (iii) the solution of the model updating problem using different methods, particularly sensitivity-based, maximum likelihood, nonprobabilistic, probabilistic, response surface, meta heuristic, and regularisation methods.

Traditionally, FEMU and the damage detection of bridges are based on modal parameters (i.e., acceleration-based methods), using natural frequencies and mode shapes [21–25]. However, modal parameters may be limited because structures under traffic loads experience much larger amplitude responses than those under ambient ones. Also, bridges often experience light/moderate nonlinear incursions, particularly at the bearing devices [26], the track–deck interface [27], and the pavement–deck interface [28], among others. To overcome these limitations, several authors included in the FEMU problem static responses (displacements and strains) [29], dynamic responses (accelerations, displacements, and strains) [30,31], or a combination of these, mainly under traffic loads. Most updated models are used for the continuous condition assessments of bridges, particularly damage identification.

Comparative studies considering data from different sensor types (accelerometers, displacement sensors, strain sensors, etc.) or other types of tests (static and dynamic) are sparse. This paper contributes to a deeper perception of the differences between acceleration- and strain-based FEMU strategies. Understanding these differences is vital as the already established technologies are re-emerging, such as using bridge weigh-in-motion (B-WIM) for SHM, as proposed in [32]. Moreover, this paper contributes to understanding how the gradual increase in the number of variables affects the FEMU results. Lastly, the error-domain model falsification (EDMF) methodology, which was adopted for FEMU, in addition to the residual minimisation methodology, proved crucial, as it allowed for gaining critical insights into the updated values of variables. Despite its success, EDMF is still underused for FEMU. Hence, this study supports using EDMF for FEMU in civil structures, such as highway bridges.

2. Materials and Methods

2.1. Description of the Viaduct

The case study, the Ravbarkomanda highway viaduct, is located in the southwestern region of Slovenia. It is over 50 years old, 560 m long, and comprises a 16-span precast I girder-type superstructure (Figure 1).

Figure 1. A view under the case study viaduct.

As shown in Figures 1 and 2, the viaduct consists of two parallel independent structures, the left carrying traffic northeast and the right one in the opposite direction. Each structure is divided into units bounded by expansion joints on both sides. Each of the two structures has four units. Precast I girders are discontinued above the piers, i.e., each girder bridges only one span, and the slab is continuous over the piers, except at expansion joint locations [33]. A detailed description of the viaduct and the established long-term monitoring that includes a B-WIM system can be found in [32].

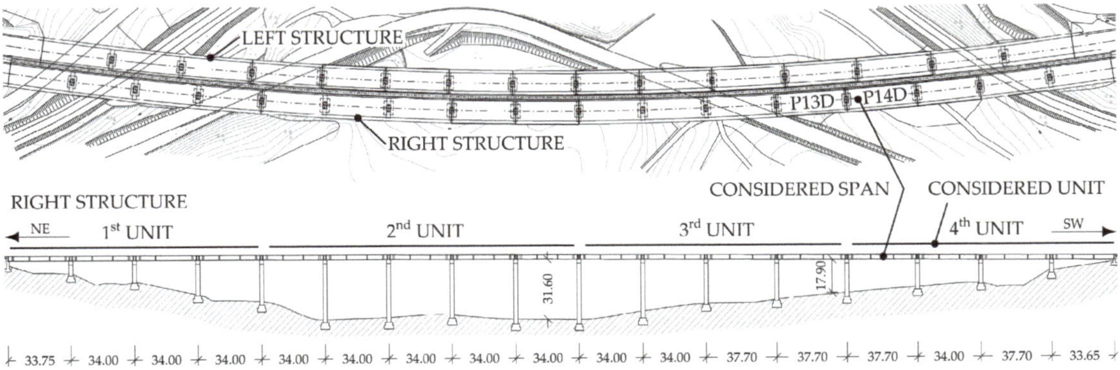

Figure 2. Plan view of both Ravbarkomanda viaduct structures and side view of the right structure with a notation of the P14D span and 4th unit, considered in this FEMU study (adapted from [32]).

This paper focuses on the fourth unit and the P14D span of the right structure, denoted as the *viaduct* throughout the paper. The paper follows the concept from a separate study [32], where strain-based FEMU was performed on the P14D span. This span was selected due to its extensive array of installed strain-gauge sensors, the largest of any span. A B-WIM system is also installed in this span to collect axle loads and spacings of all crossing vehicles.

2.2. Measurements of Strains under Passages of Calibration Vehicles

Strains were measured under crossings of three different calibration vehicles, designated V1 (two-axle rigid truck), V2, and V3. Both V2 and V3 were two-axle tractors with a three-axle semi-trailer. The calibration vehicles' passages were performed primarily to calibrate the B-WIM system installed in the P14D span. Their axle loads and gross vehicle weights (GVWs) were preweighted statically, and their axle spacings were measured manually. The results are shown in Table 1.

Table 1. Axle loads, axle spacing, and gross vehicle weights (GVWs) of the calibration vehicles.

Vehicle	1st Axle		2nd Axle		3rd Axle		4th Axle		5th Axle	
	Load [kN]	Spacing [m]	Load [kN]	Spacing [m]	Load [kN]	Spacing [m]	Load [kN]	Spacing [m]	Load [kN]	GVW [kN]
V1	67.69	3.30	85.35	1.35	88.29	/	/	/	/	241.33
V2	68.67	3.60	93.20	5.60	76.52	1.30	75.54	1.30	76.52	390.44
V3	68.67	3.30	87.31	1.35	87.31	5.17	76.52	1.33	76.52	396.32

Vehicles V1, V2, and V3 crossed the structure in the driving lane (Figure 3) 16, 17, and 18 times, respectively. Their response was measured by strain gauges installed at the mid-span of the bottom flange of the P14D span's main girders, labelled in Figure 3 as MG1, MG2, MG3, and MG4.

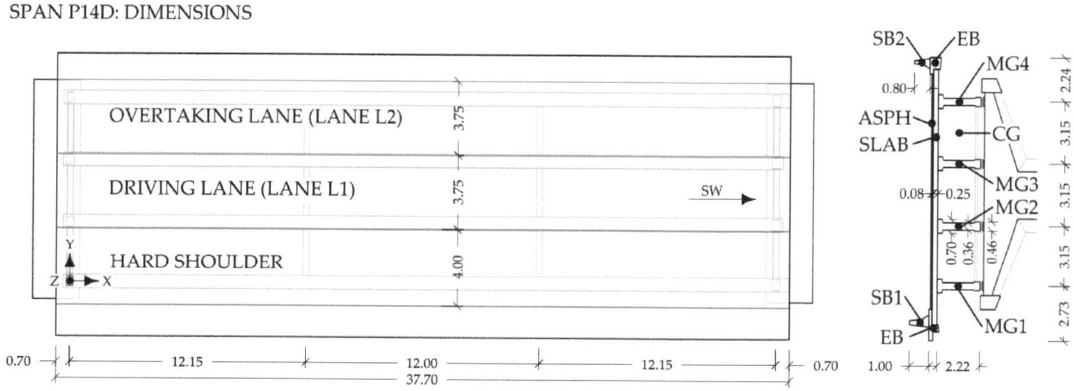

Figure 3. Plan view and cross-section of the P14D span with dimensions and notations of the structural and nonstructural elements: MG1–MG4 denote main girders, CG refers to cross-girders, SB1 and SB2 refer to safety barriers, EB refers to edge beam and SLAB denotes slab (adapted from [32]).

Each girder had 2 or 3 nearby strain-gauge sensors installed near the mid-span. The manufacturer's instructions were strictly followed in all installation stages: (concrete) surface preparation, glueing, protection, and connection of sensors. Two different types of strain gauges were used: TML PL-60-11-1LJC-F (120 Ω, half-Wheatstone type bridge, 60 mm gauge length; Tokyo Measuring Instruments Laboratory Co., Ltd., Tokyo, Japan) and Vishay C2A-06-20CLW-350 (350 Ω, half-Wheatstone type bridge, 50.8 mm gauge length; Vishay Intertechnology, Inc., Malvern, PA, USA). Signals from the girders were averaged to obtain more reliable strain responses per girder by reducing the errors due to possible uncertainties in location and faulty behaviour of the individual strain gauges. More is described in detail in Section 2.6.2 and in [32]. It is sufficient to assume that sensor SG_01 corresponds to the (average) measurements at the mid-span of girder MG1 and analogously applies to sensors SG_02, SG_03, and SG_04. Locations of sensors are shown in Figure 4, indicating that more strain-gauge sensors were installed at the same girder. Accelerometers are also shown in the figure, which is described in Section 2.3.

For the strain-based FEMU, described in Section 2.6.2, it was necessary to postprocess the strain measurements. The strain-based FEMU compared the measured strains to the FE-modelled ones under the calibration vehicles. Only the maximum values of the modelled and measured responses were compared, not the full-length signals. A separate study was performed to determine the position of all three vehicles that gave the greatest response at the strain-gauge sensor locations. Once determined, vehicles in the FE model were positioned in this location at every FEMU analysis. Such response under linear static analysis does not contain the dynamic component, and to compare it with the measured response, the latter should also be free of dynamics. The measured signals were, therefore, postprocessed with a 2 Hz low-pass filter to eliminate the dynamic component of the signal, thus obtaining the 'pseudo-static' response. A value of 2 Hz was selected based on a two-pass calculation of dynamic amplification factor (DAF) [34]. Table 2 shows the number of signals, mean, standard deviation, and coefficient of variation values for the maximum measured values in strain-gauge sensors.

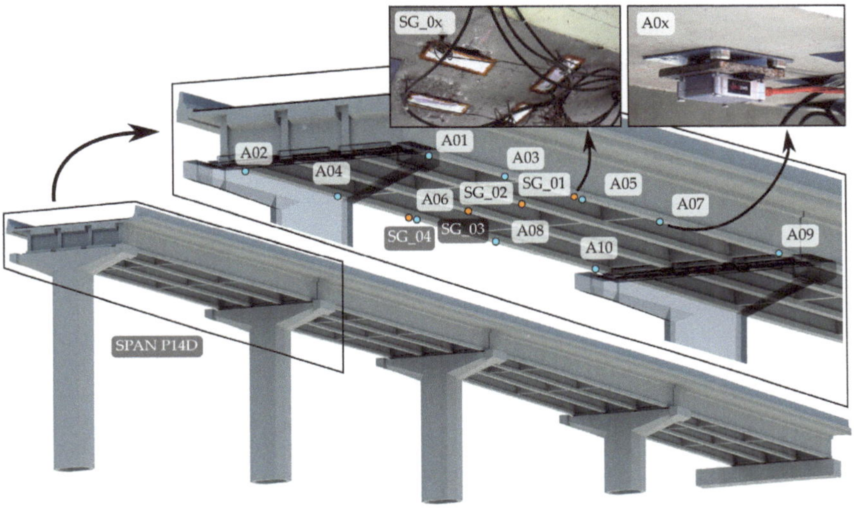

Figure 4. Render of a 4th unit with a detailed display of accelerometers and strain-gauge disposition in the P14D span.

Table 2. The number of signals (n), means, standard deviations (STDs), and coefficients of variation (CVs) for maximum measured values of calibration vehicle passages in lane L1.

n, Mean [µm/m], STD [µm/m], CV [%]		V1	V2	V3
SG_01	n	32	34	36
	Mean	19.1	29.5	31.5
	STD	0.7	0.9	1.3
	CV	3.5	2.9	4.2
SG_02	n	48	51	54
	Mean	27.1	35.4	37.9
	STD	1.3	1.2	1.4
	CV	4.8	3.4	3.7
SG_03	n	48	51	54
	Mean	27.9	35.5	36.8
	STD	1.5	1.3	1.6
	CV	5.3	3.5	4.4
SG_04	n	48	51	54
	Mean	18.2	27.2	27.4
	STD	0.9	1.2	1.6
	CV	5.2	4.6	5.7

2.3. *Ambient and Traffic-Induced Vibration Tests*

The long-term monitoring system installed on the viaduct does not include accelerometers on the superstructure. To perform the acceleration-based FEMU, additional short-term acceleration measurements were taken on the 4th unit. They were performed at 10 locations on the external main girders (MG1 and MG4) of the P14D span and on 30 more locations in the adjacent spans, namely P15D, P16D, and P17D (10 per span). Figure 5 introduces the measurement setup as a plan view of this unit, highlighting the placement of mobile and reference accelerometers. Measurements were performed in four setups; mobile sensors were moved between setups, and reference sensors remained in the same position during all setups.

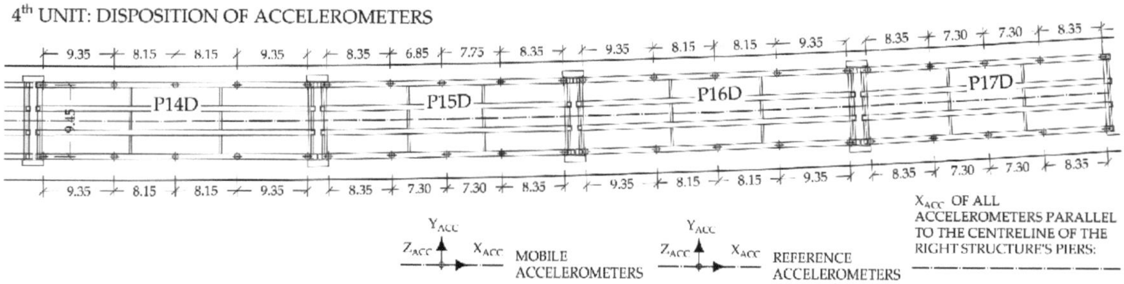

Figure 5. Plan view of the 4th unit with the disposition of the accelerometers during ambient and traffic-induced vibration tests; only Y_{ACC} and Z_{ACC} signals were used.

Measurements were taken under a partial traffic closure; the hard shoulder was closed for traffic, and the driving lane (lane L1) was closed for traffic most of the time. During the measurements, the bridge experienced no congestion. However, trucks weighing over 3.5 tons were present, with an average frequency of one truck every 30 s.

For each setup, twelve Dewesoft type IOLITEi 3xMEMS-ACC triaxial MEMS accelerometers (Dewesoft, Trbovlje, Slovenia) [35] were used for approximately 30 min at a 1000 Hz sampling frequency. Accelerometers were attached on the lower side of the bottom flange of the main girders (Figure 4) via magnets and a steel plate glued to the concrete surface. DewesoftX 2023.5 data acquisition software [36] was used for data recording. Data were imported into the ARTeMIS Modal Pro 7.2 software [37] to estimate the modal parameters. Only measurements in the Y and Z directions, according to Figure 5, were used. Basic signal processing was performed before estimation, such as linear detrending and decimation to a new frequency range of [0–100 Hz]. The operational modal analysis (OMA) frequency domain decomposition (FDD) technique was used to extract the natural frequencies and mode shapes, where the spectra resolution was set to 1024 Hz, with a 66% overlap, representing a frequency resolution of 0.098 Hz.

The results of the first test setup, with eight mobile accelerometers installed in the P14D span and four reference accelerometers in the P15D and P17D spans, are shown in Figure 6. The figure presents singular values of spectral densities. It is annotated with different coloured markers for the identified modes: 1st torsional mode (T-1), 1st and 2nd bending modes (B-1 and B-2), 1st main girder local bending mode (MG_B-1), and 3rd bending mode. All modes except T-1 appear on the first (highest) SVD line.

Table 3 provides a comprehensive look at the identified natural frequencies and corresponding mode shapes from the experimental campaign. Mode shapes are shown in general and close-up views of the P14D span. Although five modes were identified, only four were considered for the acceleration-based FEMU. As shown in Figure 6, all modes are well separated, except the T-1 and B-1 modes, which are closely spaced. The T-1 mode, which appears on the second SVD line, and as such, is not the best estimate, according to [38], was omitted from the acceleration-based FEMU.

Figure 7 presents the auto-modal assurance criterion (Auto-MAC) matrix for the experimental mode shapes. MAC provides a measure of consistency (degree of linearity) between the considered mode shapes [39], for example, the modelled mode shapes with the measured ones. Auto-MAC is a version of the MAC used to compare mode shapes with themselves [40], in this case, experimental mode shapes.

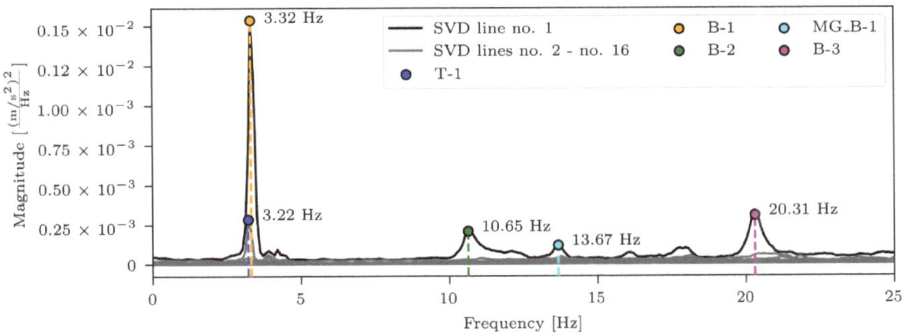

Figure 6. Singular values of spectral densities for the 1st test setup, with blue, orange, green, cyan, and magenta markers denoting 1st torsional mode (T-1), 1st bending mode (B-1), 2nd bending mode (B-2), 1st main girder local bending mode around the weak axis (MG_B-1), and 3rd bending mode (B-3).

Table 3. All identified natural frequencies and corresponding mode shapes from the experimental campaign. Red color indicates the greatest magnitude of displacements, while blue indicates the lowest.

T-1	**B-1**	**B-2**	**MG_B-1**	**B-3**
3.22 Hz	3.32 Hz	10.65 Hz	13.67 Hz	20.31 Hz
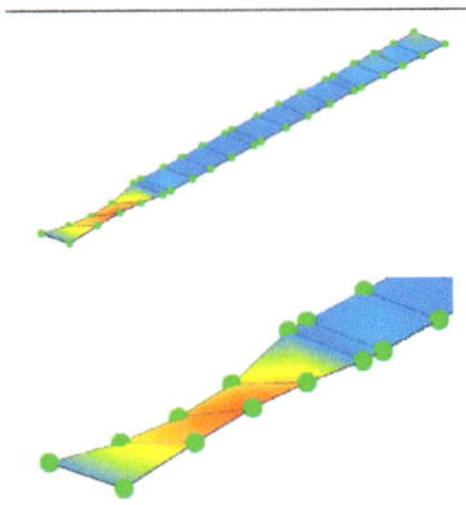	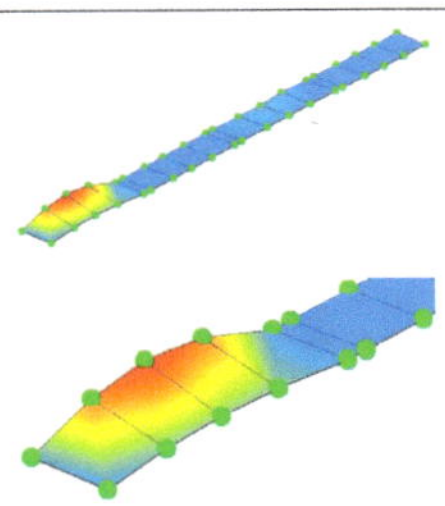	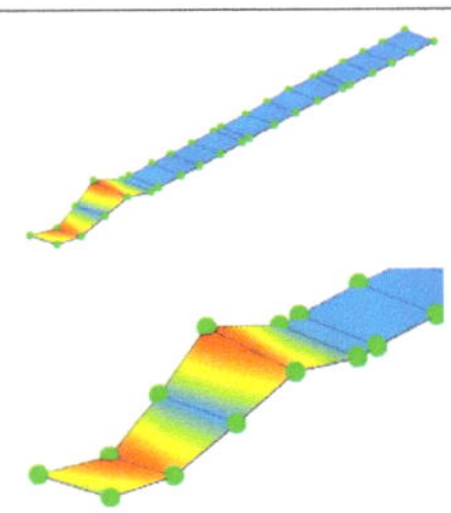	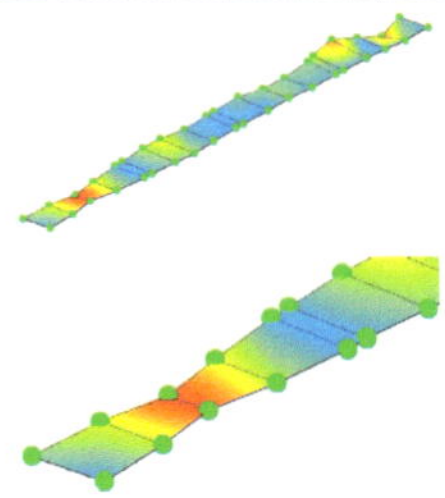	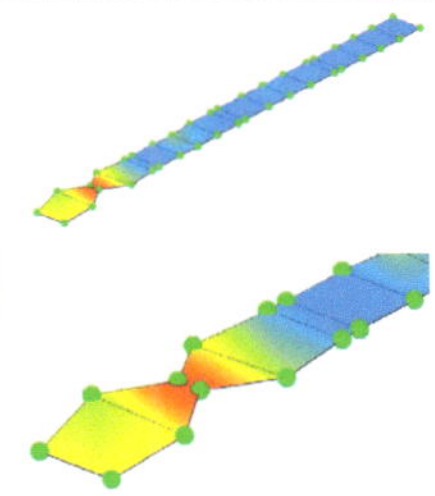

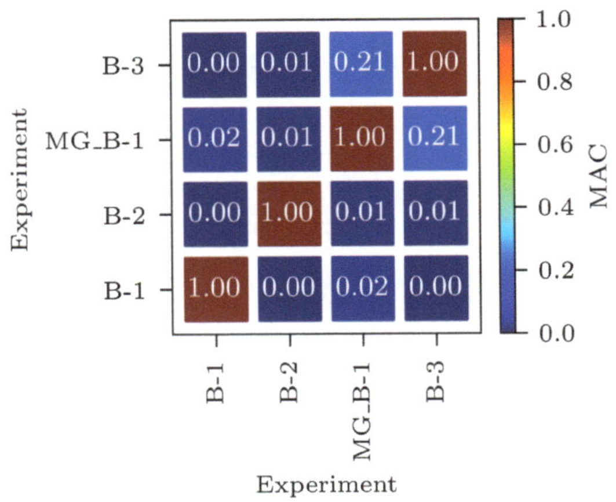

Figure 7. Auto-MAC matrix for the experimental mode shapes.

It can be seen from Figure 7 that the experimental mode shapes of most nondiagonal values are close to 0, showing a low level of consistency (linearity), except for the MG_B-1 and B-3 modes, where the auto-MAC value is 0.21. The similarity of those two mode shapes can also be seen in Table 3.

2.4. Finite Element (FE) Model

The finite element (FE) model for the analysis of the 4th viaduct unit was developed in finite element analysis (FEA) software Abaqus 2019 [41] in two stages. First, the initial model (in the following designated as M1_FULL_INIT) was created, on which preliminary studies were performed. In the second stage, a model with reduced degrees of freedom (DOFs) was created (in the following designated as M1_SUBSTR_INIT), focusing on the P14D span, as shown in Figure 8. Besides the notations of the P14D span, substructure, supports, and location of the interaction between the P14D span and substructure, Figure 8 also shows the location and notations of the structural bearings, described in Section 2.4.2.

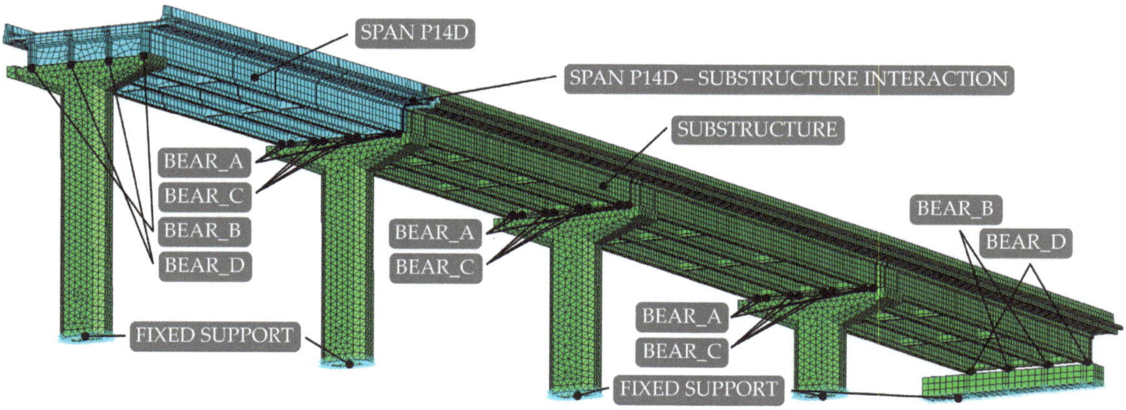

Figure 8. Initial finite element (FE) model M1_SUBSTR_INIT of the 4th unit.

The main features of the initial model M1_FULL_INIT and its assumptions to form a model with a reduced number of DOFs M1_SUBSTR_INIT are outlined in Sections 2.4.1–2.4.4.

2.4.1. Geometry and Materials

The FE model followed the geometry from original design documentation [33,42], with minor simplifications of the edge beam. All elements were modelled with 3D solids and isotropic elastic material whose properties were taken from original design documentation [33,42] (Table 4).

Table 4. Material properties of structural elements according to design documentation [33,42].

Element	Abbreviation	Young's Modulus [GPa]	Poisson Ratio	Density [t/m^3] [1]
Piers	/	34	0.20	2.500
Slab	SLAB	33	0.20	2.500
External main girders	EMG (MG1, MG4)	35	0.20	2.575
Internal main girders	IMG (MG2, MG3)	34	0.20	2.575
Cross-girders	CG	35	0.20	2.500
Safety barriers 1	SB1	33	0.20	2.500
Safety barriers 2	SB2	33	0.20	2.500
Edge beams	EB	33	0.20	2.500
Asphalt	ASPH	8	0.35	2.582

[1] Density of the main girders is increased due to the large number of prestressing tendons and mass of the equipment/installation attached to the main girders.

2.4.2. Interactions

The viaduct superstructure elements were assembled in one part, including the main girders, cross-girders, edge beams, slab, asphalt, and safety barriers. Consequently, their full interaction was assumed, and the safety barriers were treated as structural elements, fully contributing to the overall stiffness of the superstructure. A complex anchorage model to the viaduct deck would be required to model their contribution to the superstructure's stiffness accurately, or reduction factors for their stiffness would need to be included in the FEMU process. The former would increase the computing time of the FEMU process, and the latter approach can yield a wide range of results, potentially complicating the overall outcomes of the FEMU process, as already discussed in [32]. Piers are connected to the superstructure with elastomeric bearings, modelled as wires (spring-dashpot assemblies) connecting reference points on the pier–girder contact surfaces. "Cartesian + Rotation" connector sections were assigned to these wires (assemblies), and their stiffness properties were obtained from [33]. Values of translational, vertical, and rotational stiffness for all four type of bearings were [3.10×10^3 kN/m, 1.08×10^6 kN/m, 3.09×10^3 kNm] (BEAR_A); [2.43×10^3 kN/m, 8.43×10^5 kN/m, 2.32×10^3 kNm] (BEAR_B); [3.72×10^3 kN/m, 1.56×10^6 kN/m, 7.32×10^3 kNm] (BEAR_C); and [2.92×10^3 kN/m, 1.22×10^6 kN/m, 5.49×10^3 kNm] (BEAR_D). Positions of the elastomeric bearings are shown in Figure 8.

2.4.3. Boundary Conditions and Interaction with Adjacent Unit

The foundation of the piers is represented by fixing all translational degrees of freedom for the nodes on the bottom surface of the piers, as shown in Figure 8. The 4th unit interacts with the adjacent 3rd unit (Figure 2) only via a finger-type expansion joint. The adjacent span P13D, part of the 3rd unit, additionally restricts the movement of the shared pier that supports spans P13D and P14D. Therefore, in the M1_FULL_INIT model, the influence of the 3rd unit was modelled using connectors that link the locations of elastomeric bearings on the top of the pier with the ground. The stiffness properties of these connector sections were the same as the properties of the bearings they represented, except for translational stiffness in the X-direction, where the sum of the stiffness values in the X-direction of all bearings in 1st, 2nd, and 3rd unit was assumed.

2.4.4. FE Mesh

Main girders, cross-girders, edge beams, slab, asphalt layer, and safety barriers were meshed using hexahedral 20-node quadratic (C3D20R) elements with a maximum global size of 0.50 m. Piers were discretised with 10-node quadratic tetrahedral elements (C3D10) with a maximum global size of 0.50 m. The maximum global element sizes were determined through a mesh convergence study. The global element sizes were gradually reduced, and the resulting natural frequencies and MAC values from different models were compared. The comparison was made for experimentally identified natural frequencies and corresponding mode shapes. Table 5 shows natural frequencies for FE models with 0.50 m and 0.25 m global element sizes.

Table 5. Mesh convergence study of the natural frequencies for the M1_FULL_INIT FE model.

Mode	Natural Frequencies [Hz] for 0.25 m Global Element Size	Natural Frequencies [Hz] for 0.50 m Global Element Size
B-1	3.00	3.00
B-2	9.81	9.80
MG_B-1	12.83	12.78
B-3	20.39	20.37

Figure 9a shows the Auto-MAC matrix for the M1_FULL_INIT model with a 0.25 m global element size. Figure 9b displays a MAC matrix for the M1_FULL_INIT model with global element sizes of 0.50 m and 0.25 m. Due to the balance of accuracy and computational efficiency, a global element size of 0.50 m was used.

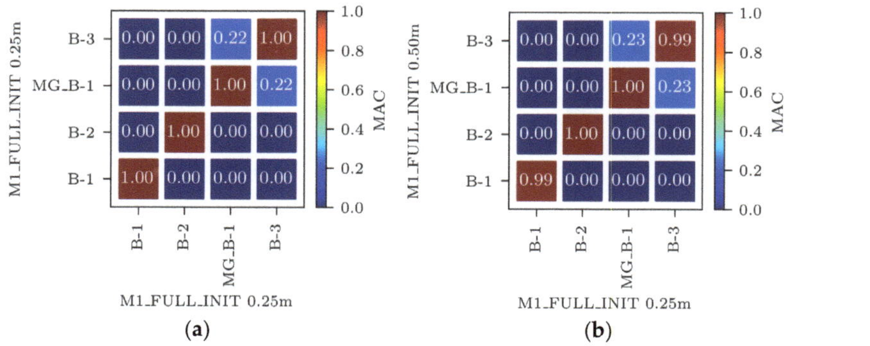

Figure 9. Mesh convergence study of the mode shapes for the M1_FULL_INIT FE model: Auto-MAC matrix for M1_FULL_INIT FE model with 0.25 m global element size (**a**) and MAC matrix for M1_FULL_INIT FE models with 0.50 m vs. 0.25 m global element size (**b**).

Even with larger finite elements, the M1_FULL_INIT model proved to be computationally intensive. To improve computational efficiency, a reduced-DOF model was created using the substructure modelling capabilities of Abaqus. In this context, "substructure" does not refer to the piers but to an entire structural component selected for separate analysis from the main structure. The P14D span was designated as the main structure, while the remaining parts of the 4th unit were modelled as a substructure (Figure 8). The substructure only contributes to the retained DOFs, including the supported nodes and nodes that interact with the main structure and provide stiffness of the substructure to the main structure during analysis. The reduced mass matrix and 90 retained modes of the substructure were computed to improve the accuracy of the main structure modal analysis (P14D span). The model with the substructure reduced the analysis time by 3.7 times compared to the M1_FULL_INIT FE model with 0.50 m global element size while maintaining the same level of result accuracy; natural frequencies of the M1_SUBSTR_INIT FE model (3.00 Hz, 9.83 Hz, 13.07 Hz, and 20.48 Hz) matched well with the M1_FULL_INIT FE model (3.00 Hz, 9.80 Hz, 12.78 Hz, and 20.37 Hz). Both models had a global element size of 0.50 m. Figure 10a shows the Auto-MAC matrix for the M1_FULL_INIT FE model, and Figure 10b displays a MAC matrix for M1_FULL_INIT and M1_SUBSTR_INIT FE models. Figure 10c shows the Auto-MAC matrix for the M1_SUBSTR_INIT FE model.

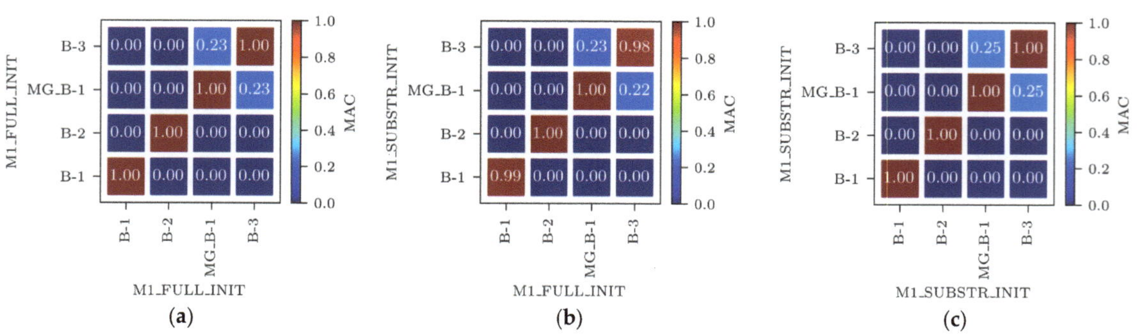

Figure 10. Auto-MAC matrix for M1_FULL_INIT FE model (**a**), MAC matrix for M1_SUBSTR_INIT vs. M1_FULL_INIT FE models (**b**), and Auto-MAC matrix for M1_SUBSTR_INIT FE model (**c**).

2.5. Comparison of the Initial FE Model M1_SUBSTR_INIT and Experiment

Table 6 compares natural frequencies and corresponding mode shapes of the M1_SUBSTR_INIT FE model and experimental values for all four modes considered within the acceleration-based FEMU: B-1, B-2, MG_B-1, and B-3. In addition, mode shapes were compared throughout the MAC matrix. From Figure 11, it is evident that the best match between the modelled and measured mode is for MG_B-1, with the MAC value amounting to 0.90. By contrast, the least similar are the B-1 mode shapes, with the MAC value of 0.68.

Table 6. Comparison of natural frequencies and corresponding mode shapes of the M1_SUBSTR_INIT FE model and experimental values. Red color indicates the greatest magnitude of displacements, while blue indicates the lowest.

	Mode	B-1	B-2	MG_B-1	B-3
FE model M1_SUBSTR_INIT	Frequency	3.00 Hz	9.83 Hz	13.07 Hz	20.48 Hz
	Mode Shape (Full)	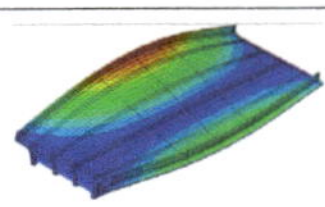	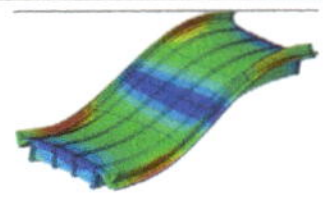	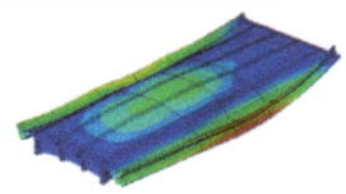	
	Mode Shape (Z Cut)	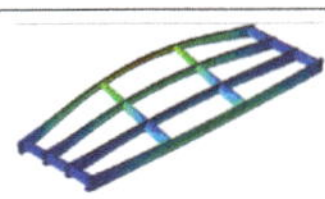	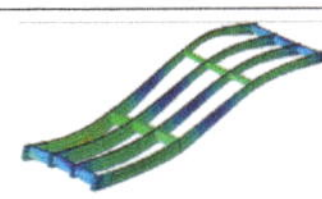	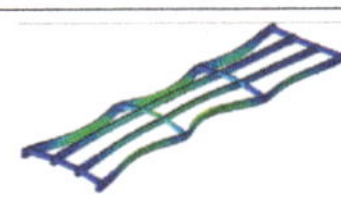	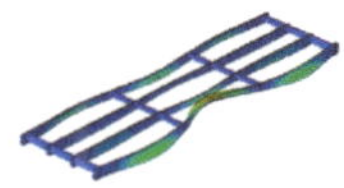
Experiment	Frequency	3.32 Hz	10.65 Hz	13.67 Hz	20.31 Hz
	Mode Shape	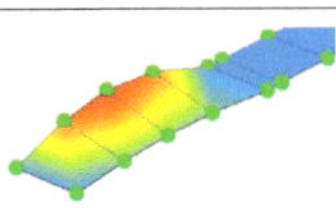	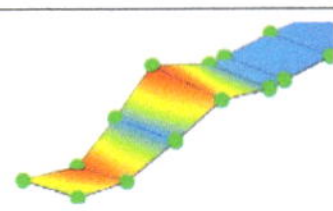	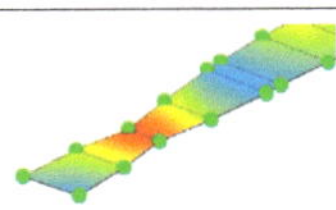	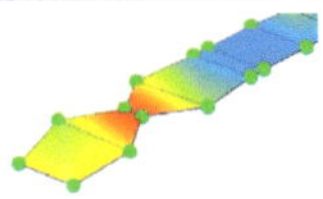

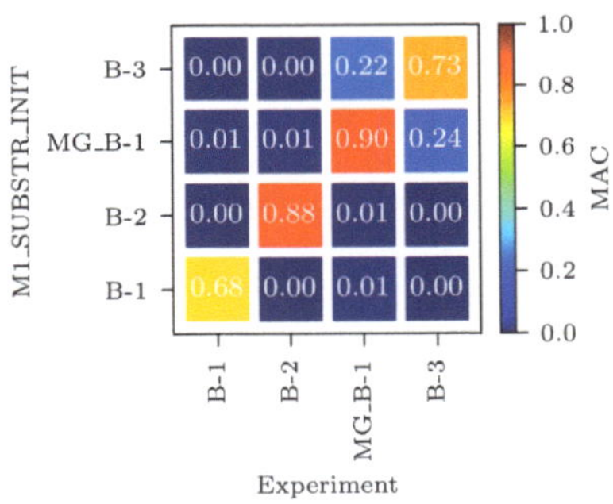

Figure 11. MAC matrix for M1_SUBSTR_INIT vs. experiment.

Comparison results for static analysis, where maximum strains under calibration vehicles were calculated and compared to the measured strains, are shown in Table 7, which compares the maximum modelled and measured strain values in sensors SG_01, SG_02, SG_03, and SG_04 (P14D span) under calibration vehicles V1, V2, and V3. In addition, for the measured strains, the STD (standard deviation) values are listed. Figure 12 graphically shows the values from Table 7.

Table 7. Maximum strains in the M1_SUBSTR_INIT FE model compared to the mean and STD (standard deviation) values of maximum measured strains in sensors SG_01, SG_02, SG_03, and SG_04 under calibration vehicles V1, V2, and V3.

Strains [μm/m]		V1		V2		V3	
		Mean	STD	Mean	STD	Mean	STD
SG_01	M1_SUSBSTR_INIT	21.0	/	31.4	/	33.4	/
	Experiment	19.1	0.7	29.5	0.9	31.5	1.3
SG_02	M1_SUSBSTR_INIT	34.1	/	44.7	/	46.6	/
	Experiment	27.1	1.3	35.4	1.2	37.9	1.4
SG_03	M1_SUSBSTR_INIT	35.2	/	45.5	/	47.0	/
	Experiment	27.9	1.5	35.5	1.3	36.8	1.6
SG_04	M1_SUSBSTR_INIT	22.4	/	32.7	/	34.2	/
	Experiment	18.2	0.9	27.2	1.2	27.4	1.6

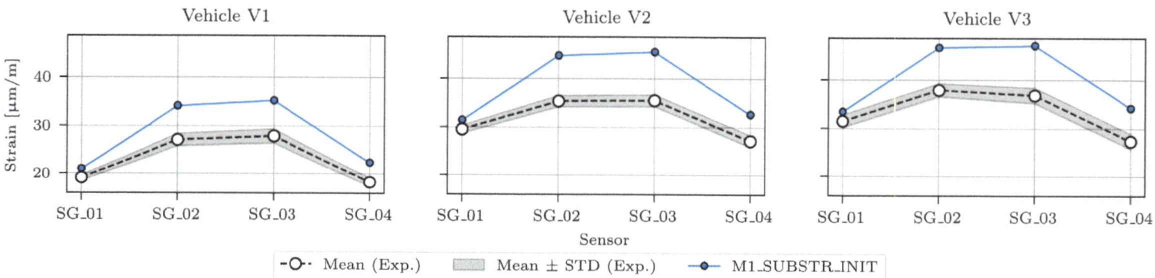

Figure 12. Maximum strains in the M1_SUBSTR_INIT FE model compared to the mean ± STD (standard deviation) values of maximum measured strains in sensors SG_01, SG_02, SG_03, and SG_04 under calibration vehicles V1, V2, and V3.

Figure 12 shows how M1_SUBSTR_INIT overestimates responses in all sensors and for all vehicles. The overestimation is the smallest in SG_01 sensor (<10%), and the most significant one in SG_02 and SG_03 sensors (>20%, <30%).

2.6. Finite Element Model Updating (FEMU): Residual Minimisation

FEMU aims to reduce the difference between the modelled and measured response. Two approaches for large-scale structures are often used for FEMU, namely Residual minimisation and Bayesian interference, the first being considered in this study. Besides the residual minimisation approach, the less common EDMF methodology [43] was also performed in this study, which is described in Section 2.7.

A function that combines the measured and modelled responses is called an "index of discrepancy" or objective function. In this section, the objective functions used for the acceleration- and strain-based FEMU analyses are formulated, and the optimisation algorithm used for the automatic nonlinear single objective optimisation is presented.

2.6.1. Objective Functions for Acceleration-Based FEMU

Three objective functions, namely J_f, J_{MAC}, and $J_{f,\mathrm{MAC}}$, were considered for acceleration-based FEMU. The J_f objective function measures the similarity of the modelled and measured natural frequencies. It is defined as follows:

$$J_f = \sum_{i=1}^{4} \left(\frac{f_{i,\mathrm{num}} - f_{i,\mathrm{exp}}}{f_{i,\mathrm{exp}}} \right)^2, \tag{1}$$

where $f_{i,\text{num}}$ and $f_{i,\text{exp}}$ are the ith matching mode pair of the natural frequencies from the FE model experiment, respectively. According to the [44], this is the "normalised" J_2 type objective function.

The J_{MAC} objective function measures the similarity of the modelled and measured mode shapes. It is defined, similarly as in [22] or [45], as follows:

$$J_{\text{MAC}} = \sum_{i=1}^{4} (1 - \text{MAC}_i)^2, \tag{2}$$

where MAC_i compares the ith mode shape of the FE model with the ith reference experimental mode shape.

The $J_{f,\text{MAC}}$ objective function combines the J_f and J_{MAC} objective functions, similar to [22]. Since J_f and J_{MAC} are of different orders of magnitude, w_f and w_{MAC} weights were considered to ensure that contribution of both to the determination of $J_{f,\text{MAC}}$ would be comparable:

$$J_{f,\text{MAC}} = w_f \cdot J_f + w_{\text{MAC}} \cdot J_{\text{MAC}}. \tag{3}$$

The value of w_f and w_{MAC} were set to 11.3 and 1.0, respectively. The value of 11.3 represents the ratio of J_{MAC} and J_f, calculated for the M1_SUBSTR_INIT FE model.

2.6.2. Objective Function for Strain-Based FEMU

The J_ε objective function is defined to measure the similarity of maximum modelled and measured strains at the mid-span of the P14D span when loaded by calibration vehicles V1, V2, and V3. It is defined as the sum of squared relative differences with standard deviation as a normalisation term. According to [44], this is the J_4-type objective function, with a minor modification, considering average responses in SG_01, SG_02, SG_03, and SG_04 sensors, as described in Section 2.2. The objective function J_ε is defined as follows:

$$J_\varepsilon = \sum_{v=1}^{n_v} \sum_{g=1}^{n_g} \frac{\left(z_{\text{num},v,g} - z_{\text{exp},v,g}\right)^2}{\text{STD}_{\text{exp},v,g}{}^2} \tag{4}$$

where $z_{\text{num},v,g}$ and $z_{\text{exp},v,g}$ are calculated as follows:

$$z_{\text{num},v,g} = \frac{1}{n_{g,s}} \sum_{s=1}^{n_{g,s}} \varepsilon_{\text{num},v,g,s} \text{ and} \tag{5}$$

$$z_{\text{exp},v,g} = \frac{1}{n_{g,s}} \sum_{s=1}^{n_{g,s}} \left(\frac{1}{n_{v,p}} \sum_{p=1}^{n_{v,p}} \varepsilon_{\text{exp},v,g,s,p} \right). \tag{6}$$

The $z_{\text{exp},v,g}$ and $\text{STD}_{\text{exp},v,g}$ values are the "experimental" mean and STD values from Table 7. Individual terms in equations are described as follows:

- g denotes the main girder index;
- n_g denotes the number of main girders considered (four in this study);
- $n_{g,s}$ denotes the number of strain gauges considered in a given girder g (two or three in this study);
- n_v denotes the number of calibration vehicles considered (three in this study);
- $n_{v,p}$ denotes the number of vehicle v passages;
- p denotes the passage index of the selected calibration vehicle;
- s denotes the strain-gauge sensor index on the selected main girder;
- $\text{STD}_{\text{exp},v,g}$ denotes the standard deviation of measured strains for main girder g and vehicle v;
- v denotes the calibration vehicle index;

- $\varepsilon_{\text{exp},v,g,s,p}$ denotes the maximum measured longitudinal strain (section) in the *s*th strain-gauge sensor on the *g*th main girder, caused by the *v*th calibration vehicle during *p*th passage;
- $\varepsilon_{\text{num},v,g,s}$ denotes the FE model longitudinal strain, oriented parallel to the X (longitudinal) direction of the viaduct, ε_{XX}, in the selected node that corresponds to the *s*th strain-gauge sensor on the *g*th main girder, caused by the *v*th calibration vehicle positioned on the location that results in the maximum strain at sensors SG_0*g*.

2.6.3. Optimisation Algorithm

In this study, the particle swarm optimisation (PSO) algorithm [46] was used to update the FE model automatically, which is one of the most commonly used algorithms in FEMU [20]. For the automatic FEMU, it is advantageous if the FEA software can interact with external programming environments such as MATLAB, Python, and Mathematica. This interaction involves preparing input files for analysis, submitting the FEA job, examining the FEA outcomes (output files), and generating new input files based on the decisions of the optimisation algorithm. In this research, the Abaqus 2019 FEA software was employed, along with Python 3.10, using the scipy.optimise.minimise [47] and pymoo [48] libraries. All parameters of the PSO algorithm were set to default (according to [48]) for all FEMU analyses, except for the population size, which was set to 100, and 20 generations were set as the stop criteria.

2.7. FEMU: Error-Domain Model Falsification (EDMF)

EDMF is a methodology for structural identification, introduced for bridge load testing in 2013 [43] and applied in 2019 [49] and recently in 2023 [50]. The falsification concept, as stated by [43], has been well known in science for centuries but was formalised only in the 1930s by Karl Popper, who stated that, in science, models cannot be fully validated by data. Instead, they can only be falsified. EDMF identifies plausible values of the FE model variables (parameters) based on experimental values from field measurements and prescribed uncertainty levels. A population of FE model instances is generated where each instance has a unique combination of variable values. Then, the FE model predictions (responses) are compared with the sensor data collected during the experiment. FE model instances where the difference between the modelled and measured responses exceeds thresholds defined based on uncertainty levels are falsified (falsified models), and the rest are designated as candidates. Updated ranges of variables are obtained by discarding variable values from falsified model instances.

As stated by [51], using thresholds for falsification enables EDMF to be robust to correlation assumptions between uncertainties; moreover, EDMF explicitly accounts for model bias based on engineering heuristics. Consequently, EDMF, when compared with traditional Bayesian model updating and residual minimisation, has been shown to provide more accurate identification and prediction when there is significant systematic uncertainty. EDMF has been gaining popularity in recent years, since only between 2015 and 2022, there were nine case studies on bridges, four on buildings, and two on geotechnical excavations reported worldwide [51].

EDMF for the considered case study was primarily utilised to verify the suspicious FEMU results from the residual minimisation, particularly the final values of variables that reached the lower and upper bound of the preset range and were not in accordance with the engineering expectation.

3. Results

3.1. Sensitivity Study

A deterministic sensitivity study was performed to understand the impact of the individual structural elements on the values of objective functions J_f and J_{MAC}. For the J_ε objective function, the sensitivity study results from the reference P14D-span-only study [32] are shown.

3.1.1. Variables

A sensitivity study for J_f and J_{MAC} was performed on the M1_SUBSTR_INIT FE model such that the variable of the selected element was set to lower and upper values. In contrast, the properties of all other elements in the model were kept constant. Variables and their lower and upper values are defined in Table 8.

Table 8. List of variables considered in the sensitivity analysis with the description of modified variables.

Element/Variable/Property	Lower Value [1]	Upper Value [1]	Description
ASPH, SB1, SB2, EB, EMG (MG1, MG4), IMG (MG2, MG3), SLAB, CG	0.75 × design	1.25 × design	Young's modulus change
BEARINGS TRANSL. STIFF.	0.75 × design	1.25 × design	Horizontal (X and Y) stiffness change
BEARINGS VERT. STIFF.	0.75 × design	1.25 × design	Vertical (Z) stiffness change
BEARINGS ROT. STIFF.	0.75 × design	1.25 × design	Rot. (around Y) stiffness change
DENSITY	0.95 × design	1.05 × design	Change in the density of elements ASPH, SB1, SB2, EB, EMG (MG1, MG4), IMG (MG2, MG3), SLAB, and CG

[1] Design values from [33,42].

For structural elements, the lower and upper values are defined as 0.75 and 1.25 times the design Young's elastic modulus values, which are shown in Table 4. For elastomeric bearings, the lower and upper values are defined as 0.75 and 1.25 times the design stiffness, as shown in Section 2.4.2. To assess the influence of density variations on the structural elements, they were simultaneously adjusted to two different levels for all elements, i.e., to 0.95 (lower value) and 1.05 (upper value) times their design values (Table 4).

3.1.2. Acceleration-Based FEMU

The results of the sensitivity study for the acceleration-based FEMU are shown separately for natural frequencies (J_f) and mode shapes (J_{MAC}). Figures 13 and 14 show the sensitivity results where the FE model objective function, either taking a lower or upper value, is first summed up over all four modes considered and then compared to the summed-up objective function of the M1_SUBSTR_INIT FE model. The results are shown in % as a relative change compared to the M1_SUBSTR_INIT FE model. Such a representation gives a general insight into which variables contribute the most to the relative change in the objective function.

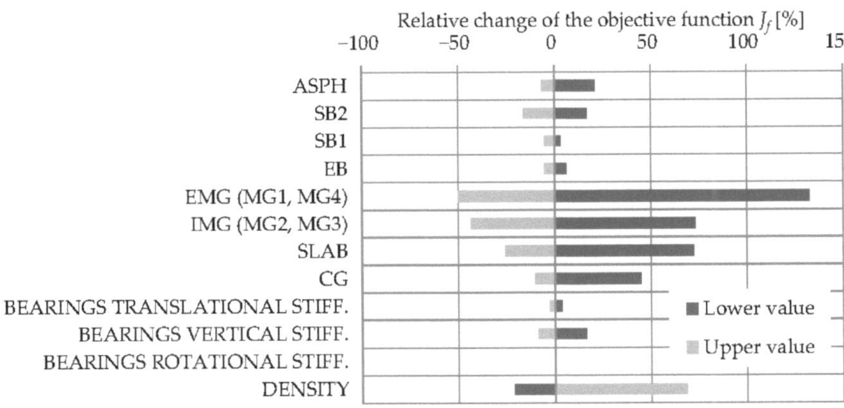

Figure 13. The sensitivity study results of the influence of structural elements Young's modulus, bearing stiffness, and density on the relative change in the objective function J_f.

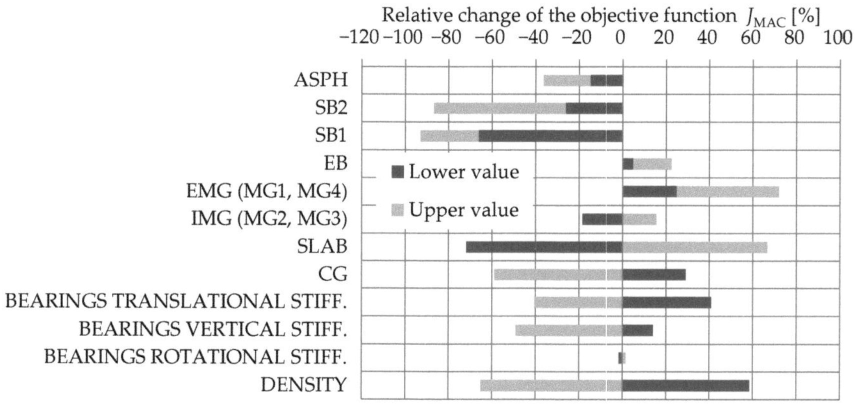

Figure 14. The sensitivity study results of the influence of structural elements Young's modulus, bearing stiffness, and density on the relative change in the objective function J_{MAC}.

Figure 13 yields the conclusion that among all variables considered, the reduction in the objective function J_f is the most sensitive to EMG, IMG, and SLAB elements' increase in Young's elastic modulus. Bearings do not have a significant impact. Figure 14, compared to Figure 13, is less concrete in suggesting which variables the objective function J_{MAC} is most sensitive to. Reducing the objective function J_{MAC} is mostly affected by SLAB and SB1 elements' decrease in Young's elastic modulus and by an increase in Young's elastic modulus in CG and SB2. An increase in translational and vertical stiffness of elastomeric bearings, as well as a decrease in their density, importantly reduces the objective function J_{MAC}.

3.1.3. Strain-Based FEMU

For the J_ε objective function, the sensitivity study results are shown from the reference study, where only the P14D span was modelled. The interested reader is referred to [32] for a detailed description. The same structural elements and bearings were checked for sensitivity as for J_f and J_{MAC}; only the density was omitted.

As seen in Figure 15, the reduction in the objective function J_ε is the most sensitive to EMG and IMG elements' increase in Young's elastic modulus. SB2, SB1, ASPH, and SLAB elements have comparable but much smaller influence.

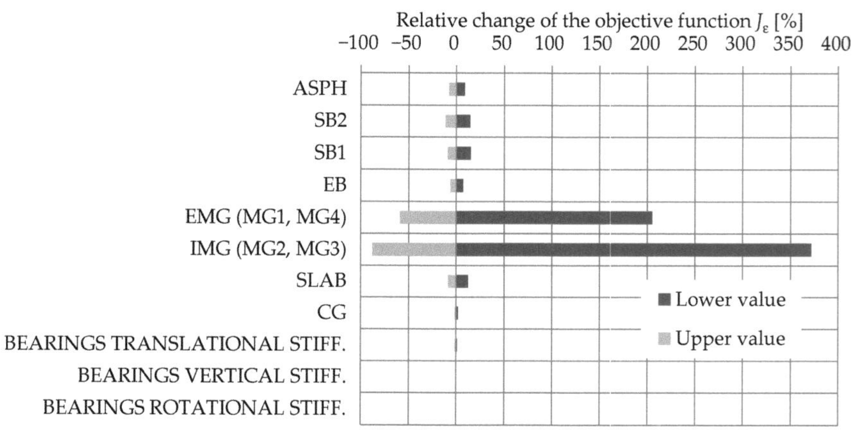

Figure 15. The strain-based sensitivity study results show the influence of structural elements Young's modulus and bearing stiffness on the objective function value J_ε (adapted from [32]).

3.1.4. Variables Selected for FEMU

Based on the sensitivity study results, it was decided to update only Young's modulus of structural elements and consider the constant design values of other properties. Furthermore, instead of updating Young's modulus of individual structural elements, a grouping was performed such that Young's modulus for all elements in the same group was updated for the same percentage/correction factor, in the following labelled as a "Young's modulus adjustment factor". Grouping was performed to observe the influence of several variables on the FEMU results. For the first FEMU studies, all structural elements were grouped. Thus, only one variable (α_{ALL}) was updated. Later, the structural elements were regrouped into the EMG+IMG (MG) group and the OTHER group, consisting of all other elements. Two variables, α_{MG} and α_{OTHER} were updated in that case. Finally, the EMG+IMG (MG) group was split into EMG and IMG groups. Thus, three variables were updated: α_{ALL}, α_{EMG}, and α_{IMG}. The variables and ranges within which the updated variables can take values are described in Table 9.

Table 9. Description of variables selected for FEMU and their range.

Variable	Description	Range	
		Res. Min.	EDMF
α_{ALL}	Young's modulus adjustment factor for ASPH, SB2, SB1, EB, EMG, IMG, SLAB, and CG	[0.9, 1.5]	/
α_{MG}	Young's modulus adjustment factor for EMG and IMG	[0.9, 1.5]	/
α_{OTHER}	Young's modulus adjustment factor for ASPH, SB2, SB1, EB, SLAB, and CG	[0.9, 1.5]	[0.10, 1.90]
α_{EMG}	Young's modulus adjustment factor for EMG	[0.9, 1.5]	[0.10, 2.00]
α_{IMG}	Young's modulus adjustment factor for IMG	[0.9, 1.5]	[0.10, 2.00]

It is important to emphasise that the goal of FEMU, as stated by [43], is not to update the model parameters to improve the agreement between predicted and measured values. Instead, model-based system identification uses physics-based models to infer parameter values. As such, the variables selected for FEMU do not represent the actual properties of the structural elements, i.e., Young's modulus (adjustment factor). Instead, they should be treated as a mixture of structural properties condensed in a single variable. This needs to be kept in mind, especially when interpreting the absolute values of updated variables.

3.2. *Updated FE Model*

3.2.1. List of Analyses

Twelve FEMU residual minimisation methodology analyses were performed. Four of them considered one variable (α_{ALL}), four considered two variables (α_{MG}, α_{OTHER}), and the last four involved three variables (α_{EMG}, α_{IMG}, and α_{OTHER}). In each group, four FEMU analyses were performed: frequency-based, MAC-based, frequency-and-MAC-based, and strain-based. All acceleration-based analyses considered B-1, B-2, MG_B-1, and B-3 modes, and all strain-based analyses considered all three calibration vehicles V1, V2, and V3. Three FEMU analyses considered EDMF methodology, all for three variables. One was acceleration-based, one was strain-based, and the last one was acceleration-and-strain-based methods. A summary of all these analyses is presented in Table 10.

Table 10. List of FEMU analyses describing variables, mode shapes/vehicles considered, and type of objective functions used.

Analysis Number	Analysis Type	Mode Shapes/Vehicles Considered	Variables	Acceleration-Based			Strain-Based (J_ε)
				Frequency-Based (J_f)	MAC-Based (J_{MAC})	Frequency-and-MAC-Based ($J_{f,MAC}$)	
1	Res. min.	B-1, B-2, MG_B-1, B-3	α_{ALL}	X			
2	Res. min.	B-1, B-2, MG_B-1, B-3	α_{ALL}		X		
3	Res. min.	B-1, B-2, MG_B-1, B-3	α_{ALL}			X	
4	Res. min.	V1, V2, V3	α_{ALL}				X

Table 10. *Cont.*

Analysis Number	Analysis Type	Mode Shapes/Vehicles Considered	Variables	Acceleration-Based			Strain-Based (J_ε)
				Frequency-Based (J_f)	MAC-Based (J_{MAC})	Frequency-and-MAC-Based ($J_{f,MAC}$)	
5	Res. min.	B-1, B-2, MG_B-1, B-3	α_{MG}, α_{OTHER}	X			
6	Res. min.	B-1, B-2, MG_B-1, B-3	α_{MG}, α_{OTHER}		X		
7	Res. min.	B-1, B-2, MG_B-1, B-3	α_{MG}, α_{OTHER}			X	
8	Res. min.	V1, V2, V3	α_{MG}, α_{OTHER}				X
9	Res. min.	B-1, B-2, MG_B-1, B-3	α_{EMG}, α_{IMG}, α_{OTHER}	X			
10	Res. min.	B-1, B-2, MG_B-1, B-3	α_{EMG}, α_{IMG}, α_{OTHER}		X		
11	Res. min.	B-1, B-2, MG_B-1, B-3	α_{EMG}, α_{IMG}, α_{OTHER}			X	
12	Res. min.	V1, V2, V3	α_{EMG}, α_{IMG}, α_{OTHER}				X
13	EDMF	B-1, B-2, MG_B-1, B-3	α_{EMG}, α_{IMG}, α_{OTHER}			X	
14	EDMF	V1, V2, V3	α_{EMG}, α_{IMG}, α_{OTHER}				X
15	EDMF	B-1, B-2, MG_B-1, B-3 & V1, V2, V3	α_{EMG}, α_{IMG}, α_{OTHER}			X	X

3.2.2. FEMU Results: Residual Minimisation

In this section, the results for all twelve FEMU analyses are presented, and prior to that, the evolution throughout the FEMU is shown for analysis number 5 (frequency-based analysis). Figure 16 illustrates the evolution of the objective function J_f including all data (**a**) and using a zoomed-in view (**b**).

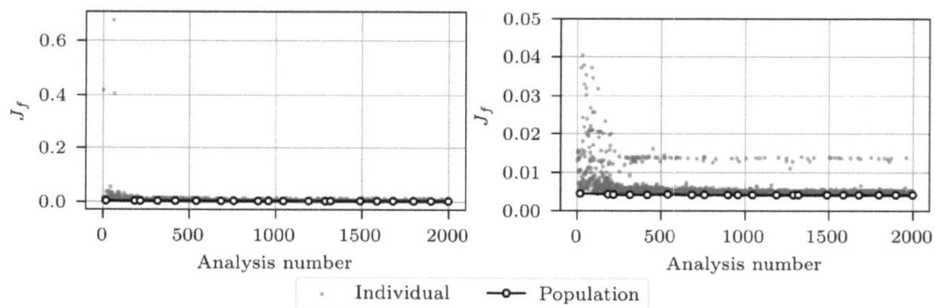

Figure 16. Evolution of the objective functions J_f for analysis number 5, including all data (**a**) and zoomed-in view (**b**).

The grey markers in Figure 16 represent the J_f values of 2000 FE analyses, and the white markers denote the minimum J_f values of each of the 20 data subsets (populations), each containing 100 results. The grey markers' scatter decreases with the number of analyses. Some analyses, despite the high sequence number, even after the 1000th analysis, give a high J_f value. This results from the incorrectly paired FE experimental modes, which could not be completely eliminated. Figure 17a,b present the evolution of α_{MG} and α_{OTHER}, respectively.

The most significant variation in α_{MG} and α_{OTHER} occurs within the first 500 analyses and finally converges towards 1.20 and 1.09, respectively. Table 11 presents the FEMU results for all analyses of 12 separately updated FE "M1_SUBSTR_UPDATE_*i*" models, where *i* represents the analysis number. For each analysis, the final updated FE model was selected as the best individual from the final data subset (the last white marker, shown in Figure 16.

For single-variable analyses (α_{ALL}), frequency-based, and frequency-and-MAC-based FEMU analyses give the same α_{ALL} value of 1.18, which matches the strain-based value of 1.21. MAC-based FEMU results differ considerably, with a value of 0.99.

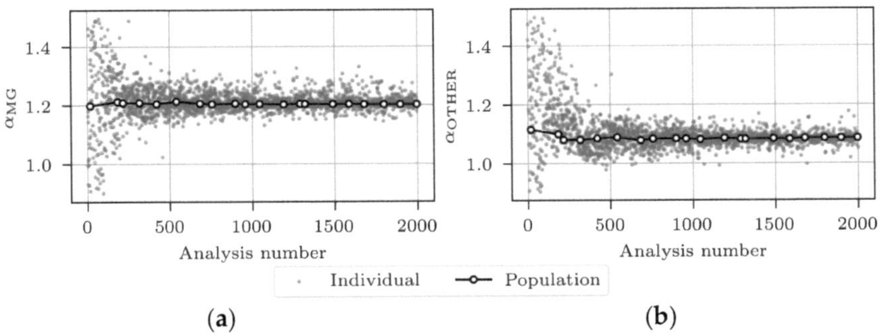

Figure 17. Evolution of α_{MG} (**a**) and α_{OTHER} (**b**) for FEMU analysis number 5.

Table 11. Summary of FEMU results: values of variables that correspond to the best match within the last population.

Analysis Number	Variables	Acceleration-Based			Strain-Based
		Frequency-Based	MAC-Based	Frequency-and-MAC-Based	
1, 2, 3, 4	α_{ALL}	1.18	0.99	1.18	1.21
5, 6, 7, 8	α_{MG}, α_{OTHER}	1.20, 1.09	0.96, 0.90	0.97, 1.33	1.17, 1.50
9, 10, 11, 12	α_{EMG}, α_{IMG}, α_{OTHER}	0.91, 1.50 1.17	0.90, 0.96, 1.02	1.36, 0.93, 1.09	0.90, 1.50, 1.01

For analyses with two variables (α_{MG} and α_{OTHER}), a good match between the frequency- and strain-based FEMU is observed for α_{MG}: The values are 1.20 and 1.17, respectively. MAC-based and frequency-and-MAC-based FEMU analyses for α_{MG} give comparable values of 0.96 and 0.99, which, however, are 20% lower than the frequency- and strain-based values. Contrary to α_{MG}, FEMU gives very different frequency- (1.09) and strain-based α_{OTHER} (1.50) results. As all objective functions are more sensitive to α_{MG} than α_{OTHER}, the better match for α_{MG} is reasonable.

For analyses with three variables (α_{EMG}, α_{IMG}, and α_{OTHER}), a good match between the frequency- and strain-based FEMU is again observed for α_{EMG}, namely 0.91 and 0.90, respectively. Additionally, MAC-based FEMU also gives a value of 0.90. Values of α_{IMG} are 1.50 for both frequency- and strain-based FEMU analyses, differing considerably from MAC-based (0.96) and frequency-and-MAC-based (0.93) FEMU.

It is evident from Table 11 for analyses with three variables that α_{EMG} reaches the lower bound (0.90), and α_{IMG} reaches the upper bound (1.50) of the preset range. One way to avoid variables reaching these bounds is to rerun FEMU analyses 9–12 with extended lower and upper bounds. The question of whether the results are reasonable in an engineering context arises when the range is too wide, i.e., there is a concern about whether a global minimum that does not reflect the physical properties of the considered problem is reached.

The following analysis pairs are expected to give comparable values of α_{OTHER}: 5–9, 6–10, 7–11 and 8–12. These values are expected to be similar because the only difference between them is the split of the MG group into the EMG and IMG groups. Elements within the OTHER group remained unchanged. The previously mentioned analysis pairs do not give comparable values of α_{OTHER}, due to the insensitivity coincidence of the objective functions to Young's modulus of the elements in the OTHER group. Splitting the OTHER group into more subsets would increase the insensitivity even more. Therefore, three variables for the considered case study represent the sensible upper bound.

When comparing the FEMU results, it is important to stress the influence of temperature during the experiments. Strain and acceleration measurements were not taken simultaneously. The latter was obtained at a later stage, with the ambient temperature roughly 3 °C above the average 15 °C recorded during the strain measurements. According to [52], such a temperature difference would cause an insignificant 1% decrease in Young's elastic modulus of concrete at higher temperatures.

Table 12 shows values of measured, initial model's (M1_SUBSTR_INIT), and updated model's (M1_SUBSTR_UPDATE_*i*) natural frequencies for frequency-based, MAC-based, and frequency-and-MAC-based FEMU analyses. As expected, the match between the modelled and measured frequencies is the best for the frequency-based FEMU (analyses 1, 5, and 9). MAC-based FEMU (analyses 2, 6, and 10) generally underestimates the first three frequencies by approximately 10%. Frequency-and-MAC-based FEMU analyses 3 and 11 give good matches for all four natural frequencies, comparable to frequency-based FEMU analyses 1 and 9, respectively. By contrast, analysis 7 gives a poor match for frequencies—in between the frequency-based analysis 5 and MAC-based analysis 6.

Figure 18 shows the MAC matrices for frequency-based, MAC-based, and frequency-and-MAC-based FEMU analyses. Nine MAC matrices are shown, where mode shapes of the FE models M1_SUBSTR_UPDATE_*i* are compared to the experimental mode shapes. The modelled and experimental mode shapes match the best for MAC-based FEMU (analyses 2, 6, and 10) and the worst for frequency-based FEMU (analyses 1, 5, and 9). The most significant difference in the MAC values can be seen for the B-1 mode shape; while frequency-based FEMU analyses give MAC values between 0.53 and 0.62, MAC-based FEMU analyses give MAC values of 0.93. Frequency-and-MAC-based FEMU analyses 7 and 11 provide a good match for all mode shapes, comparable to MAC-based FEMU analyses 2 and 8, respectively. By contrast, analysis 3 results in a poor match.

Figure 19 shows the maximum strains for the strain-based FEMU (analysis numbers 4, 8, and 12), compared to the maximum strains of the M1_SUBSTR_INIT FE model and mean ± STD (standard deviation) values of the maximum measured strains in sensors SG_01, SG_02, SG_03, and SG_04 under the calibration vehicles V1, V2, and V3. The best match was achieved for FEMU analysis number 12, and the poorest match was recorded for FEMU analysis number 4.

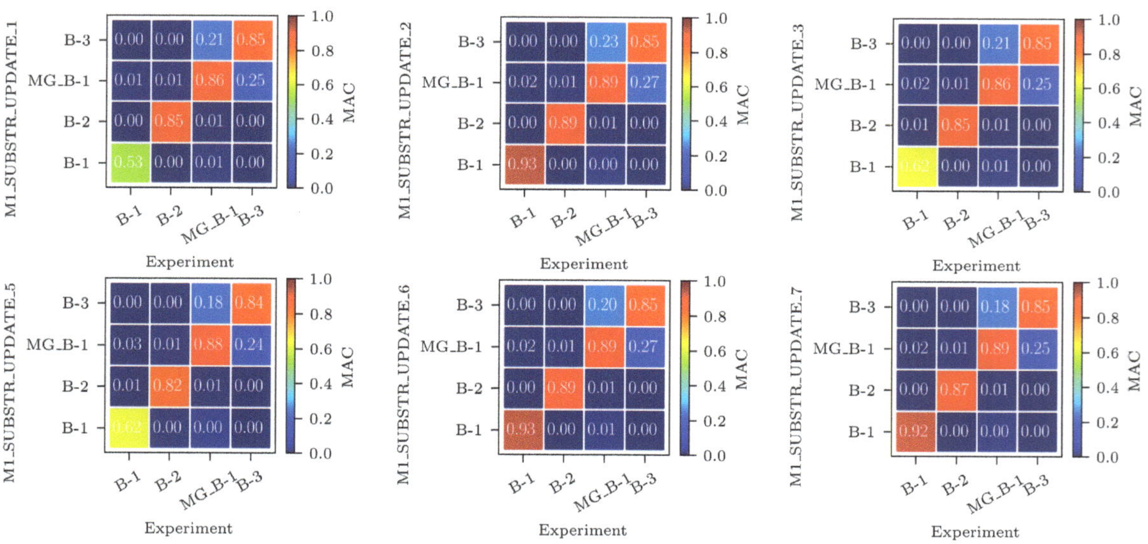

Figure 18. *Cont.*

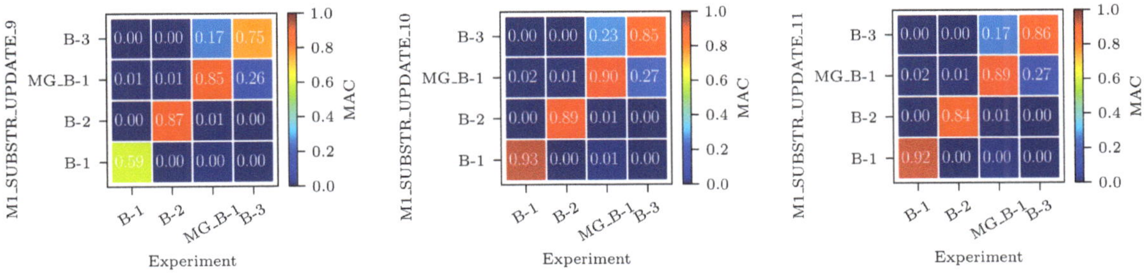

Figure 18. MAC matrices for frequency-based, MAC-based, and frequency-and-MAC-based FEMU analyses: M1_SUBSTR_UPDATE_*i* FE model vs. experiment.

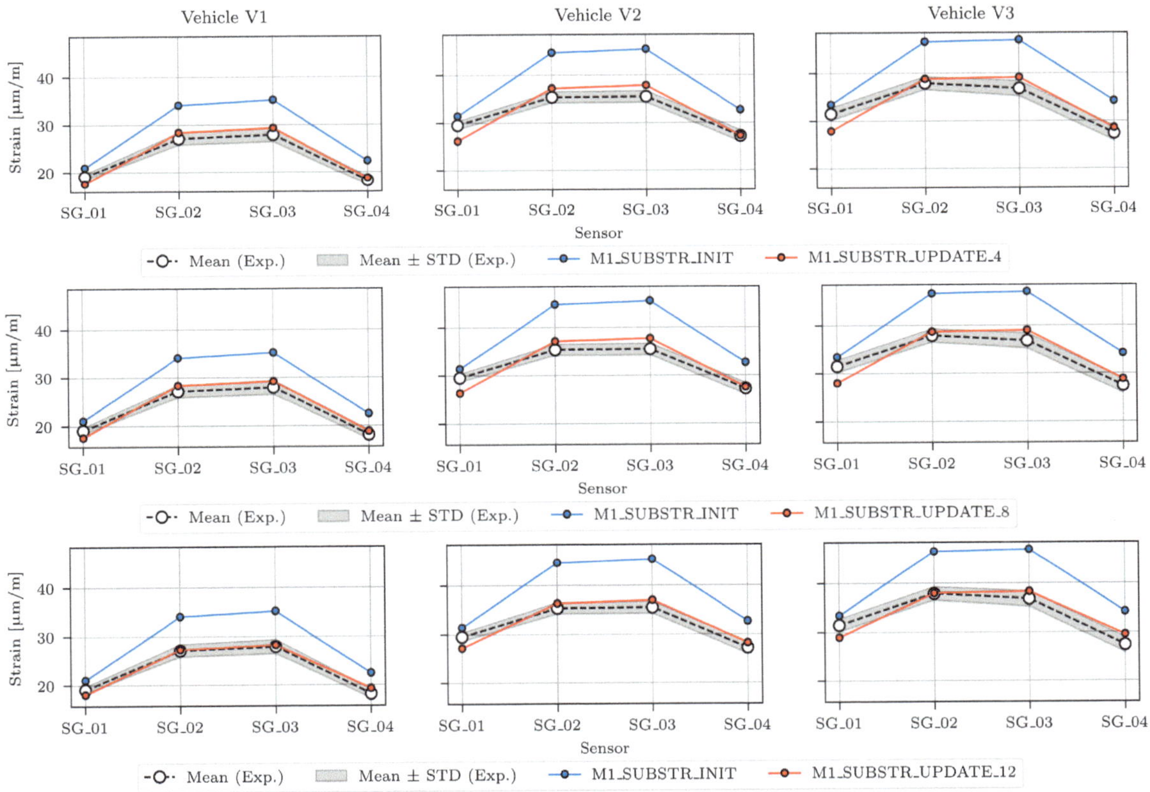

Figure 19. Maximum strains in the M1_SUBSTR_UPDATE_4, M1_SUBSTR_UPDATE_8, and M1_SUBSTR_UPDATE_12 FE models compared to the M1_SUBSTR_INIT model and mean ± STD (standard deviation) values of maximum measured strains in sensors SG_01, SG_02, SG_03, and SG_04 under calibration vehicles V1, V2, and V3.

Table 12. Results for all acceleration-based FEMU analyses: Values of measured, initial model's (M1_SUBSTR_INIT), and updated model's (M1_SUBSTR_UPDATE_*i*) natural frequencies.

Analysis Number (*i*)	Experiment [Hz]				M1_SUBSTR_INIT [Hz]				M1_SUBSTR_UPDATE_*i* [Hz]			
	B-1	B-2	MG_B-1	B-3	B-1	B-2	MG_B-1	B-3	B-1	B-2	MG_B-1	B-3
1	3.32	10.65	13.67	20.31	3.00	9.83	13.07	20.48	3.20	10.35	14.06	21.33
2									2.95	9.69	13.05	20.22
3									3.19	10.35	14.06	21.33
5									3.25	10.36	13.59	21.43
6									2.93	9.59	12.39	20.13
7									3.08	10.18	14.14	20.80
9									3.24	10.31	13.59	20.86
10									2.94	9.67	13.06	20.14
11									3.17	10.33	13.59	21.04

3.2.3. FEMU Results: EDMF

The EDMF methodology was adopted for FEMU considering the three variables, in addition to the residual minimisation analyses 9–12, to gain a more comprehensive insight into the problem. This allowed for more detailed insight into which variables and how they affect the FE model's response. As stated by [43], model simplifications are always present when modelling full-scale civil structures, and the relationship between errors is usually unquantifiable. Model simplifications usually come, among others, from the omission of load-carrying elements (in this study, safety barriers) or improper distribution of loads (in this study, the position of calibration vehicles and the filtration of the dynamic strain signal).

EDMF results in this section are shown separately for acceleration-based, strain-based, and acceleration-and-strain-based FEMU analyses. Initially, 9464 FE models with a unique combination of variable values α_{EMG}, α_{IMG}, and α_{OTHER} were calculated for static analysis (strain-based FEMU) and modal analysis (acceleration-based FEMU). The range for the variables was intentionally set to be wider than for analyses 9–12. This was carried out to show the models with physically unacceptable variable values and how EDMF methodology can help avoid them. The set of variable values was the same for α_{EMG} and α_{IMG}. Each can take 26 different values: the minimum value of 0.10 (lower bound) and the maximum value of 2.00 (upper bound). The range for α_{OTHER} was defined between the lower bound of 0.10 and the upper bound of 1.90 (14 values overall). As shown in Table 13, the intervals between values are not uniform. To optimise the number of FE analyses, the range for α_{OTHER} was, based on the sensitivity study results, sparser than for the α_{EMG} and α_{IMG}. After the FE analyses were performed, falsification thresholds were defined.

Table 13. Initial ranges of variables.

Variable	Initial Range
α_{EMG}	[0.10, 0.20, . . ., 0.90, 0.95, . . ., 1.50, 1.60, . . ., 2.00]
α_{IMG}	[0.10, 0.20, . . ., 0.90, 0.95, . . ., 1.50, 1.60, . . ., 2.00]
α_{OTHER}	[0.10, 0.30, . . ., 0.50, 0.60, . . ., 1.30, 1.50, . . ., 1.90]

For acceleration-based EDMF, falsification thresholds were defined for natural frequencies and MAC values of the mode shapes for all four modes. Threshold values were defined iteratively, initially allowing natural frequencies to deviate ±5% relative to the experimental values, and the absolute MAC values being at least 0.90, following the recommendation by [53] of a 'good correlation' between the FE model and experiment. For strain-based EDMF, the initial falsification thresholds were set to ±5% relative to the experimental values in sensors SG_01, SG_02, SG_03, and SG_04. Table 14 shows the final falsification threshold values, modified from the initial ones, to obtain enough candidates. Too narrow thresholds could lead to too few or even no candidates, and too loose ones could give too many.

Table 14. Falsification thresholds for EDMF.

Analysis Number	Source		Falsification Thresholds [1]	
			Min	Max
13	Acc.-Based	Frequencies MAC values	−5%, −5%, −5%, −5% 0.35, 0.20, 0.20, 0.20	+5%, +5%, +5%, +5% 0, 0, 0, 0
14	Strain-Based	Strains	−10%, −10%, −10%, −10%	+10%, +10%, +10%, +10%
15	Acc.-and-Strain-Based	Frequencies MAC values Strains	−5%, −5%, −5%, −5% 0.35, 0.20, 0.20, 0.20 −10%, −10%, −10%, −10%	+5%, +5%, +5%, +5% 0, 0, 0, 0 +10%, +10%, +10%, +10%

[1] For frequencies, the threshold is defined as deviation in percentage from the measured frequencies; for MAC values, it is defined as absolute deviation from 1.0; and for strains, it is defined as deviation in percentage from the measured strains.

Finally, the results of all FE models were tested for the fit within the falsification threshold bounds for the following factors:

- All frequencies and all MACs (acceleration-based EDMF, analysis number 13);
- All strains (strain-based EDMF, analysis number 14);
- All frequencies, all MACs, and all strains (acceleration-and-strain-based EDMF, analysis number 15).

Only the FE models that fit all threshold bounds were designated as acceleration-based, strain-based, or acceleration-and-strain-based candidates and were included in the candidate model set, meaning their variable values are plausible. The last step was to critically overview the candidate model sets in terms of whether the results were meaningful in an engineering context. The acceleration-based EDMF results are shown in Figure 20.

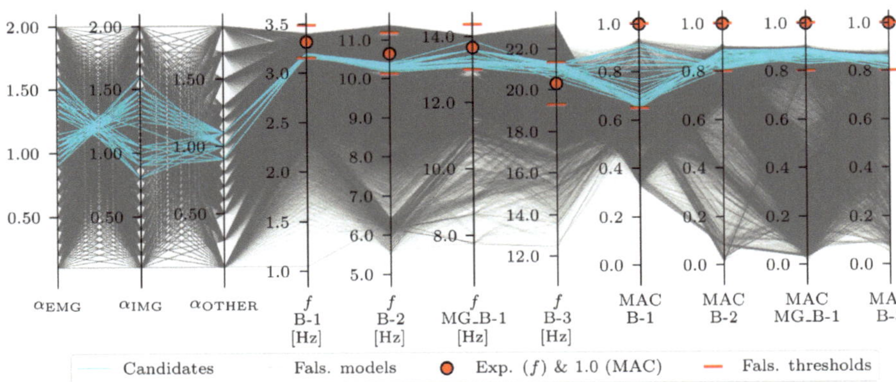

Figure 20. Acceleration-based EDMF results.

Overall, 35 candidates that fit into all threshold bounds for acceleration-based EDMF were identified (Table 15). However, not all of them were final, engineering-feasible candidates. Recalling the partially connected safety barriers, modelled as structural elements (described in Section 2.4.2 and [32]) and positioned close to the EMG elements, it was expected that this would manifest in the α_{EMG} lower than the α_{IMG}. Therefore, only the candidate model sets with $\alpha_{EMG} < \alpha_{IMG}$ were expected to be the final candidates. The strain-based EDMF results are shown in Figure 21.

Table 15. Variable ranges: initial and after EDMF.

Variable	Initial Range	Range after EDMF		
		Analysis No. 13	Analysis No. 14	Analysis No. 15
		Acceleration-Based	Strain-Based	Acceleration-and-Strain-Based
	n = 9464	n = 35	n = 199	n = 7
α_{EMG}	[0.10, 2.00]	[0.90, 1.60]	[0.40, 1.20]	[0.90, 1.10]
α_{IMG}	[0.10, 2.00]	[0.80, 1.60]	[1.05, 2.00]	[1.25, 1.50]
α_{OTHER}	[0.10, 1.90]	[0.90, 1.10]	[0.30, 1.90]	[1.00, 1.10]

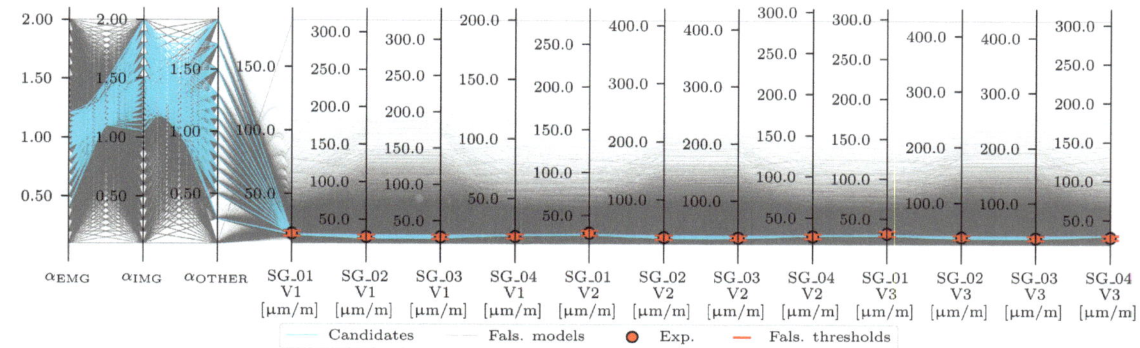

Figure 21. Strain-based EDMF results.

Overall, 199 candidates were identified for the strain-based EDMF. Most (191) satisfied the $\alpha_{EMG} < \alpha_{IMG}$ criteria. Although the number of candidates was reduced from 9464 to 199 ($\approx$2%), the range of the variables after strain-based EDMF, shown in Table 15, was still wide, especially for α_{OTHER}. Figure 22 shows the acceleration-and-strain-based EDMF results.

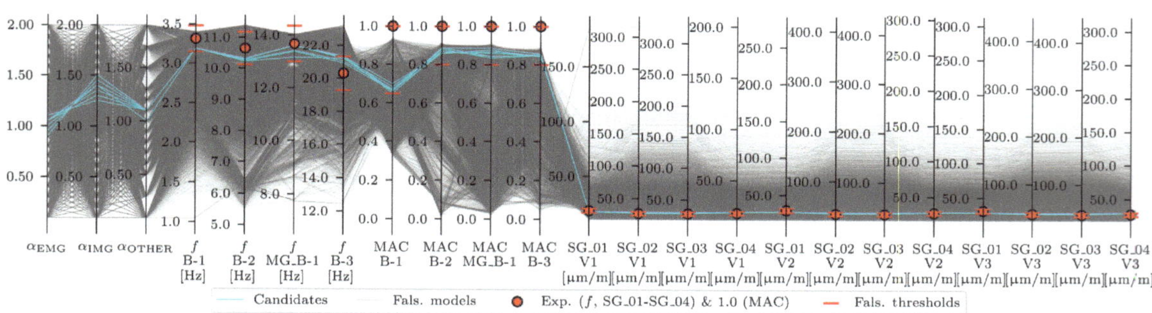

Figure 22. Acceleration-and-strain-based EDMF results.

Overall, seven candidates were identified for the acceleration-and-strain-based EDMF. All seven candidates satisfied the $\alpha_{EMG} < \alpha_{IMG}$ criteria and the updated range of the variables was narrowed significantly compared to the acceleration-based EDMF and strain-based EDMF, as shown in Table 15. Moreover, none of the variables reached the lower and upper bounds of the predefined range. As experimental values and falsified thresholds for strains are not clearly visible in Figure 22, a similar plot in Figure 23 shows a limited range of model sets with values of $\alpha_{EMG} \geq 0.7$, $\alpha_{IMG} \geq 0.7$, and $\alpha_{OTHER} \geq 0.7$.

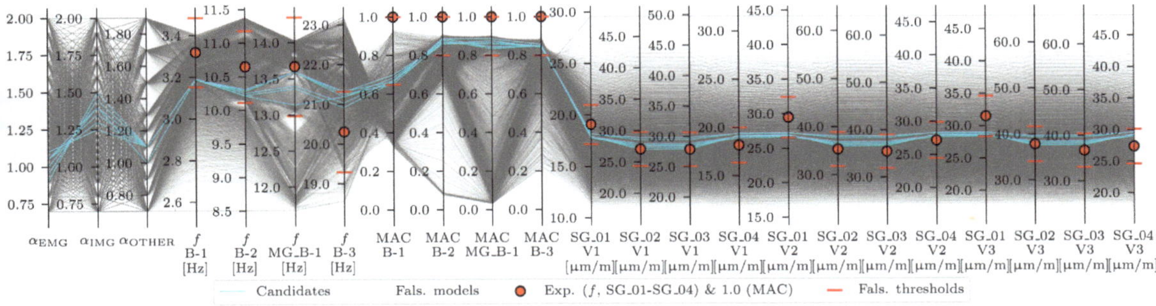

Figure 23. Acceleration-and-strain-based EDMF results, shown for a limited range of model sets with values of $\alpha_{EMG} \geq 0.7$, $\alpha_{IMG} \geq 0.7$, and $\alpha_{OTHER} \geq 0.7$.

With the provided falsification thresholds, both acceleration-based EDMF and strain-based EDMF significantly reduced the number of candidate model sets from an initial 9464 to 35 (0.4%) and 199 (2%), respectively. However, acceleration-based EDMF included engineering unacceptable candidate model sets with values of α_{EMG} greater than α_{IMG}. This was also true for a very small proportion of candidate model sets given by strain-based EDMF. Only the candidate model sets, given by the acceleration-and-strain-based EDMF, contained engineering-acceptable values of α_{EMG}, α_{IMG}, and α_{OTHER}. The deviation between ranges is mainly attributed to the safety barriers SB1 and SB2, which, although modelled as fully connected to the superstructure, are only partially connected.

Although the 3D finite elements allow for a high level of the FE model detailing, the systematic biases in the FE model remain present, resulting from the partially connected safety barriers and the unknown exact position of the calibration vehicles. EDMF methodology is computationally more demanding for the FEMU than the residual minimisation. Nevertheless, it was the key for the case study, as it allowed for combining acceleration- and strain-based FEMU studies and making an engineering decision about their updated values.

4. Conclusions

This paper presents the results of multiple FEMU studies of a highway viaduct that considered both strain responses under the traffic loading and accelerations from the traffic-induced and ambient vibration tests. The updated parameters from these two types of tests were compared. Furthermore, the residual minimisation FEMU approach was combined with the EDMF methodology. Despite being known to perform well in system identification, the latter is still underused in FEMU, compared to residual minimisation and Bayesian interference.

This study focused on updating structural parameters through Young's elastic modulus of different groups of superstructure elements, e.g., all members, main, external main, or internal main girders. A dozen FEMU analyses were performed considering residual minimisation methodology. Four of them considered one variable (α_{ALL}), four considered two variables (α_{MG}, α_{OTHER}), and the last four considered three variables (α_{EMG}, α_{IMG}, and α_{OTHER}). Four separate FEMU analyses were performed for each number of variables: frequency-based, MAC-based, frequency-and-MAC-based, and strain-based. Acceleration-based analyses considered four modes (natural frequencies and mode shapes), while strain-based analyses considered the maximum strains measured under three calibration vehicles. Frequency- and strain-based FEMU studies for the single variable α_{ALL} yielded comparable values of 1.18 and 1.21. For analyses with two variables (α_{MG} and α_{OTHER}), a good match between the frequency- and strain-based FEMU was observed for α_{MG}: 1.20 and 1.17. For analyses with three variables, α_{EMG} reached the lower bound (0.90), and α_{IMG} reached the upper bound (1.50), in frequency- and strain-based FEMU analyses. Three additional FEMU analyses for three variables, applying the EDMF methodology, yielded engineeringly sensible results for the considered problem. The last EDMF analysis which combined acceleration and strain data, proved to be crucial; initial ranges of variables were narrowed to [0.90, 1.10] for α_{EMG}, [1.25, 1.50] for α_{IMG}, and [1.00, 1.10] for α_{OTHER}.

The results of this study show that frequency- and strain-based FEMU similarly overestimated the superstructure's design bending stiffness by approximately 20%. When the main girders were separated from other elements, both methods again overestimated the design bending stiffness of the main girders by approximately 20%. When the main girders were additionally split into external and internal ones, the acceleration- and strain based EDMF overestimated the internal main girders' design bending stiffness by 25–50% No significant overestimation was obtained for the external main girders, most likely due to the partially connected safety barriers.

The key advantages of the EDMF methodology over residual minimisation are highlighted in this study, particularly its intuitiveness and the capability of combining different types of measurement within FEMU, without having to decide which one to put more weight to. Furthermore, the EDMF revealed the engineering-acceptable candidate model sets and narrowed the updated variable ranges in the FE model. This suggests that relying solely on modal parameters (frequencies and/or mode shapes) is not recommended, particularly when the FE model will serve to simulate the response under traffic loads, for example, to support bridge structural safety analyses.

The future aim is to extend the proposed FEMU approach with B-WIM results. This will involve different magnitudes of traffic loading, even the extreme ones caused by the exceptional heavy vehicles; the recorded strain responses under the crossing heavy vehicles of known axle loads and configurations; the measured modal parameters; and the measured, not theoretical, influence lines. Finally, the strain and vibration measurements can be integrated into long-term monitoring systems, providing simultaneous strains and modal parameters to allow for more reliable identification and variation of the mode shapes.

Author Contributions: Conceptualisation, D.H., D.R., A.A., A.Ž. and P.Č.; formal analysis, D.H. and P.Č.; methodology, D.H., D.R., A.A., A.Ž. and P.Č.; software, D.H., A.A. and P.Č.; investigation, D.H., D.R., A.A. and A.Ž.; resources, D.H., A.A., A.Ž. and P.Č.; data curation, D.H.; writing—original draft preparation, D.H. and P.Č; writing—review and editing, D.H., D.R., A.A., A.Ž. and P.Č.; visualisation, D.H. and A.Ž.; supervision, D.R., A.Ž. and P.Č.; project administration, A.Ž. and P.Č.; funding acquisition, A.Ž. and P.Č. All authors have read and agreed to the published version of the manuscript.

Funding: The first and fifth authors acknowledge the financial support from the Slovenian Research Agency (Young Researcher Funding Programme (ARRS No. 53694) and research core funding No. P2-0260. The third and fourth authors acknowledge the research core funding No. P2-0273 and infrastructure programme No. I0-0032. The second author acknowledges the financial support from programmatic funding—UIDP/04708/2020 with DOI 10.54499/UIDP/04708/2020 of the CONSTRUCT—Instituto de I&D em Estruturas e Construções, funded by national funds through the FCT/MCTES (PIDDAC).

Data Availability Statement: The finite element models and the data from the case study are available upon request from the corresponding author.

Acknowledgments: The authors would like to express their gratitude to the Motorway Company of the Republic of Slovenia (DARS d.d.), for allowing us to use the long-term monitoring results from the highway viaduct. The authors also acknowledge the contributions of Maja Kreslin, Mirko Kosič, and Jan Kalin associated with field measurement, data processing, and FE modelling. Company CESTEL d.o.o. is acknowledged for providing B-WIM-related support.

Conflicts of Interest: The authors declare no conflicts of interest.

References

1. Pour-Ghaz, M.; Isgor, O.B.; Ghods, P. The Effect of Temperature on the Corrosion of Steel in Concrete. Part 1: Simulated Polarization Resistance Tests and Model Development. *Corros. Sci.* **2009**, *51*, 415–425. [CrossRef]
2. Nasr, A.; Kjellström, E.; Björnsson, I.; Honfi, D.; Ivanov, O.L.; Johansson, J. Bridges in a Changing Climate: A Study of the Potential Impacts of Climate Change on Bridges and Their Possible Adaptations. *Struct. Infrastruct. Eng.* **2020**, *16*, 738–749. [CrossRef]
3. Cho, R. What Uncertainties Remain in Climate Science? Available online: https://news.climate.columbia.edu/2023/01/12/what-uncertainties-remain-in-climate-science/ (accessed on 18 December 2023).
4. Tabari, H. Climate Change Impact on Flood and Extreme Precipitation Increases with Water Availability. *Sci. Rep.* **2020**, *10*, 13768. [CrossRef]
5. Clare Nullis Economic Costs of Weather-Related Disasters Soars but Early Warnings Save Lives. Available online: https://public-old.wmo.int/en/media/press-release/economic-costs-of-weather-related-disasters-soars-early-warnings-save-lives (accessed on 18 December 2023).
6. International Transport Forum. *ITF Transport Outlook*; OECD: Paris, France, 2023.
7. Kos, M. Experimental Statistics: Road Traffic Counting 2022. Available online: https://www.stat.si/StatWeb/News/Index/11293 (accessed on 20 December 2023). (In Slovene).
8. Slovenian Infrastructure Agency Traffic Loads from 1997 Onwards. Available online: https://podatki.gov.si/dataset/pldp-karte-prometnih-obremenitev (accessed on 20 December 2023). (In Slovene)
9. Žnidarič, A.; Pakrashi, V.; OBrien, E.J.; O'Connor, A. A Review of Road Structure Data in Six European Countries. *Proc. Inst. Civ. Eng. Urban Des. Plan.* **2011**, *164*, 225–232. [CrossRef]
10. Nowak, A.S.; Latsko, O. Are Our Bridges Safe? *Bridge* **2018**, *48*, 26–30.
11. Gartner, N.; Kosec, T.; Legat, A. Monitoring the Corrosion of Steel in Concrete Exposed to a Marine Environment. *Materials* **2020**, *13*, 407. [CrossRef]
12. Page, C.L. Mechanism of Corrosion Protection in Reinforced Concrete Marine Structures. *Nature* **1975**, *258*, 514–515. [CrossRef]
13. Bertolini, L.; Elsener, B.; Pedeferri, P.; Redaelli, E.; Polder, R.B. *Corrosion of Steel in Concrete: Prevention, Diagnosis, Repair*; John Wiley & Sons: Hoboken, NJ, USA, 2013; ISBN 3527651713.
14. Scope, C.; Vogel, M.; Guenther, E. Greener, Cheaper, or More Sustainable: Reviewing Sustainability Assessments of Maintenance Strategies of Concrete Structures. *Sustain. Prod. Consum.* **2021**, *26*, 838–858. [CrossRef]
15. *Brücken an Bundesfernstraßen: Bilanz und Ausblick (Only in German)*; BMDV: Bonn, Germany, 2022.
16. Gkoumas, K.; Marques Dos Santos, F.; Van Balen, M.; Tsakalidis, A.; Ortega Hortelano, A.; Grosso, M.; Haq, A.; Pekar, F. *Research and Innovation in Bridge Maintenance, Inspection and Monitoring*; Publications Office of the European Union: Luxembourg, 2019.
17. Bigaj-van Vliet, A.; Allaix, D.L.; Köhler, J.; Scibilia, E. Standardisation in Monitoring, Safety Assessment and Maintenance of the Transport Infrastructure: Current Status and Future Perspectives. In *Proceedings of the 1st Conference of the European Association on Quality Control of Bridges and Structures, Padua, Italy, 29 August–1 September 2021*; Pellegrino, C., Faleschini, F., Zanini, M.A., Matos, J.C., Casas, J.R., Strauss, A., Eds.; Springer International Publishing: Cham, Switzerland, 2022; pp. 1152–1162.
18. James, G.; Pianigiani, G.; White, J.; Karthik, P. Genoa Bridge Collapse: The Road to Tragedy. Available online: https://www.nytimes.com/interactive/2018/09/06/world/europe/genoa-italy-bridge.html (accessed on 18 December 2023).
19. Woof, M.J. Italy Bridge Monitoring Tender. Available online: https://www.worldhighways.com/wh9/news/italy-bridge-monitoring-tender#:~:text=An%20important%20tender%20is%20underway,condition%20of%20bridges%20and%20viaducts (accessed on 18 December 2023).
20. Ereiz, S.; Duvnjak, I.; Fernando Jiménez-Alonso, J. Review of Finite Element Model Updating Methods for Structural Applications. *Structures* **2022**, *41*, 684–723. [CrossRef]
21. Xia, Z.; Li, A.; Li, J.; Shi, H.; Duan, M.; Zhou, G. Model Updating of an Existing Bridge with High-Dimensional Variables Using Modified Particle Swarm Optimization and Ambient Excitation Data. *Measurement* **2020**, *159*, 107754. [CrossRef]
22. Tran-Ngoc, H.; Khatir, S.; De Roeck, G.; Bui-Tien, T.; Nguyen-Ngoc, L.; Abdel Wahab, M. Model Updating for Nam O Bridge Using Particle Swarm Optimization Algorithm and Genetic Algorithm. *Sensors* **2018**, *18*, 4131. [CrossRef]
23. Cheng, X.X.; Song, Z.Y. Modal Experiment and Model Updating for Yingzhou Bridge. *Structures* **2021**, *32*, 746–759. [CrossRef]
24. Casas, J.R.; Moughty, J.J. Bridge Damage Detection Based on Vibration Data: Past and New Developments. *Front. Built Environ.* **2017**, *3*, 4. [CrossRef]
25. Scozzese, F.; Dall'Asta, A. Nonlinear Response Characterization of Post-Tensioned R.C. Bridges through Hilbert–Huang Transform Analysis. *Struct. Control Health Monit.* **2024**, *2024*, 5960162. [CrossRef]
26. Hester, D.; Koo, K.; Xu, Y.; Brownjohn, J.; Bocian, M. Boundary Condition Focused Finite Element Model Updating for Bridges. *Eng. Struct.* **2019**, *198*, 109514. [CrossRef]
27. Ticona Melo, L.R.; Malveiro, J.; Ribeiro, D.; Calçada, R.; Bittencourt, T. Dynamic Analysis of the Train-Bridge System Considering the Non-Linear Behaviour of the Track-Deck Interface. *Eng. Struct.* **2020**, *220*, 110980. [CrossRef]
28. Wang, X.; Zhang, C.; Sun, R. Response Analysis of Orthotropic Steel Deck Pavement Based on Interlayer Contact Bonding Condition. *Sci. Rep.* **2021**, *11*, 23692. [CrossRef]
29. Wang, H.; Li, A.; Li, J. Progressive Finite Element Model Calibration of a Long-Span Suspension Bridge Based on Ambient Vibration and Static Measurements. *Eng. Struct.* **2010**, *32*, 2546–2556. [CrossRef]

30. Meixedo, A.; Ribeiro, D.; Santos, J.; Calçada, R.; Todd, M. Progressive Numerical Model Validation of a Bowstring-Arch Railway Bridge Based on a Structural Health Monitoring System. *J. Civ. Struct. Health Monit.* **2021**, *11*, 421–449. [CrossRef]
31. Chen, S.-Z.; Wu, G.; Feng, D.-C. Damage Detection of Highway Bridges Based on Long-Gauge Strain Response under Stochastic Traffic Flow. *Mech. Syst. Signal Process.* **2019**, *127*, 551–572. [CrossRef]
32. Hekič, D.; Anžlin, A.; Kreslin, M.; Žnidarič, A.; Češarek, P. Model Updating Concept Using Bridge Weigh-in-Motion Data. *Sensors* **2023**, *23*, 2067. [CrossRef]
33. *VA0174 Ravbarkomanda Viaduct Rehabilitation Plan (in Slovene): Notebook 3/1.1-General Part, Technical Part*; Promico d.o.o.: Ljubljana, Slovenia, 2019.
34. Kalin, J.; Žnidarič, A.; Anžlin, A.; Kreslin, M. Measurements of Bridge Dynamic Amplification Factor Using Bridge Weigh-in-Motion Data. *Struct. Infrastruct. Eng.* **2022**, *18*, 1164–1176. [CrossRef]
35. DEWESoft. DEWESoft IOLITEi 3xMEMC-ACC. Available online: https://downloads.dewesoft.com/brochures/dewesoft-iolitei-3xmems-acc-datasheet-en.pdf (accessed on 26 December 2023).
36. *DEWESoft DewesoftX 2023*, Version 2023.5. Available online: https://dewesoft.com/products/dewesoftx (accessed on 26 December 2023).
37. Structural Vibration Solutions A/S ARTeMIS Modal Pro 2023. Available online: https://www.svibs.com/artemis-modal-pro/ (accessed on 26 December 2023).
38. Brincker, R.; Ventura, C. *Introduction to Operational Modal Analysis*; Wiley Online Library: Hoboken, NJ, USA, 2015; ISBN 9781119963158.
39. Allemang, R. The Modal Assurance Criterion—Twenty Years of Use and Abuse. *Sound Vib.* **2003**, *37*, 14–23.
40. Ewins, D.J. Model Validation: Correlation for Updating. *Sadhana* **2000**, *25*, 221–234. [CrossRef]
41. *Dassault Systemes SIMULIA User Assistance 2019: Abaqus/CAE User's Guide 2019*; Dassault Systemes: Waltham, MA, USA, 2019.
42. *Technical Report about the General Project of the Ravbarkomanda Viaduct (in Slovene)*; Tehnogradnja Maribor: Maribor, Slovenia, 1970.
43. Goulet, J.-A.; Smith, I.F.C. Structural Identification with Systematic Errors and Unknown Uncertainty Dependencies. *Comput. Struct.* **2013**, *128*, 251–258. [CrossRef]
44. Schlune, H.; Plos, M.; Gylltoft, K. Improved Bridge Evaluation through Finite Element Model Updating Using Static and Dynamic Measurements. *Eng. Struct.* **2009**, *31*, 1477–1485. [CrossRef]
45. Kurent, B.; Brank, B.; Ao, W.K. Model Updating of Seven-Storey Cross-Laminated Timber Building Designed on Frequency-Response-Functions-Based Modal Testing. *Struct. Infrastruct. Eng.* **2023**, *19*, 178–196. [CrossRef]
46. Eberhart, R.; Kennedy, J. A New Optimizer Using Particle Swarm Theory. In Proceedings of the MHS'95, Sixth International Symposium on Micro Machine and Human Science, Nagoya, Japan, 4–6 October 1995; pp. 39–43.
47. Virtanen, P.; Gommers, R.; Oliphant, T.E.; Haberland, M.; Reddy, T.; Cournapeau, D.; Burovski, E.; Peterson, P.; Weckesser, W.; Bright, J.; et al. {SciPy} 1.0: Fundamental Algorithms for Scientific Computing in Python. *Nat. Methods* **2020**, *17*, 261–272. [CrossRef]
48. Blank, J.; Deb, K. Pymoo: Multi-Objective Optimization in Python. *IEEE Access* **2020**, *8*, 89497–89509. [CrossRef]
49. Cao, W.-J.; Koh, C.G.; Smith, I.F.C. Enhancing Static-Load-Test Identification of Bridges Using Dynamic Data. *Eng. Struct.* **2019**, *186*, 410–420. [CrossRef]
50. Bertola, N.J.; Henriques, G.; Brühwiler, E. Assessment of the Information Gain of Several Monitoring Techniques for Bridge Structural Examination. *J. Civ. Struct. Health Monit.* **2023**, *13*, 983–1001. [CrossRef]
51. Pai, S.G.S.; Smith, I.F.C. Methodology Maps for Model-Based Sensor-Data Interpretation to Support Civil-Infrastructure Management. *Front. Built Environ.* **2022**, *8*, 801583. [CrossRef]
52. Jiao, Y.; Liu, H.; Wang, X.; Zhang, Y.; Luo, G.; Gong, Y. Temperature Effect on Mechanical Properties and Damage Identification of Concrete Structure. *Adv. Mater. Sci. Eng.* **2014**, *2014*, 191360. [CrossRef]
53. Jiménez-Alonso, J.F.; Naranjo-Perez, J.; Pavic, A.; Sáez, A. Maximum Likelihood Finite-Element Model Updating of Civil Engineering Structures Using Nature-Inspired Computational Algorithms. *Struct. Eng. Int.* **2021**, *31*, 326–338. [CrossRef]

sensors

MDPI

Article

Spatial Galloping Behavior of Iced Conductors under Multimodal Coupling

Fujiang Cui [1,2], Kaihong Zheng [1,*], Peng Liu [2] and Han Wang [2]

1. Beijing Aircraft Technology Research Institute, Commercial Aircraft Corporation of China, Ltd., Beijing 102211, China; cuifujiang@tyut.edu.cn
2. College of Aeronautics and Astronautics, Taiyuan University of Technology, Taiyuan 030024, China; liupeng01@tyut.edu.cn (P.L.); wanghan@tyut.edu.cn (H.W.)

* Correspondence: zhengkh@pku.edu.cn

Abstract: In this study, the coupled ordinary differential equations for the galloping of the first two modes in iced bundled conductors, including in-plane, out-of-plane, and torsional directions, are derived. Furthermore, through numerical analysis, the critical conditions of this modal galloping are determined in the range of wind speed–sag parameters, and the galloping patterns and variation laws in different parameter spaces are analyzed. The parameter space is then divided into five regions according to the different galloping modes. Under the multimodal coupling mechanism of galloping, the impact of single and two kinds of coupled mode galloping on the spatial nonlinear behavior is explored. The results reveal that the system exhibits an elliptical orbit motion during single mode galloping, while an "8" motion pattern emerges during coupled mode galloping. Moreover, two patterns of "8" motion are displayed under different parameter spaces. This research provides a theoretical foundation for the design of transmission lines.

Keywords: iced conductors; multimodal coupling; galloping behavior; nonlinear vibration

Citation: Cui, F.; Zheng, K.; Liu, P.; Wang, H. Spatial Galloping Behavior of Iced Conductors under Multimodal Coupling. *Sensors* **2024**, *24*, 784. https://doi.org/10.3390/s24030784

Academic Editors: Bing Li, Yongbo Li and Khandaker Noman

Received: 15 November 2023
Revised: 6 January 2024
Accepted: 12 January 2024
Published: 25 January 2024

1. Introduction

The cross-sectional shape of iced conductors changes from circular to noncircular under complex climatic conditions such as ice storm, snow, and freezing rain. Then, under the excitation of a certain wind speed, the unstable aerodynamic nonlinear load in different directions, including in-plane, out-of-plane, and torsional directions, induces galloping of the iced conductor. It is a self-excited vibration phenomenon characterized by a low frequency (approximately 0.08–3 Hz) and a large amplitude (reaching 5–300 times the diameter of the conductor) [1]. The galloping lasts for several hours and can cause serious damage to the power conductors, including tower damage, conductor fracture, insulator damage, wear of fittings and components, and phase flashover, which pose a significant threat to the safety of power conductors.

Until now, several researchers have explored the excitation mechanism of galloping, mainly focusing on the aerodynamic coefficients of iced conductors, line structure parameters, etc. Further, the galloping phenomenon has been examined from different perspectives. The three typical galloping mechanisms include vertical galloping [2], torsional galloping [3], and inertially coupled galloping [4]. In recent years, researchers from the China Electric Power Research Institute have investigated numerous domestic and foreign galloping accidents and classified galloping as a dynamic instability phenomenon. It has been reported that only unstable vibration can lead to large galloping. Accordingly, the dynamic stability mechanism of galloping has been proposed to analyze different types of galloping [5,6]. Based on these mechanisms, galloping has been explored through theoretical analysis.

Due to the geometric nonlinearity of iced conductor structures and the aerodynamic nonlinearity caused by aerodynamic loads, nonlinear coupling between multiple directions/modes of galloping is very likely to occur. Jones [7] developed a dynamic model for the coupling of in-plane and out-of-plane galloping and found that there is an inherent coupling between the equations describing in-plane and out-of-plane motions. This model demonstrated that out-of-plane motion has a significant impact on the stability of galloping. Blevins and Iwan [8] developed a dynamic model considering the coupling of torsional and in-plane motions, and they comprehensively analyzed galloping under both resonant and nonresonant conditions. This study was the first to prove that torsional motion can enhance galloping in the system. Liu et al. [9] used Hamilton's principle to establish three types of dynamic models for coupled galloping with in-plane, in-plane+torsional, and in-plane+out-of-plane+torsional motions. By analyzing the in-plane amplitude and torsional angle of galloping under various influencing factors, such as wind speed, air density, span length, damping ratio, and initial tension, they found that the coupled dynamic model of in-plane+out-of-plane+torsional motion was more accurate in evaluating the galloping characteristics. Chen and Wu [10] then clarified the generation mechanism of various galloping vibrations with different coupled motions. Matsumiya et al. [11] indicated the coupling effects on in-plane oscillation by considering the energy balance of the in-plane motion with the defined amplitudes and phase differences of the out-of-plane and torsional motions. In addition, the finite element method is widely used to simulate the galloping of iced conductors. Diana et al. [12] presented a finite-element model of quad-bundled conductors, predicting the onset speed of galloping instability and the maximum oscillation amplitudes through time-domain simulations and the proposed energy approach. The complex structural part of the system can also be reproduced, including iced eight-bundled conductors [13,14] and a tower line system [15]. Complex wind field conditions can also be simulated, such as unsteady and stochastic wind fields [16].

Most studies have focused on the first-order modes in different directions. However, Liu and Huo [17] included higher-order modes and established the coupled motion equations for the first four in-plane modes and the first torsional mode to describe the nonlinear interactions between the in-plane and torsional vibration, considering geometrical and aerodynamic nonlinearities. They found that the galloping of higher-order modes can excite the galloping of lower-order modes, and the energy transfer phenomenon between symmetric and anti-symmetric modes was also analyzed. Huo et al. [18] established the coupled motion equations for the first three in-plane modes and the first three torsional modes, and the results revealed that the torsional modes contributed to the in-plane galloping behavior. With the increase in wind speed, the lower-order in-plane modes were gradually replaced by higher-order in-plane modes. Luongo and Nayfeh [19,20] also pointed out that higher-order modes exist, and these modes can excite the galloping of lower-order modes during the vibration of suspended cables. For such suspended structures, due to the additional tension caused by motion-induced deformation, the natural frequency of the in-plane symmetric mode exceeds that of the anti-symmetric mode, resulting in a frequency crossover phenomenon [21–23]. The various correspondence relationships between frequencies provide multiple possibilities for the interaction between modes. Accordingly, in some previous studies, the first two modes of in-plane, out-of-plane, and torsional motion were considered, and the nonlinear coupling between in-plane+out-of-plane+torsional modes and high-order+low-order modes was investigated from the perspective of energy transfer, clarifying the multimode coupling galloping mechanism of iced conductors. Under the coupling effect of multiple modes, the galloping behavior of iced conductors is bound to become complex and diverse. Therefore, exploring the galloping behavior under the coupling of multiple modes is of immense theoretical significance for clearly understanding the galloping of iced conductors and formulating effective anti-galloping measures.

To this end, in this study, the multimodal coupling mechanism of galloping in iced conductors is utilized to investigate the spatial nonlinear behavior of galloping in iced bundled conductors. The rest of this paper is organized as follows. In Section 2, for the common iced bundled crescent-shaped conductors, considering geometric and aerodynamic nonlinearity, the coupled vibration ordinary differential equations for the first two modes in the in-plane, out-of-plane, and torsional directions are derived. In Section 3, through numerical analysis, the critical conditions for in-plane galloping are obtained in the wind speed–span parameter space. Combined with the multimodal coupling galloping mechanism, the influence of single-mode galloping and coupled-mode galloping on spatial galloping behavior is analyzed for different galloping patterns within the parameter domain. Finally, this study is concluded in Section 4.

2. Establishment of Dynamical Equations

Compared to a single conductor, the bundled conductors are affected by spacers, and the twisting stiffness of its sub-conductors is much higher than that of a single conductor with the same cross-section. This leads to a more irregular cross-sectional shape of the iced bundled conductors, and the aerodynamic load acting on them is more complex and prone to causing vibrations. Therefore, the common crescent-shaped iced quad-bundled conductor is taken as the research object here.

The iced quad-bundled conductors are simplified as a single conductor for analysis assuming both ends to be fixed. A detailed description of the galloping model is reported in [24]; the main details are hereafter presented. In actual engineering, the sag-to-span ratio of transmission lines is generally small, within the range from 0 to 0.1. The conductor has a slender and flexible body structure, so the influence of bending stiffness can be neglected. In addition, it is assumed that the ice is uniformly distributed along the surface of the conductor with a length of l, and the incoming wind blows perpendicular to the plane where the conductor is located with a speed U. A schematic of the spatial model is shown in Figure 1a. The initial configuration of the iced conductor under its own weight is represented by Γ_0, while the dynamic configuration of the iced conductor under aerodynamic loads is represented by Γ. $u(x,t)$, $v(x,t)$, $w(x,t)$, and $\theta(x,t)$ represent the axial (x-axis), in-plane (y-axis), out-of-plane (z-axis), and torsional (yOz plane) displacement, respectively, of a point on the iced conductor at time t with respect to the coordinate origin O. The differential element dx is studied, and a schematic of its motion is shown in Figure 1b, where $A0B0$ and $A2B2$ represent the differential element before and after structural deformation, respectively. The catenary equation of the structure at time $t = 0$ is expressed as follows:

$$y_0 = \frac{-2H}{mg}\sinh\left(\frac{mgx}{2H}\right)\sinh\left(\frac{mg(l-x)}{2H}\right) \tag{1}$$

where H is the initial horizontal tension of the iced conductor, m is the mass per unit length, and g is the gravitational acceleration.

The aerodynamic force analysis diagram of the iced conductor cross-section is shown in Figure 1c, where G is the centroid, e_r is the eccentricity, θ_0 is the initial wind angle of attack, I is the moment of inertia about point O, and the overdot denotes the derivative with respect to time. F_L, F_D, and M are the aerodynamic lift force, drag force, and torsional moment, respectively. α_0 is the angle caused by vertical velocity, α is the angle of attack, U is the mean wind speed, U_r is the relative wind speed, and D is the bare conductor diameter.

Considering the geometric and aerodynamic nonlinearity and simplifying the axial motion to the in-plane and out-of-plane directions, the coupled vibration ordinary differential equations of the first two modes, including in-plane, out-of-plane, and torsional modes, can be derived using Hamilton's principle and the Galerkin space-discretization method [24], i.e.,

$$\begin{aligned}
&\ddot{q}_{vk} + g_{vk,1}\ddot{q}_{\theta k} + 2\xi_{vk}\omega_{g,vk}\dot{q}_{vk} + \omega_{g,vk}^2 q_{vk} = g_{vk,2}q_{w2}^2 + g_{vk,3}q_{w1}q_{w2} + g_{vk,4}q_{w1}^2 \\
&\quad + g_{vk,5}q_{v2}^2 + g_{vk,6}q_{v1}q_{v2} + g_{vk,7}q_{v1}^2 + g_{vk,8}q_{v2}q_{w2}^2 + g_{vk,9}q_{v2}q_{w1}q_{w2} \\
&\quad + g_{vk,10}q_{v2}q_{w1}^2 + g_{vk,11}q_{v2}^3 + g_{vk,12}q_{v1}q_{w2}^2 + g_{vk,13}q_{v1}q_{w1}q_{w2} + g_{vk,14}q_{v1}q_{w1}^2 \\
&\quad + g_{vk,15}q_{v1}q_{v2}^2 + g_{vk,16}q_{v1}^2 q_{v2} + g_{vk,17}q_{v1}^3 + g_{vk,18}\dot{q}_{\theta 1}^2 + g_{vk,19}\dot{q}_{\theta 2}^2 + g_{vk,20}\dot{q}_{\theta 1}\dot{q}_{\theta 2} \\
&\quad + \int_0^l \left[\varphi_{vk}(x)F_{vk}\left(\dot{q}_{v1}, \dot{q}_{v2}, \dot{q}_{w1}, \dot{q}_{w2}, q_{\theta 1}, q_{\theta 2}, \dot{q}_{\theta 1}, \dot{q}_{\theta 2}\right)\right]\mathrm{d}x,\ k = 1,2 \\
&\ddot{q}_{wk} + g_{wk,1}\ddot{q}_{\theta k} + 2\xi_{wk}\omega_{g,wk}\dot{q}_{wk} + \omega_{g,wk}^2 q_{wk} = g_{wk,2}q_{v2}q_{w2} + g_{wk,3}q_{v2}q_{w1} \\
&\quad + g_{wk,4}q_{v1}q_{w2} + g_{wk,5}q_{v1}q_{w1} + g_{wk,6}q_{w2}^3 + g_{wk,7}q_{w1}q_{w2}^2 + g_{wk,8}q_{w1}^2 q_{w2} \\
&\quad + g_{wk,9}q_{w1}^3 + g_{wk,10}q_{v2}^2 q_{w2} + g_{wk,11}q_{v2}^2 q_{w1} + g_{wk,12}q_{v1}q_{v2}q_{w2} + g_{wk,13}q_{v1}q_{v2}q_{w1} \\
&\quad + g_{wk,14}q_{v1}^2 q_{w2} + g_{wk,15}q_{v1}^2 q_{w1} + g_{wk,16}\dot{q}_{\theta 1}^2 + g_{wk,17}\dot{q}_{\theta 2}^2 + g_{wk,18}\dot{q}_{\theta 1}\dot{q}_{\theta 2} \\
&\quad + \int_0^l \left[\varphi_{wk}(x)F_{wk}\left(\dot{q}_{v1}, \dot{q}_{v2}, \dot{q}_{w1}, \dot{q}_{w2}, q_{\theta 1}, q_{\theta 2}, \dot{q}_{\theta 1}, \dot{q}_{\theta 2}\right)\right]\mathrm{d}x,\ k = 1,2 \\
&\ddot{q}_{\theta k} + g_{\theta k,1}\ddot{q}_{vk} + g_{\theta k,2}\ddot{q}_{wk} + 2\xi_{\theta k}\omega_{g,\theta k}\dot{q}_{\theta k} + \omega_{g,\theta k}^2 q_{\theta k} = +g_{\theta k,3}\dot{q}_{v1}\dot{q}_{\theta 1} + g_{\theta k,4}\dot{q}_{v1}\dot{q}_{\theta 2} \\
&\quad + g_{\theta k,5}\dot{q}_{v2}\dot{q}_{\theta 1} + g_{\theta k,6}\dot{q}_{v2}\dot{q}_{\theta 2} + g_{\theta k,7}\dot{q}_{w1}\dot{q}_{\theta 1} + g_{\theta k,8}\dot{q}_{w1}\dot{q}_{\theta 2} + g_{\theta k,9}\dot{q}_{w2}\dot{q}_{\theta 1} + g_{\theta k,10}\dot{q}_{w2}\dot{q}_{\theta 2} \\
&\quad + \int_0^l \left[\varphi_{\theta k}(x)M_{\theta k}\left(\dot{q}_{v1}, \dot{q}_{v2}, \dot{q}_{w1}, \dot{q}_{w2}, q_{\theta 1}, q_{\theta 2}, \dot{q}_{\theta 1}, \dot{q}_{\theta 2}\right)\right]\mathrm{d}x,\ k = 1,2
\end{aligned} \tag{2}$$

where $q_{vk}(t)$, $q_{wk}(t)$, and $q_{\theta k}(t)$ are the generalized coordinates for in-plane, out-of-plane, and torsional motions, respectively. ξ_{vk}, ξ_{wk}, $\xi_{\theta k}$; $\omega_{g,vk}$, $\omega_{g,wk}$, $\omega_{g,\theta k}$; and φ_{wk}, φ_{vk}, and $\varphi_{\theta k}$ are the damping ratios, natural frequencies, and natural modes corresponding to the different order modal structures for in-plane, out-of-plane, and torsional directions, respectively. $g_{vi,k}(i = 1, \ldots, 22)$, $g_{wi,k}(i = 1, \ldots, 20)$, and $g_{\theta i,k}(i = 1, \ldots, 12)$ are the integral coefficients. $\int_0^l [\varphi_{wk}(x)F_{wk}]\mathrm{d}x$, $\int_0^l [\varphi_{vk}(x)F_{vk}]\mathrm{d}x$, and $\int_0^l [\varphi_{\theta k}(x)M_{\theta k}]\mathrm{d}x$ are the aerodynamic terms for in-plane, out-of-plane, and torsional directions, respectively [24].

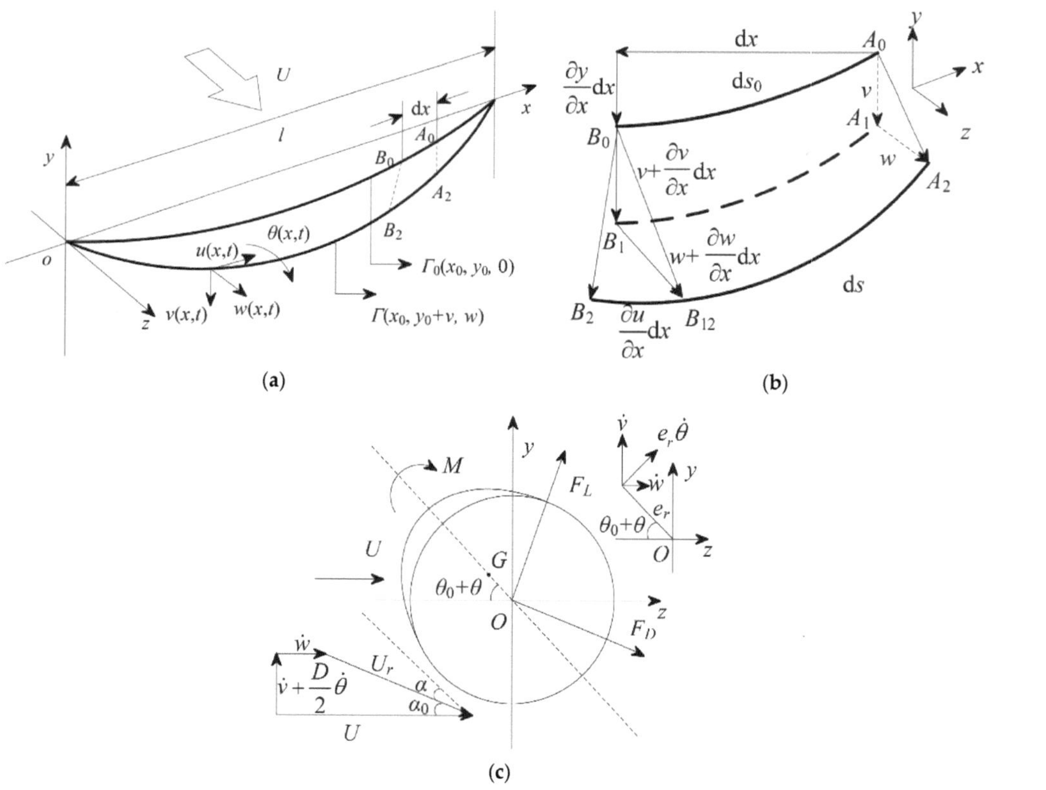

Figure 1. Galloping model of iced conductor. (**a**) Configuration. (**b**) Dynamic displacement of dx. (**c**) Aerodynamic forces acting on the chosen section of the iced conductor.

3. Numerical Analysis

The multimodal coupling mechanism of galloping is described [24] as follows: the nonlinear interaction among the in-plane, out-of-plane, and torsional modes of the same order leads to synchronized and limited galloping phenomena. The nonlinear coupling between the torsional and in-plane modes is the fundamental reason for the limited galloping characteristics. Regarding the synchronized galloping characteristics, the in-plane, out-of-plane, and torsional modes of the same order are excited simultaneously and tend to stabilize simultaneously, among which the galloping in the in-plane direction plays a dominant role. Regarding the limited galloping characteristics, the galloping is limited to a certain amplitude range. The multimodal coupling effect of galloping directly affects its spatial nonlinear behavior.

3.1. Critical Conditions of Galloping

When the span and the unit mass per unit length of the conductor are determined, the relationship between the sag (d) and the initial tension in the horizontal direction can be obtained as follows:

$$d = \frac{2H}{mg}\left[\sinh(\frac{mgl}{4H})\right]^2 \quad (3)$$

During the actual installation process of power lines, the sag of the conductors must be controlled based on different terrain sections to ensure a safe discharge distance. Therefore, sag is chosen as a bifurcation parameter to examine the critical conditions for galloping. The nonlinear dynamical equations (Equation (2)) are used to numerically analyze the unstable regions of the first two modes in different directions. Further, the critical instability conditions for the in-plane, out-of-plane, and torsional modes are explored in the wind speed (U) and sag (d) parameter plane. The common crescent-shaped iced quad-bundled conductors 4 × LGJ-400/35 are taken as an example in this study, and the selected equivalent parameters are shown in Table 1 [25].

Table 1. Equivalent parameters of iced bundled conductors.

Parameter	Value
Span l	244 m
Mass per unit length m	6.92 kg/m
Bare conductor diameter D	28.6 mm
Tensile stiffness EA	1.105×10^8 N
Torsional stiffness GJ	23,746 N·m^2/rad
Moment of inertia I	0.70065 kg·m
Inflow density ρ_a	1.29 kg/m^3
Eccentricity e_r	1.39×10^{-4} m
Initial wind attack angle θ_0	24°
In-plane damping ratio ξ_v	0.005
Out-of-plane damping ratio ξ_w	0.005
Torsional damping ratio ξ_θ	0.02

Taking in-plane motion as an example, the critical conditions and galloping region of the first two modes in the U-d parameter plane are shown in Figure 2. The blue and red lines represent the critical conditions of the first- and second-order modal galloping, respectively. According to this analysis, the following conclusions can be drawn:

(1) For the first-order modal galloping, as the sag increases, the vibration area first increases and then gradually decreases. When d > 3.8 m, the vibration area begins to decrease rapidly until the vibration disappears. However, when d > 5.6 m, the first-order modal vibration is again excited because of the nonlinear coupling between the first- and second-order modes in the in-plane direction [24]. Within the range of 5.6 m < d < 6.8 m, as the sag increases, the critical wind speeds for the upper and lower limits of the vibration increase rapidly, and the vibration area is narrow but

slowly increasing. When $d > 6.8$ m, as the sag continues to increase, the vibration area first gradually increases, reaching a maximum around $d = 8$ m, and then begins to gradually decrease. When $d > 10$ m, the first-order modal vibration is no longer excited.

(2) When the sag is low, the second-order modal galloping is not excited if the wind speed is within 20 m/s. When $d > 3.8$ m, the first-order modal galloping area begins to decrease, and the second-order modal galloping starts to be excited. The galloping area rapidly expands with the increase in sag. When $d > 5$ m, the galloping area enters a slow expansion stage. It can also be observed that the area of second-order modal galloping is significantly larger than that of first-order modal galloping, indicating that the second-order modal galloping is more likely to occur.

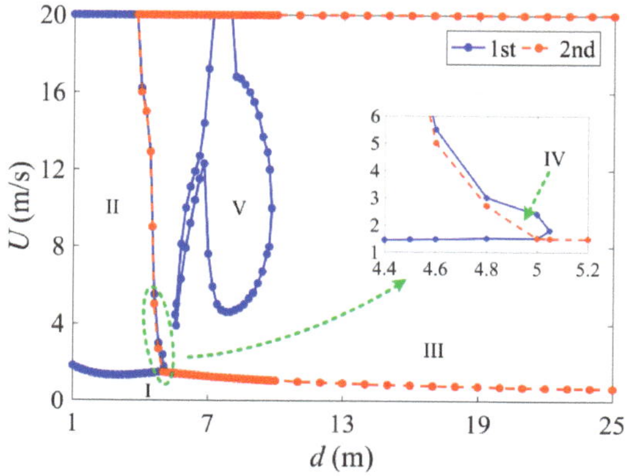

Figure 2. Critical conditions and galloping region division of first- and second-order modal galloping in the *U*-*d* parameter plane.

According to the different vibration patterns, the *U*-*d* parameter plane is divided into five regions. Region I is the stable region where no galloping occurs. Region II and Region III are both single-mode galloping regions, corresponding to single first-order modal and single second-order modal galloping, respectively. Regions IV and V have the same galloping mode, which arises from the coupling of first- and second-order modes.

3.2. Galloping Behavior

Now, the spatial galloping behavior of the five galloping regions is separately studied. The in-plane galloping of the iced conductor is a self-excited vibration, and the vibration frequency is basically the same as the natural frequency. Therefore, combined with the multimodal coupling mechanism of galloping, by comparison with the natural frequency, it can be numerically verified whether various order modes in the in-plane, out-of-plane, and torsional directions exhibit galloping and instability. The natural frequencies of the first two modes of in-plane, out-of-plane, and torsional motion at different inclinations are listed in Table 2.

Table 2. First- and second-order natural frequencies along the three directions under different sags.

Direction	Natural Frequency					
	$d = 4.3$ m		$d = 4.6$ m		$d = 6.8$ m	
	1st	2nd	1st	2nd	1st	2nd
In-plane	0.421	0.534	0.433	0.516	0.516	0.424
Out-of-plane	0.267	0.534	0.258	0.516	0.212	0.424
Torsional	0.377	0.744	0.377	0.744	0.377	0.744

In Region I, for any given nonzero initial displacement, the vibration amplitudes of the first two modes in all the directions rapidly decay to zero, as shown in Figure 3, and there is no modal vibration in this region. Keeping the sag constant, as the wind speed gradually increases, the vibrations begin to enter Region II, and the system starts to become unstable. When the wind speed increases to 14 m/s, the time history and spectral responses of the first two modes in the in-plane, out-of-plane, and torsional directions are shown in Figure 4. It can be seen that only the first-order mode in each direction is excited, and thus each direction exhibits a single first-order mode vibration (Figure 4a). According to Figure 4b–d, the frequencies of the first-order vibrations in all the directions are 0.402 Hz, which is essentially the same as the first natural frequency in the in-plane direction (see Table 2), indicating synchronous vibration characteristics, and the vibration of the first-order mode in the in-plane direction plays the dominant role.

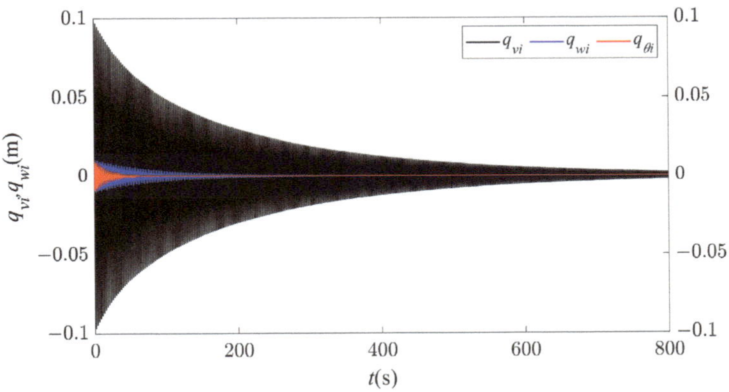

Figure 3. Time history responses of first- and second-order modal galloping along the three directions in Region I (d = 4.3 m, U = 1 m/s).

The galloping behavior within Region II is shown in Figure 5. As only the first-order modes are excited in the in-plane, out-of-plane, and torsional directions, the time history responses at 1/2 span for all the directions are stable periodic signals (Figure 5a). The galloping trajectory has only one peak in both in-plane and out-of-plane directions, exhibiting a standard first-order galloping profile (Figure 5b). When observing from the axial direction, the set of galloping trajectories in the entire span is elliptical (Figure 5c), and the galloping trajectory at 1/2 span is an inclined ellipse, showing repetitive motion on the same elliptical trajectory, as shown in Figure 5d. Therefore, the system vibrates along an inclined elliptical trajectory at all points within this region. Consequently, under the coupling effect of in-plane and out-of-plane motions, the spatial galloping trajectory has an approximately inclined elliptical spherical shape.

Keeping the sag constant and further increasing the wind speed, the galloping behavior gradually enters Region III. When the wind speed increases to 19 m/s, the time history and spectral responses of the first two modal profiles in the in-plane, out-of-plane, and torsional directions are shown in Figure 6. In this case, the first two modal profiles in the torsional direction tend to be in a stable state. However, in the in-plane and out-of-plane directions, only the second-order modal profile is excited, and the corresponding galloping frequency in each direction is 0.552 Hz (Figure 6b,c), which is basically the same as the second-order natural frequency in the in-plane direction (see Table 2).

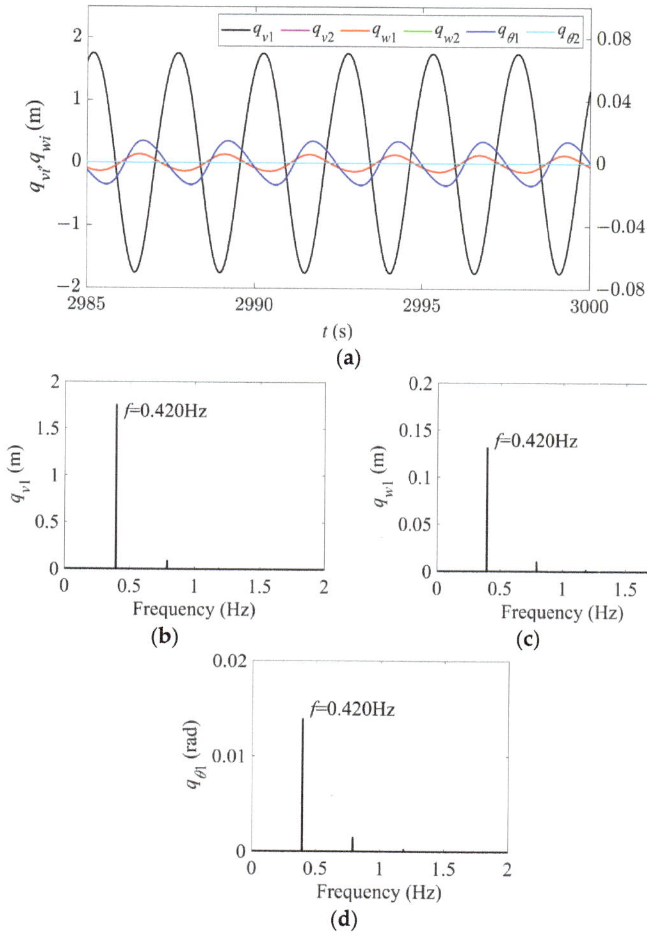

Figure 4. Time history and spectral responses of first- and second-order modal galloping along the three directions in Region II (d = 4.3 m, U = 14 m/s). (**a**) Time history responses of first- and second-order modal galloping along the three directions. (**b**) Spectral response of first-order modal galloping along the in-plane direction. (**c**) Spectral response of first-order modal galloping along the out-of-plane direction. (**d**) Spectral response of first-order modal galloping along the torsional direction.

The galloping behavior in Region III is shown in Figure 7. Since only the second-order modes are excited in both the in-plane and out-of-plane directions, the node is located at 1/2 span, and the time history responses at 1/4 span are stable periodic signals (Figure 7a). In this case, the galloping trajectories in both the in-plane and out-of-plane directions have two peaks, indicating a standard second-order galloping profile, as shown in Figure 7b. Observing from the axial direction, similar to Region II, the set of galloping trajectories for the entire iced conductor has an elliptical profile (Figure 7c). The galloping trajectory at the 1/4 span position is a tilted ellipse, and it moves cyclically on the same elliptical trajectory, as shown in Figure 7d. Therefore, it can be concluded that except for the node, all points on the conductor in Region III are vibrating along a tilted elliptical trajectory. Consequently, under the coupling of in-plane and out-of-plane motion, the spatial galloping trajectory also shows two approximately tilted elliptical spheres with a stationary node between them.

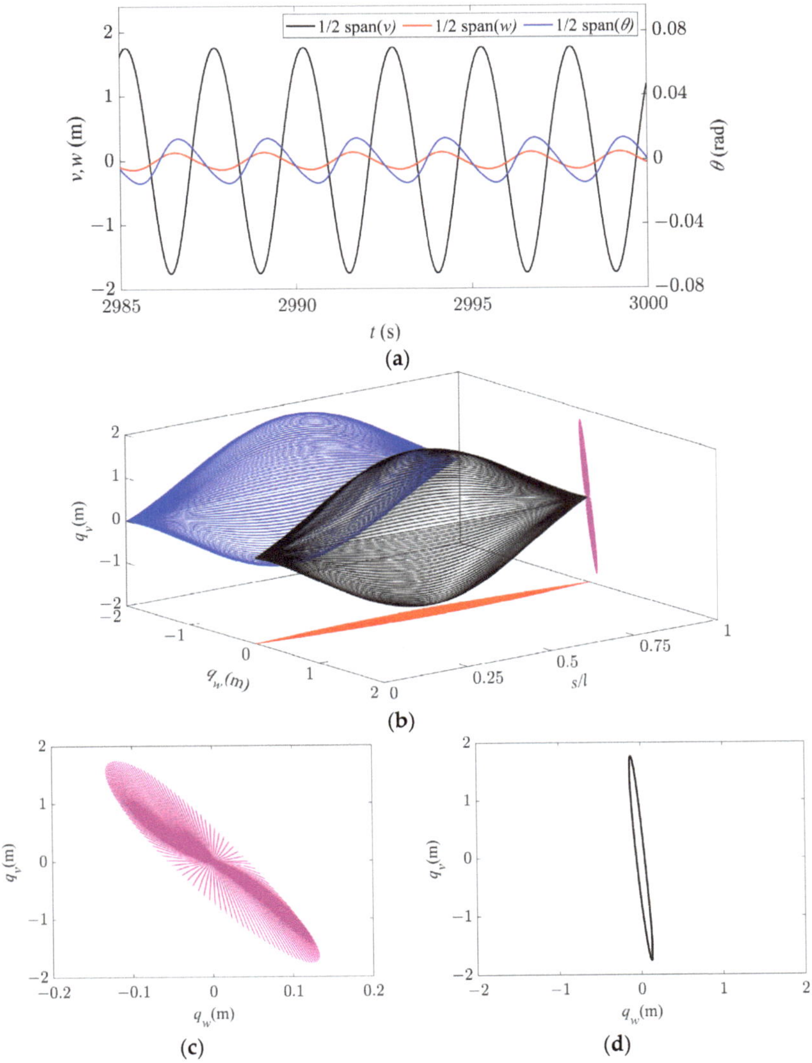

Figure 5. Galloping behavior in Region II (d = 4.3 m, U = 14 m/s). (**a**) Time history responses along the three directions at 1/2 span. (**b**) Spatial galloping trajectories along the three directions and their projections (t = 2997–3000 s). (**c**) Magnification of side view in (**b**). (**d**) Galloping orbit at 1/2 span.

Compared with Region II, in Region III, the galloping amplitude in the out-of-plane direction is larger, resulting in a greater tilt of the motion trajectory. It should be noted that in Region III, the phase difference between in-plane and out-of-plane motion is basically 180° (Figure 7a), which is larger than the difference in Region II. This results in a narrower short axis of the elliptical motion at different span positions, appearing to move along a diagonal line. Meanwhile, it can be seen that since single-mode galloping occurs in all the directions of Region II and III, except for the node, the iced conductor at each point follows its own elliptical trajectory of periodic motion.

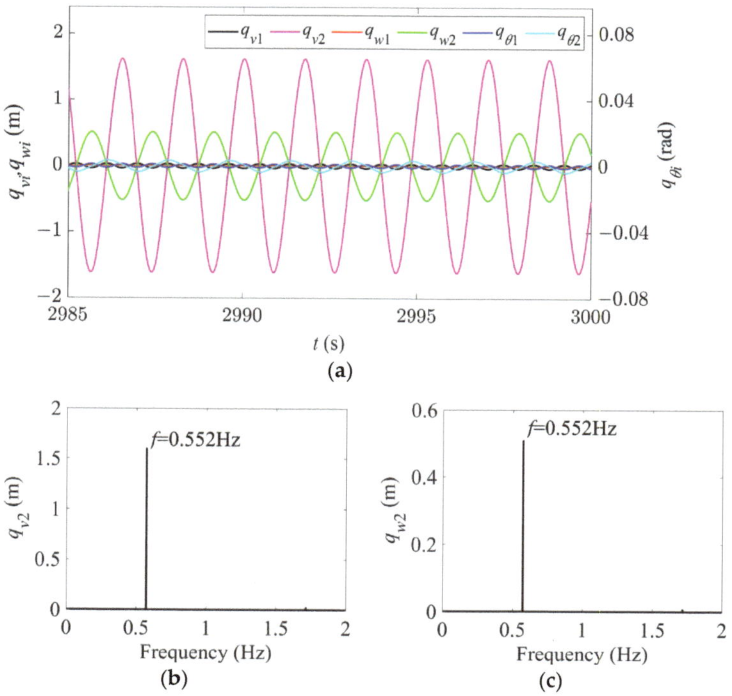

Figure 6. Time history and spectral responses of first- and second-order modal galloping along the three directions in Region III (d = 4.3 m, U = 19 m/s). (**a**) Time history responses of first- and second-order modal galloping along the three directions. (**b**) Spectral response of second-order modal galloping along the in-plane direction. (**c**) Spectral response of second-order modal galloping along the out-of-plane direction.

Until now, the galloping behavior in different regions with single-mode galloping has been analyzed. Next, the galloping behavior in Region IV and V, which both have coupled-mode galloping, are examined. It can be seen in Figure 2 that Region IV is the transition stage between the in-plane first-order and in-plane second-order modal galloping, with a narrow galloping area. When d = 4.6 m and U = 5.5 m/s, the time history and spectral response of the first two modes in the in-plane, out-of-plane, and torsional directions are shown in Figure 8. In this case, the first mode with a galloping frequency of 0.429 Hz (Figure 8b) is only excited in the in-plane direction, which is basically the same as the first in-plane natural frequency in Tables 1 and 2. The second mode is excited in both the in-plane and out-of-plane directions, and their galloping amplitudes are basically the same. Their galloping frequency is also 0.519 Hz (Figure 8c,d), which is basically the same as the second in-plane natural frequency in Table 1. The first two modes in the torsional direction are essentially stable.

The galloping behavior in Region IV is shown in Figure 9. Because the node of the second-order mode is located at 1/2 span, only the in-plane first-order mode component exists at 1/2 span (Figure 9a). At 1/4 span, due to the coupling between the two in-plane modes, the time history response is no longer a stable periodic signal and beats occur. The out-of-plane galloping only involves a single second-order mode and the response remains a stable periodic signal (Figure 9b). The galloping trajectory and its projections in one period of the entire span in different directions are shown in Figure 9c. Due to the coupling effect of the modes, there is no fixed peak and node in the in-plane galloping trajectory, but the overall galloping trajectory exhibits an anti-symmetric profile. In the out-of-plane direction, the galloping trajectory has two peaks and one node, exhibiting a standard second-order galloping profile.

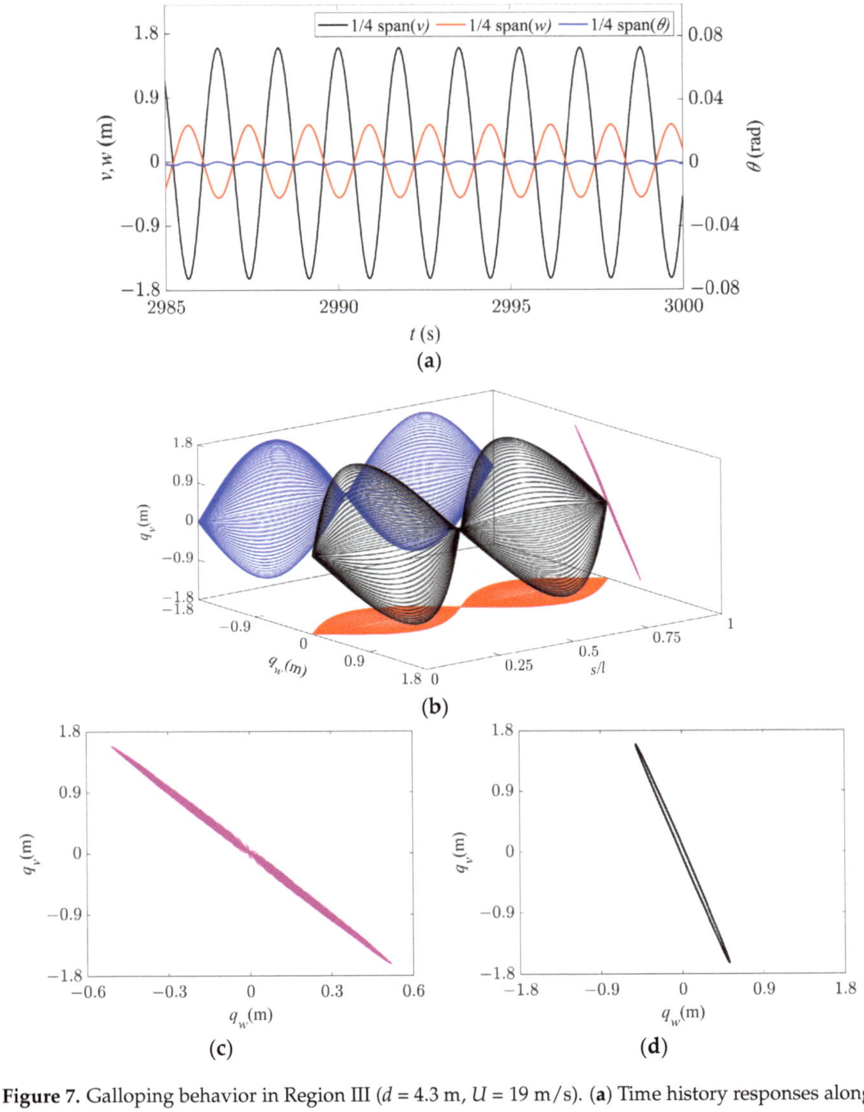

Figure 7. Galloping behavior in Region III (d = 4.3 m, U = 19 m/s). (**a**) Time history responses along the three directions at 1/4 span. (**b**) Spatial galloping trajectories along the three directions and their projections (t = 2996–2998 s). (**c**) Magnification of side view in (**b**). (**d**) Galloping orbit at 1/4 span.

Compared with the single-mode galloping regions, the trajectory in the axial direction is no longer a single ellipse, but rather a superposition of multiple inclined ellipses (Figure 9d). The conductor moves along an elliptical path at 1/2 span, but its galloping trajectory is no longer the same elliptical track, as shown in Figure 9e. However, due to the coupling between the modes, the conductor no longer moves along a single galloping trajectory at 1/4 span, as shown in Figure 9f,g. The conductor first moves in an inclined ellipse along the blue line, then in an inclined "8" profile along the black line, and then exhibits an inclined elliptical motion along the red line. Therefore, it can be inferred that in this case, the galloping trajectory exhibits a mixed motion pattern of ellipse and "8" profile due to the coupling effect of multiple modes. Hence, the spatial trajectory of galloping is anti-symmetrical and has no fixed profile.

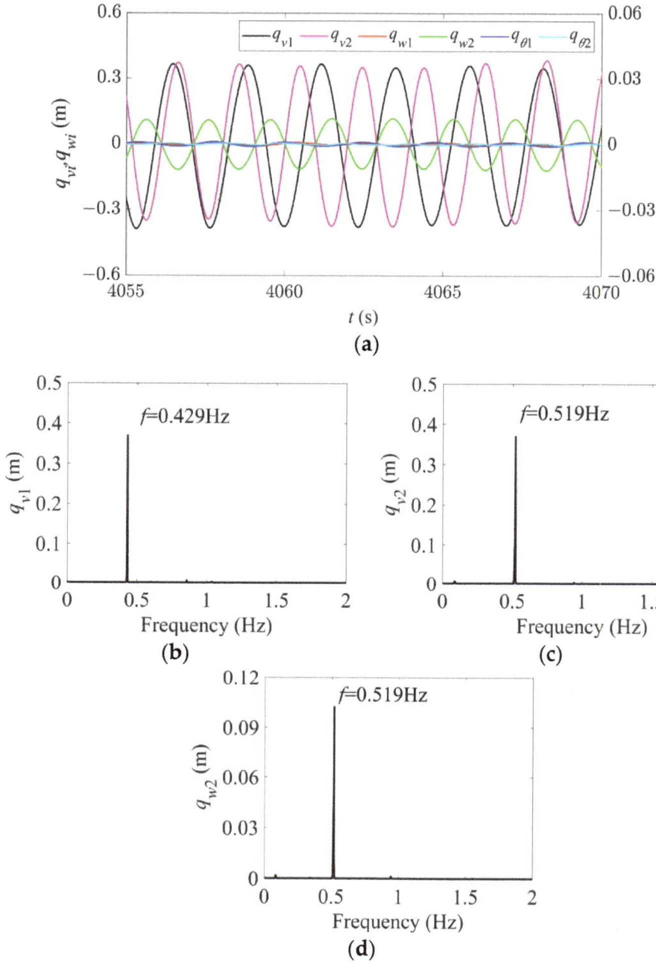

Figure 8. Time history and spectral responses of first- and second-order modal galloping along the three directions in Region IV (d = 4.6 m, U = 5.5 m/s). (**a**) Time history responses of first- and second-order modal galloping along the three directions. (**b**) Spectral response of first-order modal galloping along the in-plane direction. (**c**) Spectral response of second-order modal galloping along the in-plane direction. (**d**) Spectral response of second-order modal galloping along the out-of-plane direction.

As the parameters continue to change, the galloping enters Region V. When d = 6.8 m and U = 14 m/s, the time history and spectral responses of the first two modes in the in-plane, out-of-plane, and torsional directions are shown in Figure 10. It can be seen that all the modes except the second torsional mode are excited and there are dense sidebands at each galloping frequency. Among them, the galloping frequency of the first in-plane mode is the same as that of the second in-plane mode, both of which are dominated by 0.473 Hz (Figure 10b,c). Compared to Table 2, it is found that this frequency is the average of the first and second intrinsic frequencies of the in-plane mode, which may be due to the nonlinear coupling between the first and second in-plane modes, leading to frequent energy exchange between them. In addition, the dominant frequency of the first-order mode in the out-of-plane and torsional directions is 0.237 Hz (Figure 10d–f)). This frequency is half of the in-plane first-order mode frequency due to the synchronization among the three first-order modes [24]. The dominant frequency of the out-of-plane second-order mode is 0.473 Hz (Figure 10e), caused by the synchronization of the three second-order modes [24].

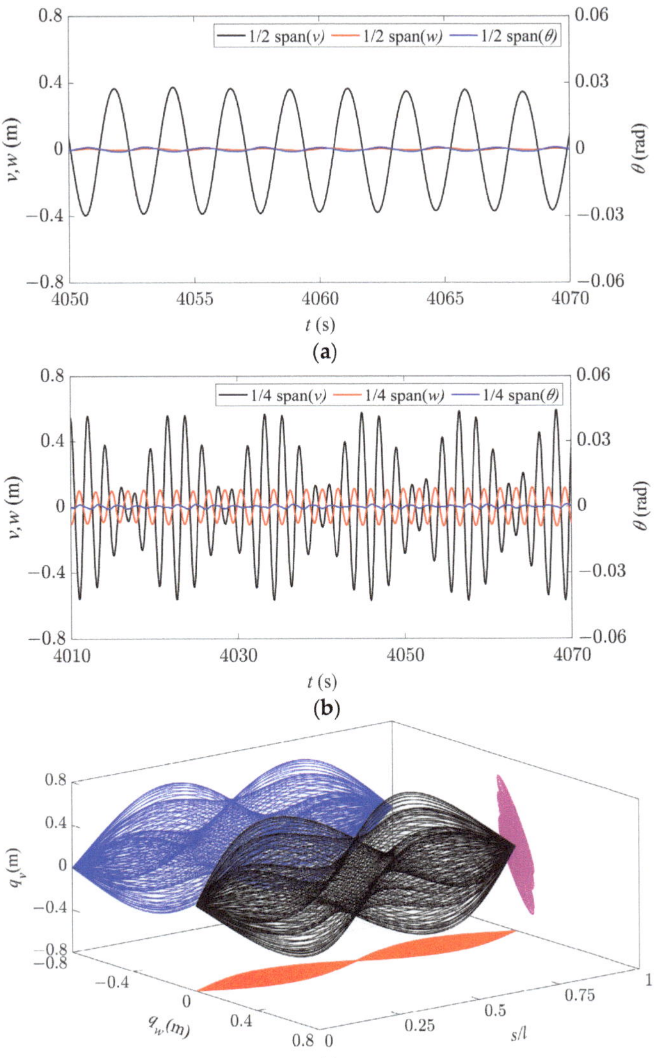

Figure 9. *Cont.*

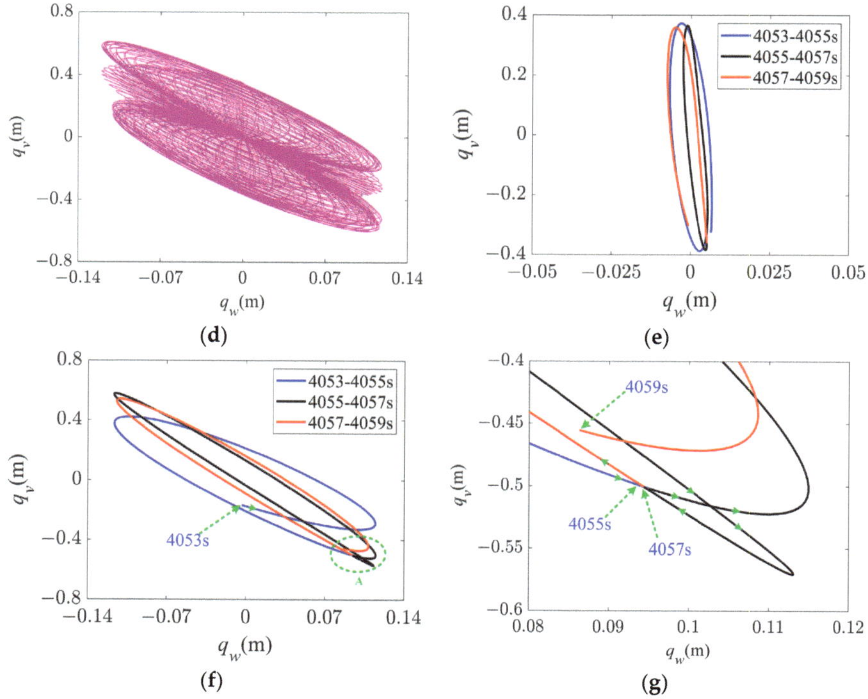

Figure 9. Galloping behavior in Region IV (d = 4.6 m, U = 5.5 m/s). (**a**) Time history responses along the three directions at 1/2 span. (**b**) Time history responses along the three directions at 1/4 span. (**c**) Spatial galloping trajectory along the three directions and the projections (t = 4052–4060 s). (**d**) Magnification of side view in (**c**). (**e**) Galloping orbit at 1/2 span. (**f**) Galloping orbit at 1/4 span. (**g**) Enlarged view of section A in (**f**).

The galloping behavior of the system in Region V is shown in Figure 11. Although there is a node of the second-order mode at 1/2 span, the time history responses are no longer stable periodic signals at this location due to strong coupling between the in-plane first- and second-order modes, resulting in beat phenomenon (Figure 11a). A similar phenomenon occurs at 1/4 span with the coupling between different order modes (Figure 11b). The galloping trajectory and its projection over a single cycle in different directions are shown in Figure 11c. Due to nonlinear coupling between the modes, there are no fixed peaks and nodes in both in-plane and out-of-plane directions, but the overall trajectory has an anti-symmetric profile.

From the axial direction, the galloping trajectories of the entire iced conductor in Region III are more complex than those in Region IV (Figure 11d). However, in this case, the galloping trajectories of each point on the iced conductor have an approximately horizontal "8" shape. Figure 11e shows the galloping trajectory at 1/2 span, where the conductor continuously moves in an "8" pattern without any transition, and the trajectory does not repeat at each cycle. The same is true for the galloping trajectory at 1/4 span (Figure 11f). In this case, the spatial trajectory is also roughly anti-symmetric but without a fixed profile. Moreover, compared to the other galloping regions, it has a wider spatial range, especially in the out-of-plane direction, which can exacerbate wear and tear of the conductor, leading to strand and line breakage as well as damage to the insulator strings and their connections at both ends of the conductor. Therefore, it is necessary to avoid the routing of transmission lines in such situations.

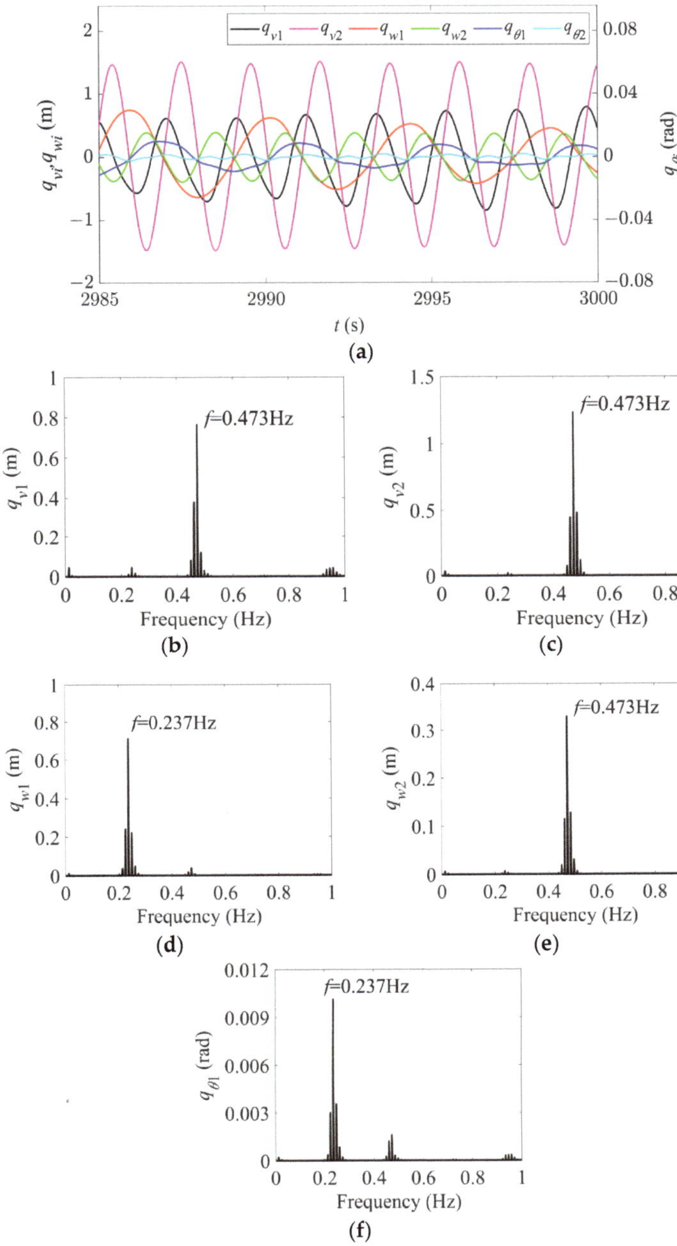

Figure 10. Time history and spectral responses of first- and second-order modal galloping along the three directions in region V (d = 6.8 m, U = 14 m/s). (**a**) Time history responses of first- and second-order modal galloping along the three directions. (**b**) Spectral response of first-order modal galloping along the in-plane direction. (**c**) Spectral response of second-order modal galloping along the in-plane direction. (**d**) Spectral response of first-order modal galloping along the out-of-plane direction. (**e**) Spectral response of second-order modal galloping along the out-of-plane direction. (**f**) Spectral response of first-order modal galloping along the torsional direction.

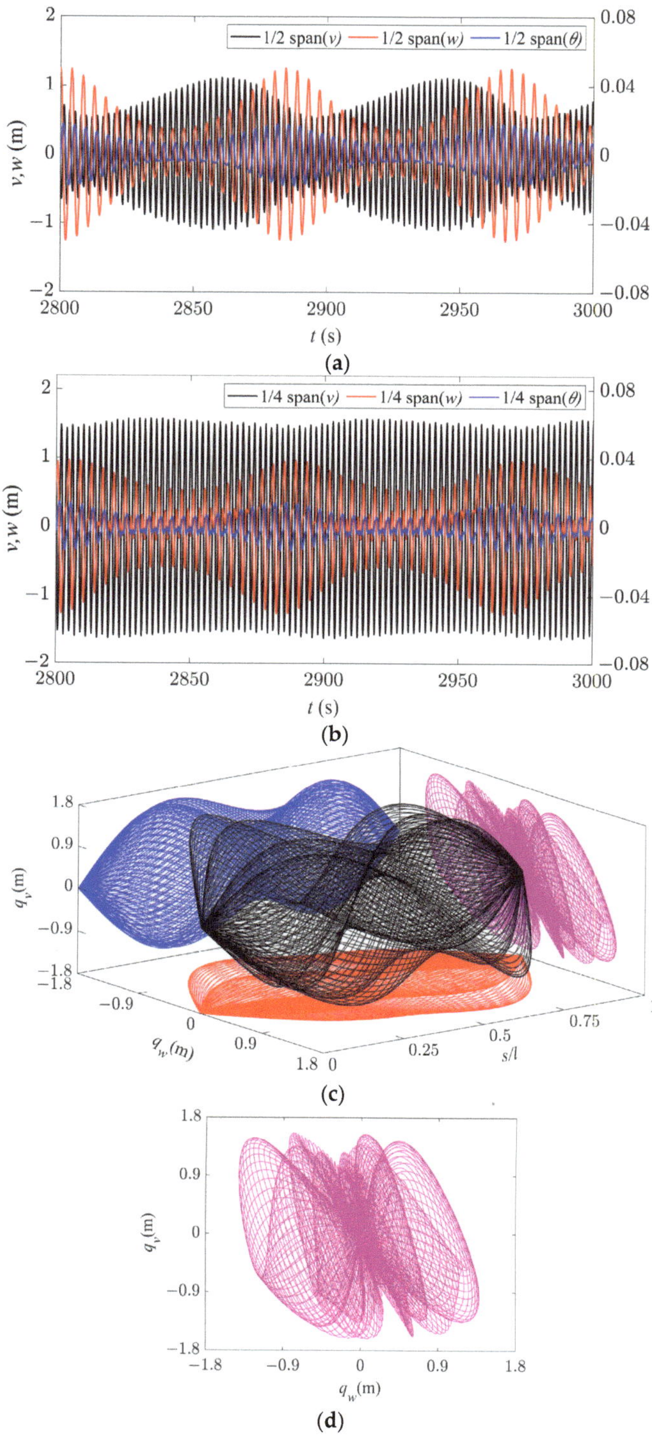

Figure 11. *Cont.*

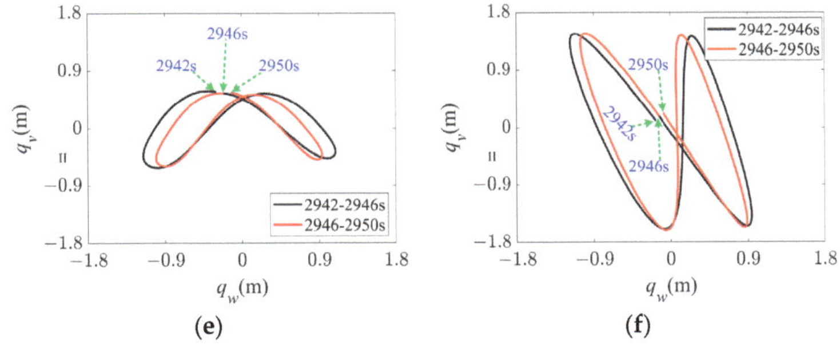

Figure 11. Galloping behavior in Region V (d = 6.8 m, U = 14 m/s). (**a**) Time history responses along the three directions at 1/2 span. (**b**) Time history responses along the three directions at 1/4 span. (**c**) Spatial galloping trajectories along the three directions and the projections (t = 2942–2970 s). (**d**) Magnification of side view in (**c**). (**e**) Galloping orbit at 1/2 span. (**f**) Galloping orbit at 1/4 span.

In summary, from the axial direction, as the wind speed increases, the galloping trajectories gradually change from an elliptical shape to an approximately "8" shape. The results are basically consistent with those obtained by the finite element method [13,15] and experiments [26].

4. Conclusions

In this study, a simplified dynamic model was established for analyzing the galloping of iced quad-bundled conductors in the in-plane, out-of-plane, and torsional directions. Using numerical analysis in the wind speed–span parameter space, the critical conditions for the first- and second-order modal galloping were investigated, and the galloping modes and their variations in different parameter space were analyzed. The parameter space was divided into five regions based on different galloping patterns, and the spatial nonlinear behavior of system galloping in different regions was discussed under the existence of the multimodal coupling mechanism. The main results of this study are summarized as follows:

(1) As the wind speed and span increased, the galloping process of the system underwent the following sequential stages: first-order mode, coupling of first- and second-order modes, second-order mode, coupling of first- and second-order modes, and second-order mode. Further, first-order mode galloping was excited twice. Moreover, the area of the second-order mode galloping was significantly larger than that of the first-order mode galloping, making it easier to occur in practical situations.

(2) When the system was in a single-modal galloping state in all the directions, it exhibited a stable periodic motion. Under the coupled effect of in-plane and out-of-plane motion, except at the nodes, all the points on the iced conductor moved along a continuous, overlapping, and inclined elliptical orbit. When the system was in a single first-order mode galloping state, the spatial trajectory of the galloping motion was an approximately inclined elliptical sphere. When the system was in a single second-order mode galloping state, the spatial trajectory of galloping was approximately two inclined elliptical spheres with an immobile node in the center.

(3) When the system was in a coupled-mode galloping state in certain directions, the spatial trajectory of galloping was basically anti-symmetric but had no fixed profile. During the first excitation stage of the first-order mode, the iced conductor moved along a continuous, inclined, elliptical orbit at the 1/2 span position, while at the other points, there was a mixed motion pattern of inclined elliptical and "8" profiles. During the second excitation stage of the first-order mode, all the points on the iced conductor vibrated along a continuous, approximately horizontal "8" profile.

Author Contributions: Conceptualization, F.C. and K.Z.; methodology, F.C., P.L. and H.W.; software, K.Z., F.C. and P.L.; validation, F.C., K.Z., P.L. and H.W.; formal analysis, F.C. and K.Z.; investigation, P.L. and H.W.; resources, F.C.; data curation, K.Z.; writing—original draft preparation, F.C.; writing—review and editing, K.Z., P.L. and H.W.; visualization, F.C.; supervision, K.Z., P.L. and H.W.; project administration, K.Z.; funding acquisition, P.L. and H.W. All authors have read and agreed to the published version of the manuscript.

Funding: This research received no external funding.

Institutional Review Board Statement: Not applicable.

Informed Consent Statement: Not applicable.

Data Availability Statement: Where data is unavailable due to privacy restrictions.

Acknowledgments: The authors gratefully acknowledge the support of the Foundation Research (Free Exploration) Youth Program in Shanxi (20210302124385), Scientific and Technological Innovation Programs of Higher Education Institutions in Shanxi (2021L069), National Natural Science Foundation of China (12201450 and 12102290), and Major Scientific and Technological Special Project in Shanxi Province (202101120401007).

Conflicts of Interest: The authors declare no conflict of interest.

References

1. EPRI. *Conductor Reference Book: Wind-Induced Conductor Motion*; Electric Power Research Institute: Palo Alto, CA, USA, 1979; pp. 4–5.
2. Hartog, J.P.D. Conductor vibration due to sleet. *Trans. Am. Inst. Electr. Eng.* **1932**, *51*, 1074–1076. [CrossRef]
3. Nigol, O.; Buchan, P.G. Conductor galloping part II—Torsional mechanism. *IEEE Trans. Power Appar. Syst.* **1981**, *100*, 708–720. [CrossRef]
4. Yu, P.; Shah, A.H.; Popplewell, N. Inertially coupled galloping of iced conductors. *J. Appl. Mech.* **1992**, *59*, 140–145. [CrossRef]
5. You, C. Stability mechanism of conductor galloping and its application on transmission line. *Electr. Equip.* **2004**, *5*, 13–17.
6. Zhu, K.; You, C.; Zhao, Y. Study and control on galloping of transmission lines. *Electr. Power Constr.* **2004**, *25*, 18–21.
7. Jones, K. Coupled vertical and horizontal galloping. *J. Eng. Mech.* **1992**, *118*, 92–107. [CrossRef]
8. Blevins, R.; Iwan, W. The galloping response of a two-degree-of-freedom system. *J. Appl. Mech.* **1974**, *41*, 1113–1118. [CrossRef]
9. Liu, B.; Zhu, K.; Sun, X.; Huo, B.; Liu, X. A contrast on conductor galloping amplitude calculated by three mathematical models with different DOFs. *Shock Vib.* **2014**, *2014*, 781304. [CrossRef]
10. Chen, X.; Wu, Y. Explicit closed-form solutions of the initiation conditions for 3DOF galloping or flutter. *J. Wind Eng. Ind. Aerodyn.* **2021**, *219*, 104787. [CrossRef]
11. Matsumiya, H.; Yagi, T.; Macdonald, J.H.G. Effects of aerodynamic coupling and nonlinear behaviour on galloping of ice-accreted conductors. *J. Fluids Struct.* **2021**, *106*, 103366. [CrossRef]
12. Diana, G.; Manenti, A.; Melzi, S. Energy method to compute the maximum amplitudes of oscillation due to galloping of iced bundled conductors. *IEEE Trans. Power Deliv.* **2021**, *36*, 2804–2813. [CrossRef]
13. Zhou, L.; Yan, B.; Zhang, L.; Zhou, S. Study on galloping behavior of iced eight bundle conductor transmission lines. *J. Sound Vib.* **2016**, *362*, 85–110. [CrossRef]
14. Shunli, D.; Mengqi, C.; Bowen, T.; Junhao, L.; Linshu, Z.; Chuan, W.; Hanjie, H.; Jun, L. Numerical simulation of galloping characteristics of multi-span iced eight-bundle conductors. *Front. Energy Res.* **2022**, *9*, 888327. [CrossRef]
15. Tian, B.; Cai, M.; Zhou, L.; Huang, H.; Ding, S.; Liang, J.; Hu, M. Numerical simulation of galloping characteristics of multi-span iced eight-bundle conductors tower line system. *Buildings* **2022**, *12*, 1893. [CrossRef]
16. Chen, J.; Sun, G.; Guo, X.; Peng, Y. Galloping behaviors of ice-coated conductors under steady, unsteady and stochastic wind fields. *Cold Reg. Sci. Technol.* **2022**, *200*, 103583. [CrossRef]
17. Liu, X.; Huo, B. Nonlinear vibration and multimodal interaction analysis of conductor with thin ice accretions. *Int. J. Appl. Mech.* **2015**, *07*, 1540007. [CrossRef]
18. Huo, B.; Li, X.; Yang, S. Galloping of Iced Conductors Considering Multi-Torsional Modes and Experimental Validation on a Continuous Model. *IEEE Trans. Power Deliv.* **2022**, *37*, 3016–3026. [CrossRef]
19. Luongo, A.; Zulli, D.; Piccardo, G. Analytical and numerical approaches to nonlinear galloping of internally resonant suspended cables. *J. Sound Vib.* **2008**, *315*, 375–393. [CrossRef]
20. Nayfeh, A.; Arafat, H.; Chin, C.; Lacarbonara, W. Multimode interactions in suspended cables. *J. Vib. Control* **2002**, *8*, 337–387. [CrossRef]
21. Irvine, H.; Caughey, T. The Linear theory of free vibrations of a suspended cable. *Proc. R. Soc. A Math. Phys. Eng. Sci.* **1974**, *341*, 299–315.
22. Luongo, A.; Piccardo, G. Linear instability mechanisms for coupled translational galloping. *J. Sound Vib.* **2005**, *288*, 1027–1047. [CrossRef]

23. Barbieri, R.; Barbieri, N.; Junior, O. Dynamical analysis of transmission line cables. Part 3—Nonlinear theory. *Mech. Syst. Signal Process.* **2008**, *22*, 992–1007. [CrossRef]
24. Cui, F.; Zhang, S.; Huo, B.; Liu, X.; Zhou, A. Analysis of stability and modal interaction for multi-modal coupling galloping of iced conductors. *Commun. Nonlinear Sci. Numer. Simul.* **2021**, *2021*, 105910. [CrossRef]
25. Lou, W.; Yang, L.; Huang, M.F.; Yang, X. Two-parameter bifurcation and stability analysis for nonlinear galloping of iced conductors. *J. Eng. Mech.* **2014**, *140*, 04014081. [CrossRef]
26. Chen, Z.; Cai, W.; Su, J.; Nan, B.; Zeng, C.; Su, N. Aerodynamic force and aeroelastic response characteristics analyses for the galloping of ice-covered four-split transmission lines in oblique flows. *Sustainability* **2022**, *14*, 16650. [CrossRef]

sensors

MDPI

Article

Robust Interval Prediction of Intermittent Demand for Spare Parts Based on Tensor Optimization

Kairong Hong [1], Yingying Ren [1], Fengyuan Li [1], Wentao Mao [2,*] and Xiang Gao [2]

1 China Railway Tunnel Group, Zhengzhou 450001, China
2 School of Computer and Information Engineering, Henan Normal University, Xinxiang 453007, China
* Correspondence: maowt@htu.edu.cn; Tel.: +86-150-3730-1821

Abstract: Demand for spare parts, which is triggered by element failure, project schedule and reliability demand, etc., is a kind of sensing data to the aftermarket service of large manufacturing enterprises. Prediction of the demand for spare parts plays a crucial role in inventory management and lifecycle quality management for the aftermarket service of large-scale manufacturing enterprises. In real-life applications, however, demand for spare parts occurs randomly and fluctuates greatly, and the demand sequence shows obvious intermittent distribution characteristics. Additionally, due to factors such as reporting mistakes made by personnel or environmental changes, the actual data of the demand for spare parts are prone to abnormal variations. It is thus hard to capture the evolutional pattern of the demand for spare parts by traditional time series forecasting methods. The reliability of prediction results is also reduced. To address these concerns, this paper proposes a tensor optimization-based robust interval prediction method of intermittent time series for the aftersales demand for spare parts. First, using the advantages of tensor decomposition to effectively mine intrinsic information from raw data, a sequence-smoothing network based on tensor decomposition and a stacked autoencoder is proposed. Tucker decomposition is applied to the hidden features of the encoder, and the obtained core tensor is reconstructed through the decoder, thus allowing us to smooth outliers in the original demand sequence. An alternating optimization algorithm is further designed to find the optimal sequence feature representation and tensor decomposition factors for the extraction of the evolutionary trend of the intermittent series. Second, an adaptive interval prediction algorithm with a dynamic update mechanism is designed to obtain point prediction values and prediction intervals for the demand sequence, thereby improving the reliability of the forecast. The proposed method is validated using the actual aftersales data from a large engineering manufacturing enterprise in China. The experimental results demonstrate that, compared with typical time series prediction methods, the proposed method can effectively grab the evolutionary trend of various intermittent series and improve the accuracy of predictions made with small-sample intermittent series. Moreover, the proposed method provides a reliable elastic prediction interval when distortion occurs in the prediction results, offering a new solution for intelligent planning decisions related to spare parts in practical maintenance.

Keywords: demand prediction; intermittent time series; tensor decomposition; interval prediction; time series forecasting

Citation: Hong, K.; Ren, Y.; Li, F.; Mao, W.; Gao, X. Robust Interval Prediction of Intermittent Demand for Spare Parts Based on Tensor Optimization. *Sensors* **2023**, *23*, 7182. https://doi.org/10.3390/s23167182

Academic Editors: Yongbo Li, Bing Li and Khandaker Noman

Received: 8 July 2023
Revised: 1 August 2023
Accepted: 4 August 2023
Published: 15 August 2023

1. Introduction

In complicated equipment manufacturing enterprises such as shield tunneling, rail transportation, and wind energy, the cost of maintaining an inventory of spare parts generally accounts for more than 60% of inventory costs [1]. Qualified inventory optimization [2] and flexible scheduling of parts [3] are critical to improve the efficiency of aftermarket service in lifecycle product management [4]. Due to various uncertain factors, such as element failure, project schedule, safe inventory level, etc., there will be an inventory shortage of spare parts in warehouses which, in turn, triggers the demand for spare parts. The demand for spare parts can then serve as a kind of sensing data to monitor the maintenance efficiency and evaluate the aftermarket service quality. Accurate prediction of the demand for spare parts plays a supporting role in intelligent inventory optimization. However, in practical operations, parts planning is often linked to new online projects or associated with the unavailability of spare parts, resulting in sporadic demand for spare parts. The data distribution of the demand for spare parts therefore has intermittent characteristics. Precisely predicting the demand of intermittent time series is challenging in spare parts management for manufacturing enterprises.

Due to the intermittent characteristics of these demand data, predicting demand relies heavily on time series prediction. Currently, time series prediction methods can be divided into three categories [5]: (1) statistical methods (e.g., exponential smoothing [6] and moving average [7]); (2) machine learning methods (e.g., support vector regression (SVR) [8], random forests (RF) [9], and LightGBM (light gradient boosting machine) [10,11]); and (3) deep learning methods (e.g., recurrent neural networks (RNN) [12] and long short-term memory (LSTM) [13]). These methods are often applicable to time series with strong periodicity and apparent trends. However, there are some challenges involved in effectively extracting evolutionary patterns from time series with strong randomness, poor continuity, and, especially, small sample sizes, leading to low prediction accuracy for intermittent series. To solve this problem, Croston [14] improved the exponential smoothing algorithm by decomposing intermittent time series into zero-interval and demand sequences. The exponential smoothing algorithm was then applied to each sequence separately, and the results were weighted to improve prediction performance. Syntetos et al. [15] further improved the Croston algorithm and designed the Syntetos–Boylan approximation (SBA) method. The SBA method introduced a bias term $(1 - \alpha/2)$ to mitigate the uncertainty of intermittent distributions. In addition, some studies proposed different metrics, such as the average demand interval (ADI) and the square of the coefficient of variation (CV2) [16], to explore intermittent characteristics to better extract intrinsic information from demand sequences [17]. Another typical approach is to perform hierarchical clustering on demand sequences [18], which divides the original sequences with weak overall patterns into multiple clusters with more significant patterns. Then, different regression algorithms can be applied to the clusters for prediction. Shi et al. [19] proposed the block Hankel tensor-autoregressive integrated moving average (BHT-ARIMA) model, which uses tensor decomposition [20] to extract the intrinsic correlations among multidimensional small-sample sequences. Although the aforementioned methods have achieved certain results, they still have some limitations. First, they are mostly based on the assumption that all sequence demands have predictive value and disregard the interference of abnormal values. In fact, in actual business, due to special events such as natural disasters, emergencies, market fluctuations, and other factors, some spare parts have abnormal demand patterns, which require manual analysis for recognition. Second, demand prediction results are random and unreliable. Operations still want to obtain trustworthy results regarding demand prediction, but the existing methods fail to make valid decisions when the prediction results are distorted.

There is also a similar concept, i.e., abnormal demand forecasting, which refers to the process of forecasting and analyzing abnormal demands that may occur in the future. Guo et al. [21] utilized the passenger flow characteristics extracted by SVR into LSTM to predict abnormal passenger flow. Liu et al. [22] combined statistical learning and linear regression to build a model of the relationship between price discounts and demand for medical consumables. Li et al. [23] proposed a multi-scale radial basis function (MSRBF) network to predict subway passenger flow under special events. Nguyen et al. [24] combined LSTM with SVM to detect outliers in a demand sequence, and then used an LSTM neural network to predict the occurrence of outliers. Li et al. [25] proposed a combination model of time series and regression analysis, plus dummy variables, to predict special factors related to passenger flow. These methods aim to predict the occurrence of abnormal but meaningful events in many fields. However, in the situation described by this paper, abnormal demands are commonly found due to reporting mistakes made by personnel or environmental changes. Such abnormal demands are harmful to the prediction of the intermittent time series, since the evolutionary pattern of the demand for spare parts is disturbed. There is no value in predicting abnormal demands if such demands have no predictability. This paper is, therefore, solely devoted to predicting normal demands against the disturbances induced by abnormal demands, instead of predicting the abnormal demands.

To solve the problems mentioned above, this paper employs a tensor decomposition technique to smooth the abnormal demands in demand sequences. Tensor Tucker decomposition decomposes a tensor into a set of factor matrices and one core tensor that can describe the intrinsic information from raw data. With the decomposed forms, tensor Tucker decomposition brings the potential to extract the evolutionary trend from intermittent time series. The concept of a prediction interval is further introduced to tackle the problem of high uncertainty and less reliability in the prediction results. The methodology is as follows: First, a tensor autoencoder network is constructed, which performs tensor Tucker decomposition on the output features extracted from the hidden layers of the autoencoder. Second, the core tensor is decoded and reconstructed with an alternating optimization strategy to obtain the optimal feature representation of intermittent time series. Third, an adaptive prediction interval (API) algorithm is developed with a dynamical update mechanism to obtain point prediction values and prediction intervals, thus improving the reliability of the prediction results. Finally, the performance of the proposed method is validated using a set of real-life aftersales data from a large engineering manufacturing enterprise from China.

The contributions of this paper can be summarized as follows:

(1) An intermittent series smoothing algorithm is proposed. By integrating tensor Tucker decomposition into a stacked autoencoder network with an alternately optimizing scheme, the proposed algorithm extracts the evolutionary trend of the intermittent time series under irregular noise interference. Compared with existing time series anomaly detection methods, the proposed algorithm does not require any pre-fixed detection thresholds, and can adaptively identify outliers in the series. It is highly suitable for smoothing the outliers in the intermittent time series. Moreover, the proposed algorithm is universally applicable, e.g., the stacked autoencoder can be easily replaced by other deep models;

(2) An adaptive prediction interval algorithm is designed. Different from the existing point prediction methods using a deterministic prediction model, this algorithm incorporates the prediction intervals with the point prediction, which can match the intermittent characteristics of demand sequence for spare parts. This algorithm provides an effective solution to address the uncertainty in the demand for spare parts. According to the literature survey, there is no related research about interval prediction, specifically for demand prediction.

2. Background

2.1. Multi-Way Delay Embedding Transform

The multi-way delay embedding transform (MDT) [26] is capable of embedding low-order data into a high-dimensional space, which can be used to construct Hankel matrices or block Hankel tensors. The tensor obtained by MDT possesses the characteristics of low rank, which can smooth the original data to make them easier to train. $v = (v_1, \ldots, v_L)^T \in R^L$ denotes a vector that is transformed into a Hankel matrix with a delay of τ through MDT. This process is referred to Hankelization of the vector. The transform process is shown in Equation (1).

$$\mathcal{H}_\tau(vs.) := \begin{pmatrix} v_1 & v_2 & \cdots & v_{L-\tau+1} \\ v_2 & v_3 & \cdots & v_{L-\tau+2} \\ \vdots & \vdots & \ddots & \vdots \\ v_\tau & v_{\tau+1} & \cdots & v_L \end{pmatrix} \in R^{\tau\times(L-\tau+1)} \frac{-b \pm \sqrt{b^2-4ac}}{2a} \quad (1)$$

First, the duplication matrix $S \in \{0,1\}^{\tau \times (L-\tau+1) \times L}$ is constructed with a delay of τ, as shown in Equation (2).

$$S^T = \begin{pmatrix} I_\tau & & & \\ & I_\tau & & \\ & & \ddots & \\ & & & I_\tau \end{pmatrix}_{\tau \times \tau}^{T} \quad (2)$$

The vector v is transformed into a Hankel matrix denoted by $\mathcal{H}_\tau(vs.)$. The duplication matrix S is essentially a linear transformation. The vectorization expansion is shown in Equation (3):

$$vec(\mathcal{H}_\tau(vs.)) = Sv, \quad Sv \in R^{\tau \times (L-\tau+1)} \quad (3)$$

where $vec(\cdot)$ expands the matrix along the column direction. The Hankel matrix through delay embedding can be shown in Equation (4).

$$\begin{aligned} &\mathcal{H}_\tau(vs.) = fold_{(L,\tau)}(Sv) := v_H, \\ &fold_{(L,\tau)} : R^{\tau \times (L-\tau+1)} \rightarrow R^{\tau \times (L-\tau+1)} \end{aligned} \quad (4)$$

where $fold_{(L,\tau)}$ folds a vector into a matrix.

The inverse transform of multi-way delay embedding for vectors can convert the data from a high-dimensional space to a lower-dimensional target space. This can be calculated using Equations (5) and (6).

$$\mathcal{H}_\tau^{-1}(V_H) = S^\dagger vec(V_H) \quad (5)$$

$$S^\dagger := \left(S^T S\right)^{-1} S^T \quad (6)$$

where † is the Moore–Penrose inverse matrix [27].

2.2. Tensor Decomposition

Tensor decomposition aims to decompose high-order tensor data into low-rank matrices or vectors [28]. It is commonly applied in data compression, dimensionality reduction, feature extraction, etc. Tensor Tucker decomposition decomposes an Nth-order tensor $\chi \in \mathbb{R}^{I_1 \times I_2 \times \cdots \times I_N}$ into the product of a core tensor $\varsigma_t \in \mathbb{R}^{J_1 \times J_2 \times \cdots \times J_N}$ and N factor matrices $U^{(n)} \in \mathbb{R}^{I_n \times J_n}$, as shown in Equation (7). The factor matrices obtained from Tucker decomposition represent the principal components of the tensor's modular expansion, while the core tensor captures the correlations between these components.

$$\chi = \varsigma \times {}_1U^{(1)} \times {}_2U^{(2)} \cdots \times {}_NU^{(N)} \quad (7)$$

where $\varsigma \times {}_1U^{(n)}$ is the n-mode product of the modular (n) expansion of tensor S and matrix $U^{(n)} \in \mathbb{R}^{I_n \times J_n}$:

$$\begin{aligned} &\left[\varsigma \times U^{(n)}\right]_{j_1 \cdots j_{n-1} i_n j_{n+1} \cdots j_N} = \sum_{j_n=1}^{J_n} g_{j_1 \cdots j_{n-1} i_n j_{n+1} \cdots j_N} u_{i_n j_n} \\ &\varsigma \times U^{(n)} \in \mathbb{R}^{J_1 \times J_2 \times \cdots \times J_N} \end{aligned} \quad (8)$$

With the above equation, one data point in the tensor can be expanded by Equation (9) as:

$$x_{i_1 i_2 \cdots i_N} = \sum_{j_1, j_2, \cdots, j_N} g_{j_1 \cdots j_N} u^{(1)}_{i_1 j_1} u^{(2)}_{i_2 j_2} \cdots u^{(3)}_{i_3 j_3} \quad (9)$$

Figure 1 illustrates a three-order tensor decomposed using Tucker decomposition, resulting in the product of a smaller core tensor and three factor matrices.

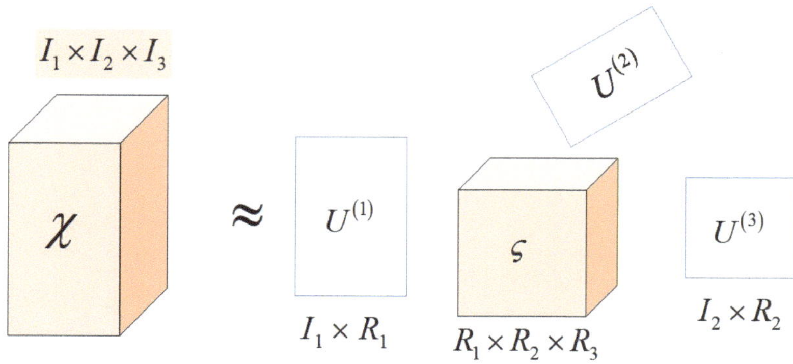

Figure 1. Illustration of three-order tensor Tucker decomposition.

2.3. LightGBM

LightGBM, developed by Microsoft engineers [10], is a gradient-boosting framework based on decision trees. By utilizing feature parallelism during its training process, LightGBM assigns discrete features to multiple bins and constructs decision trees using histogram algorithms, which provides a quick and efficient training mechanism for LightGBM. Additionally, LightGBM employs a sparse feature algorithm to significantly reduce memory consumption, which is suitable for training with extremely large-scale data. Moreover, LightGBM utilizes a leaf-wise tree growth strategy, which leads to faster convergence and higher accuracy compared to traditional level-wise strategies.

3. Proposed Method

In this section, a tensor optimization-based robust interval prediction method of intermittent time series is presented. This method consists of two parts: (1) a sequence smoothing network based on tensor decomposition and a stacked autoencoder, which aims to smooth anomalous demand values in the original sequence and to extract the evolutionary trend of demand as well; and (2) an adaptive prediction interval algorithm, which aims to construct a reliable prediction interval to avoid the oversupply risk or inventory shortage caused by inaccurate predictions. In this method, the role of tensor Tucker decomposition is: (1) extracting core tensors from the demand sequence for representing the evolutionary trend; and (2) smoothing the outliers in the sequence. The negative interference of anomalous demands can then be effectively reduced under an unsupervised mode. Moreover, training a LightGBM model with core tensors can better mine the trend information from the demand sequence. Tensor Tucker decomposition is believed suitable for intermittent time series forecasting with small samples.

The flowchart of the proposed method is shown in Figure 2. First, the hidden features are extracted from the original data using a stacked autoencoder, then tensor Tucker decomposition is performed on the hidden features to obtain the core tensors. An alternating optimization scheme is designed to obtain the optimal core tensors by alternately updating the autoencoder parameters and tensor factor matrices. Second, an adaptive interval prediction algorithm is constructed. The interval is calculated using the predicted values and prediction residuals from LightGBM estimators. Finally, a dynamic update mechanism is used to adjust the width of prediction interval. The detailed implementation will be presented as follows.

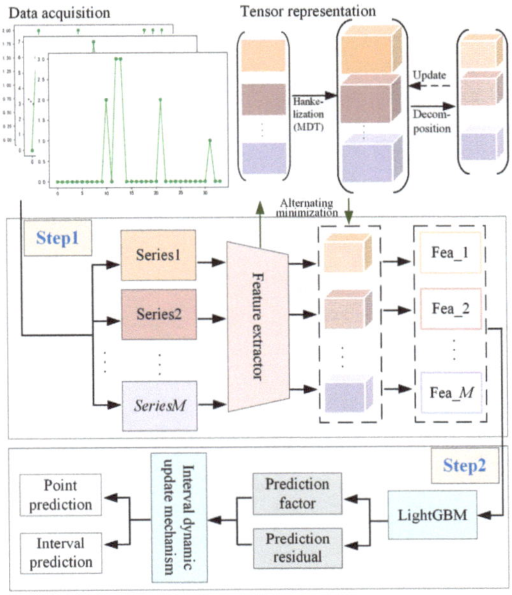

Figure 2. Flowchart of the proposed method.

3.1. Sequence Smoothing Network

Denote the demand sequence by $X = \{x_1, x_2, ..., x_m\} \in R^{m \times n}$, where m indicates the sample number and n represents the time dimension. The sequence can be encoded and mapped into a hidden layer, as shown in Equation (10).

$$h = f(Wx + b) \tag{10}$$

where h represents the hidden layer features, $f(\cdot)$ is the activation function of the encoding layer, and $W \in \mathbb{R}^r$ and $b \in \mathbb{R}^r$ are the weight matrix and bias vector of the encoding layer, respectively. The obtained feature set is denoted by $H = \{h_1, h_2, \cdots, h_m\} \in R^{m \times l}$, where l represents the dimension of the deep features.

In order to adequately represent the temporal information between samples in the feature set H, an MDT with the operations of multi-linear duplication and multi-way folding is employed to transform the original sequence to a three-order tensor. This process can also reduce noise disturbance. Denote by $H = \{h_1, h_2, \cdots, h_m\} \in R^{m \times l}$ the tensor of $\mathbb{X} = \{\mathcal{X}_1, ..., \mathcal{X}_{m-\tau+1}\} \subseteq \mathbb{R}^{l \times \tau \times (m-\tau+1)}$, then the MDT for $\mathbb{X}$ can be defined as:

$$\mathcal{X}_i = \mathcal{H}_\tau(H) = Fold_{(m,\tau)}(H \times {}_1S_1 \times \cdots \times {}_{m-\tau+1}S_{m-\tau+1}) \tag{11}$$

where τ and m represent the time window size and sample length, respectively, and S is a duplication matrix. In this paper, τ is set to 6.

With the tensors constructed by MDT, the Tucker decomposition technique is introduced to obtain the three-order core tensor $\mathcal{G} = \{g_1^1, ..., g_{m-\tau+1}^l\} \in R^{l \times \tau \times (m-\tau+1)}$ that represents the essential information of the sequence:

$$\begin{aligned} &\mathcal{G} = \mathbb{X} \times {}_1{U^{(1)}}^{\mathrm{T}} \times {}_2{U^{(2)}}^{\mathrm{T}} \times \cdots \times {}_V{U^{(v)}}^{\mathrm{T}} \\ &s.t. \quad {U^{(v)}}^{\mathrm{T}} U^{(v)} = I, v = 1, \cdots, V \end{aligned} \tag{12}$$

where $\{U^{(v)}\}_{v=1}^V$ is the projection matrix and can maximally preserve the temporal continuity between core tensors, where the superscript v represents the tensor dimension, and $m - \tau + 1$ represents the sample length after reconstruction. Equation (12) means that the decomposition result consists of a core tensor and a series of factor matrices. Obviously, the core tensor $\mathcal{G}$ contains the intrinsic information of an evolutionary trend. By optimizing $\{U^{(v)}\}_{v=1}^{V}$, the inherent correlation between feature sequences can be sufficiently captured, while noise interference can also be reduced. The loss of tensor reconstruction can be calculated as:

$$\mathcal{L}_{Tensor} = \min\left\|\mathbb{X} - \hat{\mathbb{X}}\right\|_F^2 \tag{13}$$

where $\mathbb{X}$ is the original tensor, and $\hat{\mathbb{X}} = \mathcal{G} \times {}_1U^{(1)} \times {}_2U^{(2)} \times \cdots \times {}_HU^{(H)}$ is the tensor reconstructed from the factor matrix $\{U^{(v)}\}_{v=1}^{V}$.

To implement the decoding operation, it is necessary to transform the core tensor $\mathcal{G}$ back to the original input space. Here, the inverse MDT transform [6] is applied to $\mathcal{G}$ by reversing the transformation along the time dimension to obtain a second-order core tensor $\mathcal{G}' = \{g_1', \ldots, g_m'\} \in R^{m\times l}$, as shown in Equation (14). The core tensor $\mathcal{G}'$ is then used as the input of decoding to update the network.

$$\mathcal{G}' = \mathcal{H}(\mathcal{G}) = Unfold_{(m,\tau)}(\mathcal{G}) \times {}_1S_1{}^{\dagger} \times \cdots \times {}_{m-\tau+1}S_{m-\tau+1}^{\dagger} \tag{14}$$

where $\dagger$ is the Moore–Penrose pseudo-inverse.

Through decoding $\mathcal{G}'$, the reconstructed data $\hat{x}$ can be obtained as follows:

$$\hat{x} = f^*(W^*g' + b^*) \tag{15}$$

where $f^*(\cdot)$ is the activation function of the decoding layer, $W^* \in \mathbb{R}^{n\times r}$ is the weight matrix of the decoding layer, and $b^* \in \mathbb{R}^n$ is the bias vector of the decoding layer. Consequently, the reconstruction loss of the autoencoder can be calculated as:

$$\mathcal{L}_{AE} = \frac{1}{m}\sum_{i=1}^{m}\frac{1}{2}\|\hat{x} - x\|^2 + \frac{\lambda}{2}(\|W\|_F{}^2 + \|W^*\|_F{}^2) \tag{16}$$

where λ is the weight decay parameter and $\|\cdot\|$ is the Frobenius norm.

Based on the aforementioned analysis, the whole loss function is:

$$\mathcal{L}_{loss} = \sum_{i=1}^{M}\mathcal{L}_{AE} + \eta\mathcal{L}_{Tensor} \tag{17}$$

Minimizing Equation (17) can smooth anomalous demands in the demand sequence and extract the evolutionary trend of the demand for spare parts. The key idea of this process is to reduce the significant deviations of anomalous demands, making them suitable for intermittent sequence anomaly detection. The crucial aspect of this process lies in utilizing the core tensor to represent the evolutionary trend. As Equation (12) indicates, $\{U^{(v)}\}_{v=1}^{V}$ is randomly initialized. Therefore, the optimization of tensor decomposition in minimizing Equation (17) is required to obtain the optimal representation of the core tensors.

3.2. *Alternating Optimization Scheme*

Minimization of Equation (17) includes the optimization of $\mathcal{L}_{AE}$ and $\mathcal{L}_{Tensor}$. The $\mathcal{L}_{AE}$ can be solved using a stochastic gradient descent (SGD) strategy [29]. However, SGD cannot be directly applied to tensor decomposition. Therefore, this paper adopts an alternating optimization strategy: fix $\{U^{(v)}\}_{v=1}^{V}$, and update W, W^*; then fix the updated W, W^*, and update $\{U^{(v)}\}_{v=1}^{V}$. These two steps are performed alternately until convergence. It should be noted that, since the number of tensor optimization iterations is typically smaller than the number of W, W^* updates, updating $\{U^{(v)}\}_{v=1}^{V}$ is set to stop when the difference between two consecutive tensor optimizations of $\{U^{(v)}\}_{v=1}^{V}$ is less than a specific threshold. W, W^* is updated until the model converges. With the initialized W and W^*, the specific optimization process is as follows:

(1) Fix $\{U^{(v)}\}_{v=1}^{V}$, and update W, W^*;

1. Encoding stage: the partial derivatives of $\mathcal{L}_{AE}$ with respect to the parameters are:

$$\begin{cases} \frac{\partial \mathcal{L}_{AE}}{\partial W} = \frac{2}{M}\sum_{m=1}^{M}(\hat{x}^{(m)} - x^{(m)}) \cdot \frac{\partial(\hat{x}^{(m)} - x^{(m)})^T}{\partial W} + 2\lambda \frac{\partial(\|W\|_F{}^2 + \|W^*\|_F{}^2)}{\partial W} \\ \frac{\partial \mathcal{L}_{AE}}{\partial b} = \frac{2}{M}\sum_{m=1}^{M}(\hat{x}^{(m)} - x^{(m)}) \cdot \frac{\partial(\hat{x}^{(m)} - x^{(m)})^T}{\partial b} \end{cases} \tag{18}$$

The partial derivatives corresponding to the error term on each sample can be obtained by:

$$\begin{cases} \frac{\partial(\hat{x}^{(m)} - x^{(m)})}{\partial W} = \sum_{m=1}^{M} \frac{\partial(\hat{x}^{(m)})^T}{\partial W} = f' \cdot diag(X^{(m)}) \in \mathbb{R}^{1\times r} \\ \frac{\partial(\hat{x}^{(m)} - x^{(m)})}{\partial b} = \sum_{m=1}^{M} \frac{\partial(\hat{x}^{(m)})^T}{\partial b} = f' \cdot 1_r \in \mathbb{R}^{1\times 1} \end{cases} \tag{19}$$

where "$\cdot$" represents the dot product operator, $diag(\cdot)$ is the diagonal matrix, 1_r is the unit column vector of size r, and f' is the derivative of the activation function;

2. Decoding stage: the partial derivatives of $\mathcal{L}_{AE}$ with respect to the parameters are:

$$\begin{cases} \frac{\partial \mathcal{L}_{AE}}{\partial W^*} = \frac{2}{M}\sum_{m=1}^{M}(\hat{x}^{(m)} - x^{(m)}) \cdot \frac{\partial(\hat{x}^{(m)} - x^{(m)})^T}{\partial W^*} + 2\lambda \frac{\partial(\|W\|_F{}^2 + \|W^*\|_F{}^2)}{\partial W^*} \\ \frac{\partial \mathcal{L}_{AE}}{\partial b^*} = \frac{2}{M}\sum_{m=1}^{M}(\hat{x}^{(m)} - x^{(m)}) \cdot \frac{\partial(\hat{x}^{(m)} - x^{(m)})^T}{\partial b^*} \end{cases} \tag{20}$$

For easy analysis, the error propagation term (i.e., the derivative of the error term with respect to the hidden layer output) is briefly denoted by:

$$\delta_{H^{(m)}} = \left(\frac{\partial(\hat{x}^{(m)} - x^{(m)})}{\partial H^{(m)}} \right) \tag{21}$$

Furthermore, following the chain rule, we have:

$$\begin{cases} \frac{\partial(\hat{x}^{(m)} - x^{(m)})}{\partial W^*} = \delta_{H^{(m)}} \cdot \frac{\partial H^{(m)}}{\partial W^*} = \delta_{H^{(m)}} \cdot (f' \odot diag(H^{(m)}))^T \\ \frac{\partial(\hat{x}^{(m)} - x^{(m)})}{\partial b^*} = \delta_{H^{(m)}} \cdot \frac{\partial H^{(m)}}{\partial b^*} = \delta_{H^{(m)}} \cdot (f' \odot 1_r)^T \end{cases} \tag{22}$$

where "$\cdot$" and $diag(\cdot)$ have the same meaning as in the encoding stage;

3. Substitute Equations (19) and (22) into Equations (18) and (20), respectively, and the model parameters can be updated as:

$$\begin{cases} W^{(l+1)} = W^{(l)} - \xi \cdot \frac{\partial \mathcal{L}_{AE}}{\partial W}\Big|_{W=W^{(l)}} \\ b^{(l+1)} = b^{(l)} - \xi \cdot \frac{\partial \mathcal{L}_{AE}}{\partial b}\Big|_{b=b^{(l)}} \\ W^{*(l+1)} = W^{*(l)} - \xi \cdot \frac{\partial \mathcal{L}_{AE}}{\partial W^*}\Big|_{W^*=W^{*(l)}} \\ b^{*(l+1)} = b^{*(l)} - \xi \cdot \frac{\partial \mathcal{L}_{AE}}{\partial b^*}\Big|_{b^*=b^{*(l)}} \end{cases} \tag{23}$$

where ξ is the learning rate.

(2) Fix W, W^* and update $\{U^{(v)}\}_{v=1}^{V}$.

Updating $\{U^{(v)}\}_{v=1}^{V}$ can be achieved by minimizing the tensor reconstruction error as shown in Equation (13). Since $\{U^{(v)}\}_{v=1}^{V}$ is an orthogonal matrix, Equation (13) can be rewritten as follows:

$$\begin{aligned}\left\|\mathbb{X}-\hat{\mathbb{X}}\right\|_F^2 &= \left\|vec(\mathbb{X})-(U^{(V)}\otimes U^{(V-1)}\otimes\cdots\otimes U^{(1)})\cdot vec(\mathcal{G})\right\|_F^2\\ &= \left\|vec(\mathbb{X})-(U^{(V)}\otimes U^{(V-1)}\otimes\cdots\otimes U^{(1)})\cdot(U^{(V)}\otimes U^{(V-1)}\otimes\cdots\otimes U^{(1)})^T\cdot vec(\mathbb{X})\right\|_F^2\\ &= \|vec(\mathbb{X})\|_F^2-\left\|U^T\cdot vec(\mathbb{X})\right\|_F^2\end{aligned} \tag{24}$$

where $U^T = U^{(H)}\otimes U^{(H-1)}\otimes\cdots\otimes U^{(1)}$. To minimize Equation (24), we should maximize the following equation:

$$\left\|U^T\cdot vec(\mathbb{X})\right\|_F^2 = \begin{cases}\left\|U^{(1)^T}\cdot A^{(1)}\cdot(U^{(H)}\otimes U^{(H-1)}\otimes\cdots\otimes U^{(2)})\right\|_F^2\\ \left\|U^{(2)^T}\cdot A^{(2)}\cdot(U^{(H)}\otimes U^{(H-1)}\otimes\cdots\otimes U^{(1)})\right\|_F^2\\ \cdots\\ \left\|U^{(H)^T}\cdot A^{(H)}\cdot(U^{(H-1)}\otimes U^{(H-2)}\otimes\cdots\otimes U^{(1)})\right\|_F^2\end{cases} \tag{25}$$

where $A^{(i)}$ is the unfolding matrix of the tensor $\mathbb{X}$ along the i-th dimension. The alternating least squares method [30] can be used to solve it, where each factor matrix in different directions is fixed sequentially to find the least squares solution. This process is shown in Equation (26):

$$U_{k+1}^{(v)} = \underset{\{U^{(v)}\}_{v=1}^{V}}{\arg\min}\left\|\hat{\mathbb{X}}-\mathcal{G}\times_M U_k^{(V)}\times_{M-1}\cdots\times_1 U_k^{(1)}\right\|_F^2 \tag{26}$$

3.3. Adaptive Prediction Intervals (API)

After obtaining the optimal feature representation, this paper designs the API algorithm to generate the point prediction values and reliable prediction intervals. The details of the API algorithm are presented as follows. The API algorithm consists of two stages: training and prediction. At the training stage, a fixed number of LightGBM estimators are fitted from a subset of training data. Then, the predicted value from all the LightGBM estimators is aggregated using the leave-one-out (LOO) strategy to generate LOO prediction factors and residuals. At the prediction stage, the API algorithm calculates the LOO predicted value by averaging the prediction factors from each test sample. With the predicted values, the center of the prediction interval is determined. The prediction intervals are then established using the LOO residuals. The width of prediction interval is updated using a dynamic updating strategy. The specific implementation steps are as follows:

First, a LightGBM model f is trained using the training samples $\{(x_t, y_t)\}_{t=1}^{T}$, and the prediction interval at the t-th time step is calculated as:

$$\hat{C}_t^{\alpha} = [\hat{f}_{-t}(x_t)+q_{\hat{\beta}}, \hat{f}_{-t}(x_t)+q_{(1-\alpha+\hat{\beta})}] \tag{27}$$

where α is the significance factor, $\hat{f}_{-t}$ represents the t-th estimator of f, $q_{\hat{\beta}}$ is the $\hat{\beta}$-quantile on $\{\hat{\varsigma}_i\}_{i=t-1}^{t-T}$, and $q_{(1-\alpha+\hat{\beta})}$ is the $(1-\alpha+\hat{\beta})$-quantile on $\{\hat{\varsigma}_i\}_{i=t-1}^{t-T}$. The LOO prediction residual $\hat{\varsigma}_i$ and $\hat{\beta}$ are defined as follows:

$$\hat{\varsigma}_i = y_i - \hat{f}_{-t}(x_t) \tag{28}$$

$$\hat{\beta} = \underset{\beta\in[0,\alpha]}{\arg\min}(\hat{f}_{-t}(x_t) + q_{(1-\alpha+\hat{\beta})} - \hat{f}_{-t}(x_t) + q_{\hat{\beta}}) \tag{29}$$

Second, the interval center is $\hat{f}_{-t}(x_t)$, and the interval width is the difference between the $(1-\alpha+\hat{\beta})$ and $\hat{\beta}$-quantiles on the past T residuals.

However, the obtained interval values cannot adequately address the problem of large fluctuations of demands for different spare parts. This section proposes an adaptive update mechanism to improve the reliability of the prediction interval. The specific step is as follows: First, each demand sequence is divided into a zero-interval sequence (i.e., the gap between two subsequent demands occur) and a non-zero sequence (i.e., the real demand values). The squared coefficient of variation is then calculated for the two sequences , which are denoted by cv_1 and cv_2, respectively. A larger coefficient indicates greater sequence fluctuation, thus requiring a larger interval width for prediction, and vice versa. Second, initialize the interval width parameter $\alpha = 0.1$ and update α by:

$$\alpha = \alpha + cv_1 + cv_2 \tag{30}$$

The interval update mechanism, as shown in Equation (30), can improve the rationality and reliability of the prediction interval for demand sequences with different volatility and intermittency characteristics. Obviously, different values of interval widths will directly affect the decision of inventory management, e.g., spare part coverage rate. The spare part coverage rate, reflected by the interval coverage rate in this paper, is defined as the ratio of the number of demands covered by the interval to the total number of demands in the sequence. The mechanism begins by setting an initial interval with a larger width, and then reduces the width to meet the fluctuations in the demand sequence. The update process is stopped when the interval width reaches a certain threshold. In this study, the threshold of the interval width is determined just by the interval coverage rate. The setting of the coverage rate relies heavily on business logic. Different kinds of maintenance tasks or enterprises have different requirement for the setting. In this experiment, we received help from the maintenance engineers from our cooperative enterprise and set the lower limit of coverage rate to be 60%. We also observe that, with this threshold, the prediction results become stable, which indicates that the threshold runs well.

4. Validation Results and Analysis

4.1. Experimental Settings

The experimental data for validation are the real-life demand data of spare parts from a large engineering manufacturing enterprise from China. The data set contains 75 sequences of different spare parts, in which each sequence covers 34 months from November 2018 to August 2021. The training data are the demand values in the first 33 months, while the data of the last month are for the test. The enterprise usually has one month in advance to make inventory plans and implement the allocation. Therefore, in this experiment we mainly focus on the prediction value of the last month in the whole sequence. To visualize the intermittent characteristic of the demand data, we randomly select two spare parts and illustrate their demand sequences in Figure 3.

For a fair comparison, we introduce six representative methods of time series forecasting in this experiment. These six methods are applicable for different kinds of time series. It is worth noting that BHT-ARIMA, which also adopts tensor decomposition and joint optimization strategy, can be viewed as the SOTA method for intermittent time series forecasting. See Table 1 for detailed information.

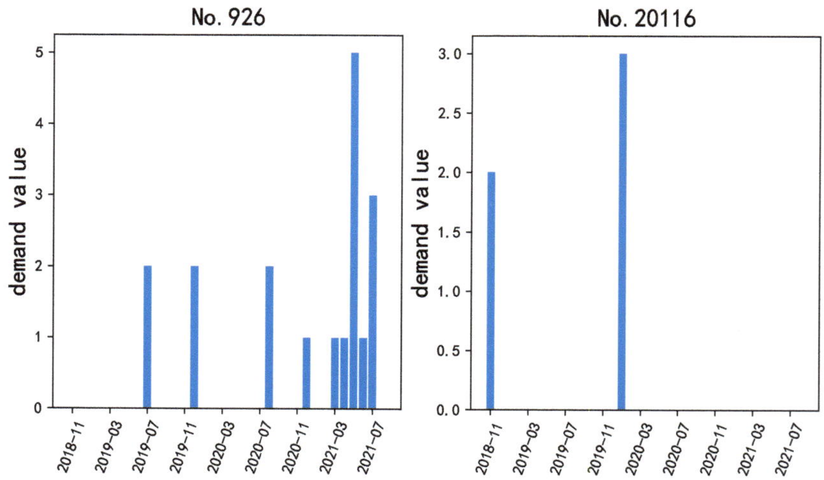

Figure 3. Example of demands for different spare parts used in this experiment. Due to commercial confidentiality requirements, only the part index, instead of its specific name, is given here.

Table 1. Six prediction methods for comparison.

Method Name	Type	Implementation
Croston [14] BHT-ARIMA [19]	Intermittent time series forecasting	Exponential smoothing Joint optimization with ARIMA and tensor decomposition
ARIMA [31]	One-dimensional time series forecasting	Autoregressive modeling with moving average
Random Forest [9] LightGBM [10]	Multidimensional time series forecasting	Ensemble prediction Decision trees using histogram algorithm
LSTM [13]	Temporal deep learning method	Modeling long-term dependencies with memory units

In the experiment, the parameters α and β in the Croston method are set in the range of [0.13, 0.26], and the step size is set to 0.07. The parameters p and q of BHT-ARIMA are set to 1 and 3, respectively. For ARIMA, the parameters p and q are set to 2 and 3, respectively. For Random Forest, a grid search strategy is adopted to select the optimal parameters, with the parameters determined as max_depth = 5, n_estimators = 30, learning = 0.05, and num_leaves = 20. The LSTM hidden layer is set to 20, and the learning rate is 0.001.

4.2. Result Analysis

In this experiment, the demand data of the first 33 months are used as training and the demand data of the 34th month is for the test. The predicted values and prediction intervals obtained by the proposed method are shown in Figure 4.

As suggested by the enterprise's engineers, a prediction result within a range of ±30% around the real values can be regarded correct. From Figure 4, our method works well when dealing with high-demand data. The reason is that, besides the unsupervised feature extraction capability of the autoencoder network, the core tensors can represent the evolutionary trend of the raw demand data well. More importantly, when the prediction results are heavily biased, the prediction intervals obtained by the proposed method can effectively cover the true value. It is clear that the prediction results by the proposed method can improve the decision-making ability of enterprises in inventory management.

Figure 5 shows the comparison of different prediction methods on the demand sequences with different degrees of intermittent distribution.

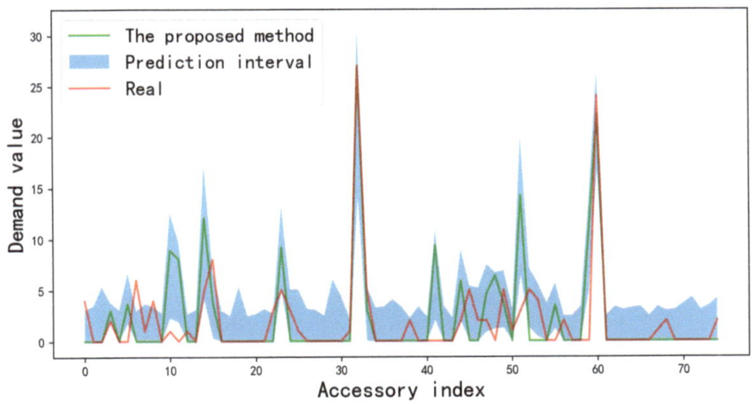

Figure 4. Prediction results of the proposed method on a total of 75 demand sequences.

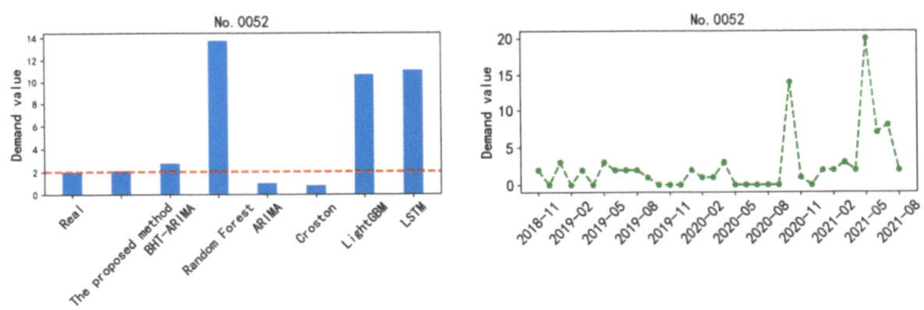

(**a**)

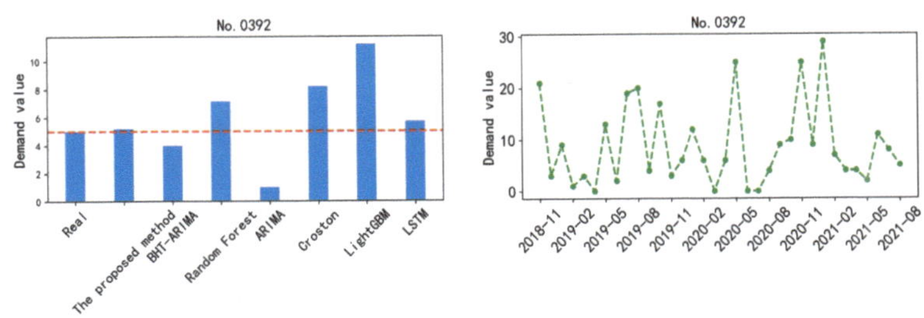

(**b**)

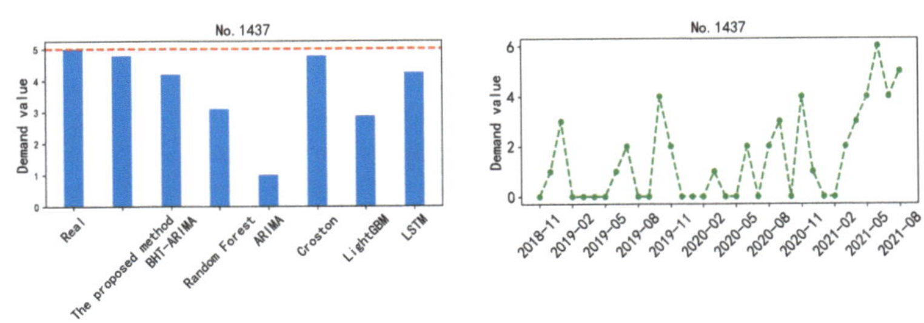

(**c**)

Figure 5. Comparative results of different prediction methods on the demand sequences for the spare parts, respectively, indexed by (**a**) No. 0052, (**b**) No. 0392, and (**c**) No. 1437. The demands for the three spare parts are randomly selected for validation and appear different levels of intermittent distribution characteristic. The left column is the prediction results, while the right column is the raw sequence for reference.

From Figure 5, the zero-interval of the No.1437 sequence is relatively stable, so the Croston method can achieve good results. However, the Croston method performs worse than the other methods on the sequences No.0392 and No.0052 with large demand fluctuations. It indicates that Croston is only suitable for the intermittent series with a stationary zero-interval. The other conventional methods also have similar effects. These methods all have certain limitations and are only applicable to certain types of intermittent time series. Our method outperforms the other methods. The key part of our method is the introduction of tensor decomposition for smoothing anomalous demands, which is beneficial for the effective extraction of evolutionary trends from intermittent time series. With no surprise, our method shows superiority in the forecasting of intermittent time series with different distribution characteristics.

We further evaluate the effectiveness of the designed prediction interval. As stated above, a reliable prediction interval is able to provide a reference for inventory management, even if the point prediction is heavily biased or distorted. Figure 6 shows the prediction interval and predicted value of the demand sequences No. 0392, 0602, 0348, and 3082.

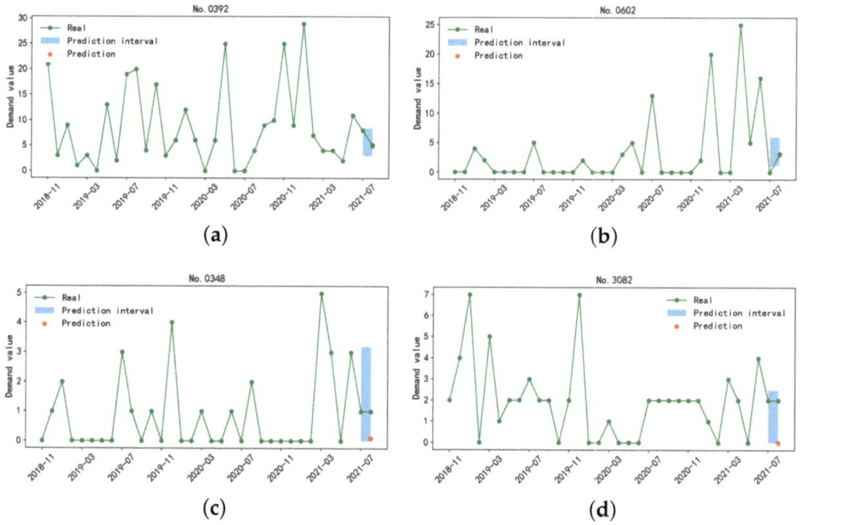

Figure 6. Prediction results of four parts sequences (**a**–**d**) with different intermittent distribution characteristics.

From Figure 6, the point prediction values of the sequences No. 0392 and No. 0602 are more accurate than the others. The prediction intervals also cover the real values, while the interval range remains small. On the contrary, the demand number in the sequences No. 0348 and No. 3082 is smaller, while the deviation of point prediction results is rather large. The prediction interval can cover the real value, so the enterprise can obtain precise information for the spare parts in this interval and make a more reliable plan on inventory management.

For numerical comparison, we introduce three commonly-used error metrics: MAE, RMSE, and RMSSE. The prediction errors obtained by different methods are listed in Table 2. Our method obtains the lowest prediction error in terms of all three metrics. With these results, our method is believed to provide a new reliable solution for enterprises to realize inventory management and parts scheduling.

The mainstream method in this field is the demand prediction method based on deep learning models. One advantage of the proposed method, compared to the current mainstream deep learning model, is introduction of tensor decomposition. By applying tensor decomposition, the core tensor can be extracted from the demand sequences. This effectively mitigates the impact of outliers in the demand sequences and diminishes their interference on the prediction results. Additionally, by utilizing LightGBM prediction with the core tensor, the proposed method can better capture the demand trend information in the demand sequences. The proposed method is then more suitable than deep learning methods for predicting intermittent time series with small samples.

We must point out the potential disadvantages of tensor Tucker decomposition. Actually, the process of tensor decomposition is rather computationally expensive. In the experiment, each round of tensor decomposition needs about 0.6 s, while the alternating optimization needs 3.07 s on average. The total cost of the proposed method is 30 s per round. In contrast, LSTM takes an average of 2.15 s per round, and the other shallow models like Croston, Random Forest, and ARIMA needs much less time. The high cost raises the potential risks of the proposed method for applications that require high real-time performance.

Table 2. Numerical comparison of different methods in terms of three error metrics on all 75 demand sequences.

Algorithm	MAE	RMSE	RMSSE
Random Forest	1.85	2.77	0.79
ARIMA	1.99	4.40	0.73
Croston	1.77	2.79	0.84
LightGBM	1.85	2.80	0.78
LSTM	1.74	3.20	0.91
BHT-ARIMA	1.67	2.73	0.71
Our method	1.64	2.57	0.58

5. Conclusions

In this paper, a new tensor optimization-based robust interval prediction method of intermittent time series is proposed to forecast the demand for spare parts. With the introduction of tensor decomposition, the proposed method can smooth the anomalous demand while preserving the intrinsic information of evolutionary trend from the demand sequences. Moreover, to tackle the distorted or biased prediction results of intermittent time series, the API algorithm is designed to transform traditional point prediction to adaptive interval prediction. The proposed method is able to provide a trustworthy interval for the prediction results, and can accurately reflect the uncertainty of the prediction results.

This study is fundamental to develop the effect of inventory management. As a typical extended application, the proposed method is critical to update the current safety stock model to a dynamic version by integrating the evolutionary trend of demand for spare parts. For instance, the prediction results from this paper can adjust the upper bound of safety stock to match the fluctuations in the actual demands. We believe this operation can decrease the cost level while keeping maintenance.

Another interesting issue is the correlation among the demand sequences for different spare parts. The demand for two spare parts is probably correlated due to many factors such as project cycle, climatic influence, failure probability, etc. As proven by many prior works, the correlation information between two time series is valuable. Especially for intermittent time series prediction, analyzing the correlation among some sequences is believed to improve the predictability, which is critical for intermittent time series. The joint prediction of two or more sequences is able to enhance the prediction performance in terms of accuracy and numerical stability. We plan to explore the utilization of correlation information in the future work. A feasible implementation is running adaptive clustering algorithms with profile coefficients to determine the proper sequence clusters before the prediction, which is expected to improve the predictability. We can further adopt multi-output regression or multi-task learning algorithms to achieve joint prediction on the sequences in the same cluster. We plan to explore the utilization of correlation information in the future work.

In our future work, transfer learning will be studied for the prediction of demand for spare parts. In actual enterprises, the demand data are usually insufficient, especially for newly deployed equipment. Considering the distinction between different types of equipment, we plan to build a transfer learning model to incorporate the evolutionary information of the demand from available equipment. The reliability of the prediction interval is also an interesting work. It is necessary to find a reliable method to converge the prediction interval into a prediction point with a high confidence level.

Author Contributions: Conceptualization, K.H. and F.L.; Methodology, W.M.; Software, X.G.; Validation, Y.R. and W.M.; Formal analysis, W.M.; Investigation, W.M.; Resources, K.H.; Data curation, W.M.; Writing—original draft preparation, X.G.; Writing—review and editing, W.M.; Visualization, Y.R.; Supervision, K.H.; Funding acquisition, K.H. All authors have read and agreed to the published version of the manuscript.

Funding: This work was supported in part by the National Key R&D Program of China under Grant 2020YFB1712105, in part by the National Natural Science Foundation of China under Grant U1704158, and in part by the Key Technology Research Development Joint Foundation of Henan Province under Grant 225101610001.

Institutional Review Board Statement: Not applicable.

Informed Consent Statement: Not applicable.

Data Availability Statement: Not applicable.

Conflicts of Interest: The authors declare no conflicts of interest.

References

1. Bao, Y.; Wang, W.; Zou, H. SVR-based method forecasting intermittent demand for service parts inventories. In *Rough Sets, Fuzzy Sets, Data Mining, and Granular Computing: Proceedings of the 10th International Conference, RSFDGrC 2005, Regina, SK, Canada, 31 August–3 September 2005*; Proceedings, Part II 10; Springer: Berlin/Heidelberg, Germany 2005; pp. 604–613.
2. Van Horenbeek, A.; Buré, J.; Cattrysse, D.; Pintelon, L.; Vansteenwegen, P. Joint maintenance and inventory optimization systems: A review. *Int. J. Prod. Econ.* **2013**, *143*, 499–508.
3. Moore, J.R., Jr. Forecasting and scheduling for past-model replacement parts. *Manag. Sci.* **1971**, *18*, B-200–B-213. [CrossRef]
4. Saaksvuori, A.; Immonen, A. *Product Lifecycle Management Systems*; Springer: Cham, Switerland, 2008.
5. Karthikeswaren, R.; Kayathwal, K.; Dhama, G.; Arora, A. A survey on classical and deep learning based intermittent time series forecasting methods. In Proceedings of the 2021 International Joint Conference on Neural Networks (IJCNN), Shenzhen, China, 18–22 July 2021; pp. 1–7.
6. Fu, G.; Zheng, Y.; Zhou, L.; Lu, C.; Zhang, L.; Wang, X.; Wang, T. Look-ahead prediction of spindle thermal errors with on-machine measurement and the cubic exponential smoothing-unscented Kalman filtering-based temperature prediction model of the machine tools. *Measurement* **2023**, *210*, 112536. [CrossRef]
7. Chen, Y.; Zhao, H.; Yu, L. Demand forecasting in automotive aftermarket based on ARMA model. In Proceedings of the 2010 International Conference on Management and Service Science, Wuhan, China, 24–26 August 2010; pp. 1–4.
8. Karmy, J.P.; Maldonado, S. Hierarchical time series forecasting via support vector regression in the European travel retail industry. *Expert Syst. Appl.* **2019**, *137*, 59–73. [CrossRef]
9. Van Steenbergen, R.; Mes, M.R. Forecasting demand profiles of new products. *Decis. Support Syst.* **2020**, *139*, 113401. [CrossRef]
10. Suenaga, D.; Takase, Y.; Abe, T.; Orita, G.; Ando, S. Prediction accuracy of Random Forest, XGBoost, LightGBM, and artificial neural network for shear resistance of post-installed anchors. In *Structures*; Elsevier: Amsterdam, The Netherlands, 2023; Volume 50; pp. 1252–1263.
11. Cao, Y.; Gui, L. Multi-step wind power forecasting model using LSTM networks, similar time series and LightGBM. In Proceedings of the 2018 5th International Conference on Systems and Informatics (ICSAI), Nanjing, China, 10–12 November 2018; pp. 192–197.
12. Cao, D.; Chan, M.; Ng, S. Modeling and Forecasting of nanoFeCu Treated Sewage Quality Using Recurrent Neural Network (RNN). *Computation* **2023**, *11*, 39. [CrossRef]
13. Abbasimehr, H.; Shabani, M.; Yousefi, M. An optimized model using LSTM network for demand forecasting. *Comput. Ind. Eng.* **2020**, *143*, 106435. [CrossRef]
14. Croston, J.D. Forecasting and stock control for intermittent demands. *J. Oper. Res. Soc.* **1972**, *23*, 289–303. [CrossRef]
15. Syntetos, A.A.; Boylan, J.E. The accuracy of intermittent demand estimates. *Int. J. Forecast.* **2005**, *21*, 303–314. [CrossRef]
16. Mor, R.S.; Nagar, J.; Bhardwaj, A. A comparative study of forecasting methods for sporadic demand in an auto service station. *Int. J. Bus. Forecast. Mark. Intell.* **2019**, *5*, 56–70. [CrossRef]

17. Fu, W.; Chien, C.F.; Lin, Z.H. A hybrid forecasting framework with neural network and time-series method for intermittent demand in semiconductor supply chain. In *Advances in Production Management Systems. Smart Manufacturing for Industry 4.0, Proceedings of the IFIP WG 5.7 International Conference, APMS 2018, Seoul, Republic of Korea, 26–30 August 2018*; Proceedings, Part II; Springer: Cham, Switerland 2018; pp. 65–72.
18. Xu, S.; Chan, H.K.; Ch' ng, E.; Tan, K.H. A comparison of forecasting methods for medical device demand using trend-based clustering scheme. *J. Data Inf. Manag.* **2020**, *2*, 85–94. [CrossRef]
19. Shi, Q.; Yin, J.; Cai, J.; Cichocki, A.; Yokota, T.; Chen, L.; Yuan, M.; Zeng, J. Block Hankel tensor ARIMA for multiple short time series forecasting. In Proceedings of the AAAI Conference on Artificial Intelligence, New York, NY, USA, 7–12 February 2020; Volume 34; pp. 5758–5766.
20. Zhou, X.Y.; Tang, T.; Zhang, S.Q.; Cui, Y.T. Missing Information Reconstruction for Multi-aspect SAR Image Occlusion. *J. Signal Process.* **2021**, *37*, 1569–1580.
21. Guo, J.; Xie, Z.; Qin, Y.; Jia, L.; Wang, Y. Short-Term Abnormal Passenger Flow Prediction Based on the Fusion of SVR and LSTM. *IEEE Access* **2019**, *7*, 42946–42955. [CrossRef]
22. Liu, P.; Ming, W.; Huang, C. Intelligent modeling of abnormal demand forecasting for medical consumables in smart city. *Environ. Technol. Innov.* **2020**, *20*, 101069. [CrossRef]
23. Li, Y.; Wang, X.; Sun, S.; Ma, X.; Lu, G. Forecasting short-term subway passenger flow under special events scenarios using multiscale radial basis function networks. *Transp. Res. Part C Emerg. Technol.* **2017**, *77*, 306–328. [CrossRef]
24. Nguyen, H.D.; Tran, K.P.; Thomassey, S.; Hamad, M. Forecasting and Anomaly Detection approaches using LSTM and LSTM Autoencoder techniques with the applications in supply chain management. *Int. J. Inf. Manag.* **2021**, *57*, 102282. [CrossRef]
25. Li, B. Urban rail transit normal and abnormal short-term passenger flow forecasting method. *J. Transp. Syst. Eng. Inf. Technol.* **2017**, *17*, 127.
26. Yokota, T.; Hontani, H. Tensor completion with shift-invariant cosine bases. In Proceedings of the 2018 Asia-Pacific Signal and Information Processing Association Annual Summit and Conference (APSIPA ASC), Honolulu, HI, USA, 12–15 November 2018; pp. 1325–1333.
27. Courrieu, P. Fast computation of Moore-Penrose inverse matrices. *arXiv* **2008**, arXiv:0804.4809.
28. Yokota, T.; Erem, B.; Guler, S.; Warfield, S.K.; Hontani, H. Missing slice recovery for tensors using a low-rank model in embedded space. In Proceedings of the IEEE Conference on Computer Vision and Pattern Recognition, Salt Lake City, UT, USA, 18–23 June 2018; pp. 8251–8259.
29. Mao, W.; Liu, J.; Chen, J.; Liang, X. An Interpretable Deep Transfer Learning-based Remaining Useful Life Prediction Approach for Bearings with Selective Degradation Knowledge Fusion. *IEEE Trans. Instrum. Meas.* **2022**, *71*, 3508616. [CrossRef]
30. Comon, P.; Luciani, X.; De Almeida, A.L. Tensor decompositions, alternating least squares and other tales. *J. Chemom. J. Chemom. Soc.* **2009**, *23*, 393–405. [CrossRef]
31. Box, G.E.; Jenkins, G.M.; Reinsel, G.C.; Ljung, G.M. *Time Series Analysis: Forecasting and Control*; John Wiley & Sons: Hoboken, NJ, USA, 2015.

sensors

MDPI

Article

Wavelet-Based Output-Only Damage Detection of Composite Structures

Rims Janeliukstis * and Deniss Mironovs

Institute of Materials and Structures, Riga Technical University, LV-1048 Riga, Latvia; deniss.mironovs@rtu.lv
* Correspondence: rimasj479@gmail.com

Abstract: Health monitoring of structures operating in ambient environments is performed through operational modal analysis, where the identified modal parameters, such as resonant frequencies, damping ratios and operation deflection shapes, characterize the state of structural integrity. The current study shows that, first, time-frequency methods, such as continuous wavelet transform, can be used to identify these parameters and may even provide a large amount of such data, increasing the reliability of structural health monitoring systems. Second, the identified resonant frequencies and damping ratios are used as features in a damage-detection scheme, utilizing the kernel density estimate (KDE) of an underlying probability distribution of features. The Euclidean distance between the centroids of the KDEs, at reference and in various other cases of structural integrity, is used as an indicator of deviation from reference. Validation of the algorithm was carried out in a vast experimental campaign on glass fibre-reinforced polymer samples with a cylindrical shell structure subjected to varying degrees of damage. The proposed damage indicator, when compared with the well-known Mahalanobis distance metric, yielded comparable damage detection accuracy, while at the same time being not only simpler to calculate but also able to capture the severity of damage.

Keywords: statistical damage detection; wavelet transform; modal features; composite structure

Citation: Janeliukstis, R.; Mironovs, D. Wavelet-Based Output-Only Damage Detection of Composite Structures. *Sensors* **2023**, *23*, 6121. https://doi.org/10.3390/s23136121

Academic Editors: Yongbo Li, Bing Li and Khandaker Noman

Received: 30 May 2023
Revised: 20 June 2023
Accepted: 23 June 2023
Published: 3 July 2023

1. Introduction

Structures made of conventional materials are being extensively replaced by fibre-reinforced polymer (FRP) composite materials in the aerospace, automotive, energy and other industries owing to their superior specific strength [1], negligible thermal expansion, as well as fatigue and corrosion resistance compared to metals. The safety and reliability of structures, such as aircraft fuselages, helicopter blades, as well as wind turbine blades, is ensured by surveys using non-destructive testing (NDT) methods or planned maintenance. Many NDT approaches require manual inspection, whereas planned maintenance often assumes taking the structure out of operation, which increases downtime and increases costs. What is more is that sometimes the inspection is unnecessary due to lack of damage. In order to detect the onset and progression of existing structural damage in a timely manner and reduce the maintenance costs, effective structural health monitoring (SHM) solutions are crucial.

Output-only SHM involves the detection and possible characterization of damage by analysing only the response signals collected from sensors mounted on the structure. Structural excitation can be provided if conditions for operational modal analysis (OMA) are fulfilled [2]. In practice, such is the case of ambient excitation of a stochastic nature, for example, that is caused by wind, sea waves or traffic loads. Specialized signal-processing algorithms can be applied to these output-only responses to extract essential information on the status of structural integrity. This is normally achieved through a statistical pattern recognition framework, where damage-sensitive features (DSFs) are extracted from the sensor measurements [3]. The DSFs in OMA are typically modal parameters—resonant frequencies, damping ratios and mode shapes. Although several techniques of modal parameter identification exist, such as stochastic subspace identification [4] and least-squares complex exponential [5], etc., it is possible to extract only a limited amount of modal parameter observations and only from repeated measurements. For a statistical damage-detection approach using a significantly larger amount of data, traditional OMA modal parameter estimation techniques can be substituted by time-frequency analysis.

Time-frequency analysis methods, such as short-time Fourier transform, Wigner–Ville distribution and Hilbert–Huang transform and wavelet transform (WT) [6], among others, have become established techniques for analysing transient signals, which are nonstationary in nature. In the case of OMA, free structural vibrations induced by ambient excitations are transient signals with a finite energy localized in time and frequency. Another merit of time-frequency techniques is that they have an ability to decompose a composite signal consisting of several modes of vibration (degrees of freedom) into individual modes [7]. This is normally achieved by finding wavelet ridges—high energy curves in a time-frequency plane tracing that allow for system identification via the extraction of modal parameters. Staszewski was the first to demonstrate the WT can be used as a tool for structural modal parameter identification [8]. Wavelet transform has been used in modal parameter estimation for real structures, such as long-span cable-stay bridges and suspension bridges using continuous wavelet transform in [9]; a cable-stay bridge in Taiwan using wavelet packet transform in [10]; a 600 m tall building in China using a combination of empirical wavelet transform and Hilbert transform in [11]; and a pedestrian overpass in the USA using multisynchrosqueezing transform, a variation of wavelet transform that yields a more concentrated estimate of modal parameters at the cost of higher computational complexity in [12]. Authors in [13] proposed an output-only modal identification and structural damage detection technique based on time-frequency techniques, including wavelets. The above studies have focused on modal parameter estimation via wavelet transform. However, it has been demonstrated in [14] that the time-frequency approach with continuous wavelet transform (CWT) allows for the extraction of numerous instances of modal parameters which, when employed in statistical pattern recognition schemes, are more beneficial, since more data are available. This leads to a larger dataset and, therefore, issues of model overfitting and underfitting can be solved.

After the extraction of DSFs, an anomaly detection algorithm is employed to identify outliers supposedly originating from damage or changes in environmental conditions [15]. A popular class of methods of anomaly detection is based on dissimilarities between a reference structural state and a potentially anomalous state. The Mahalanobis distance (MD) metric has been successfully used for such purposes [16–18], owing to its ability to detect outliers in a multidimensional feature space with an arbitrary number of extracted DSFs. However, for MD to be used, DSFs have to follow a normal distribution. On the other hand, kernel density estimation (KDE) is an approach used to identify the underlying probability density function without any assumptions regarding a probability distribution of data. KDE was used for structural damage detection in [19,20].

Building on the concept introduced in [14], the aim of the present study was to develop a structural damage detection algorithm using continuous wavelet transform as an alternative modal parameter estimation technique. The use of CWT enables a statistical approach to damage detection using KDE. The underlying tasks of the study were to estimate the probability density function of the extracted modal parameters and calculate the probability centroids for modal features. Finally, Euclidean distances between centroids at a reference point and various states of damage were to be calculated, explored as a potential damage indicator and compared with a Mahalanobis distance in terms of accuracy.

2. Modal Identification

The current study is based on the concepts described in detail in [14]. Wavelet ridges are hidden constituent elements of a finite-energy signal revealed through wavelet decomposition of a said signal. Ridges contain all of the essential information on structural modal parameters of a structure whose impulse response is available. Wavelet phase can be used to extract numerous observations of resonant frequencies and damping ratios for each mode of structural vibration. These instances comprise a dataset that is representative of a structural condition in the current state. Each new structural state, for example, the occurrence of damage is associated with changes in modal parameter values. This concept can be utilized in machine-learning-aided structural damage detection and, in a broader sense, structural health monitoring (SHM).

A quick recap of the methodology from [14] is given as follows:

- CWT on the recorded response signals $y(t)$ is carried out using analytical Morlet function and storing complex-valued CWT coefficients in a matrix form:

$$\mathbf{W}_y = \begin{bmatrix} \mathrm{Re}(W_y(s_1, b_1)) - i\mathrm{Im}(W_y(s_1, b_1)) & \cdots & \mathrm{Re}(W_y(s_1, b_L)) - i\mathrm{Im}(W_y(s_1, b_L)) \\ \vdots & \ddots & \vdots \\ \mathrm{Re}(W_y(s_n, b_1)) - i\mathrm{Im}(W_y(s_n, b_1)) & \cdots & \mathrm{Re}(W_y(s_n, b_L)) - i\mathrm{Im}(W_y(s_n, b_L)) \end{bmatrix}, \quad (1)$$

where $s = \begin{bmatrix} s_1 & \ldots & s_n \end{bmatrix}^T$ is the vector of scale factors, $b = \begin{bmatrix} b_1 & \ldots & b_L \end{bmatrix}^T$ is the vector of translation parameters, and L is the signal length.
- Wavelet ridges (in terms of scale parameters s_i^*) of each response signal are found by, firstly, finding the s and b parameters (denoted by s^* and b^*) corresponding to the maximum value of modulus of CWT coefficients and, secondly, testing the ridge condition at a fixed parameter b^* (time instant when vibration amplitude is maximum).

$$\frac{\mathrm{d}}{\mathrm{d}s} W_y(s,\ b^*) = 0. \quad (2)$$

The conversion from scale parameter to frequency is performed through the following relation:

$$f = \frac{1}{2\pi} \times \frac{\omega_0}{s}, \quad (3)$$

where ω_0 is the central frequency of wavelet function. It is essentially a pseudofrequency or a frequency that the wavelet function would have if it was a harmonic function.

- Damped natural frequencies are calculated from the derivative of phase between the real and imaginary parts of CWT coefficients along the wavelet ridge line with respect to time. The wavelet ridge line is defined at the ridge scales s_i^* along the whole time span of free vibrations starting from the time instant b^*.

$$\frac{\mathrm{d}}{\mathrm{d}t} \mathrm{Arg}(W_y(s_i^*, b^* : b_L)) = \omega_d \frac{\mathrm{d}}{\mathrm{d}s} W_y(s,\ b^*) = 0. \quad (4)$$

Damping ratios are extracted in the following substeps:

The moduli of CWT coefficients are extracted along the wavelet ridge, and its natural logarithm is calculated. It is denoted as $\ln\left|W_y(s_i^*, b^* : b_L)\right|$.

By plotting $\ln\left|W_y(s_i^*, b^* : b_L)\right|$ versus the time axis, a straight line is obtained for most of the time span of response because the modulus of CWT coefficients decays exponentially with time. This straight line is fit with a linear function, and the slope parameter is extracted. This slope parameter is equal to

$$\frac{\mathrm{d}}{\mathrm{d}t} \ln\left|W_y(s_i^*, b^* : b_L)\right| = -\xi_i \times \omega_n = \text{slope}, \quad (5)$$

where ξ is the damping ratio, and ω_n is the undamped natural frequency.

The relationship between the damped and undamped natural frequencies is well-known from structural dynamics as $\omega_d = \omega_n \times \sqrt{1 - \xi^2}$. Hence, in practice, the damping ratio is obtained as

$$\xi = \pm \sqrt{\frac{\text{slope}^2}{\omega_d^2 + \text{slope}^2}}. \quad (6)$$

3. Damage Detection Algorithm

SHM systems estimate the health state of structures during operation to ensure their safety and economic efficiency. Such structures can be industrial, transport or energy equipment with structural elements, for example, wind turbines or wind generators and their elements—tower and blades. An SHM system's sensors network will be connected to the structure, read the vibration data and send it to the workstation (computer). Then, an operator estimates the modal parameters of the structure. The resulting parameters together with the possible external operational factors, such as static loads, temperature or speed of rotation (blades), are input to a modal passport [21,22], where the past measurements are stored. Modal parameters in a reference or intact structural state together with their deviations due to operational factors form a "signature" of a structure that is unique to this structure or for structures of this type. As next step, a specialized algorithm analyses the modal passport and recognizes a particular structure. By analysing modal parameter changes with damage, one can infer on the severity of the damage, which allows for further planning of the agenda of structural serviceability—repair, replacement or resuming operation if damage is not significant.

The damage detection algorithm proposed for an SHM system originates from an anomaly detection field. It has three distinct phases, namely, Phase I: signal collection, Phase II: feature extraction, and Phase III: statistical control, as shown in Figure 1.

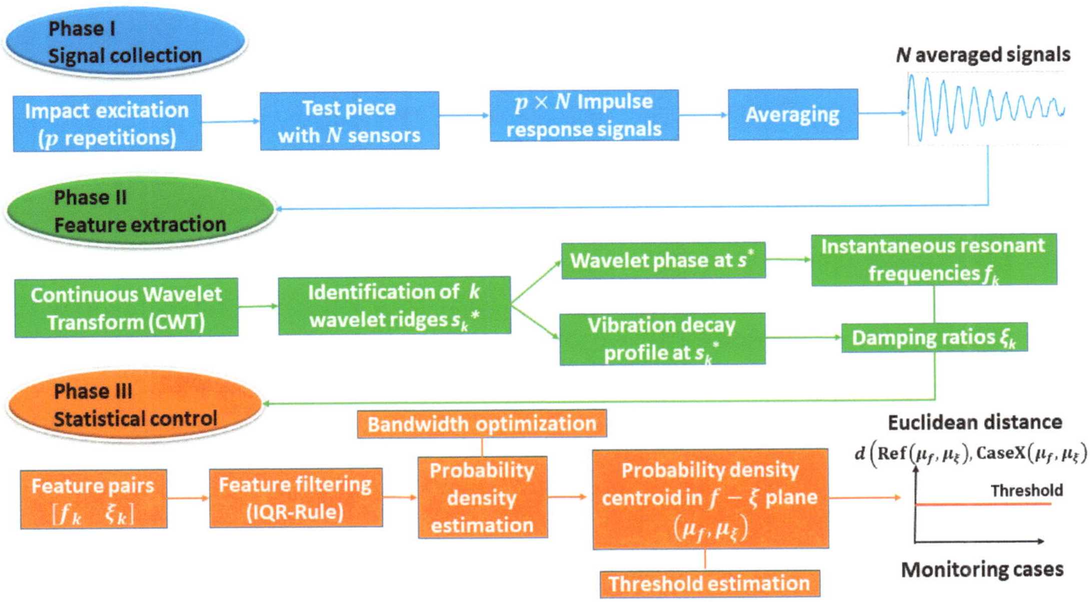

Figure 1. Anomaly detection algorithm.

3.1. Phase I—Signal Collection

Vibration signals as a response to structural excitation are measured with N sensors connected to measurement channels Ch1 to ChN. In the case of numerous instances of impact excitation (p excitation of the structure during the measurement session), any individual free-vibration decay profile is isolated from the whole signal. Afterwards, p, these individual vibration profiles, are averaged to obtain N averaged vibration responses.

3.2. Phase II—Feature Extraction

In Phase II, a time-frequency analysis of the averaged responses is performed using CWT. CWT analysis involves the identification of wavelet ridges from the ridge condition in Equation (2) in the time domain signal analysed. Wavelet ridges represent the oscillatory modes, which comprise the components of vibration decay signals. Subsequently, this ridge information from the ridges identified at $r = 1, \ldots, R$ is used to identify the resonant frequencies f_r from wavelet phase (Equation (4)), and the decay profile of CWT coefficients in time domain is used to extract the damping ratios ξ_r (Equation (6)). CWT analysis involves operating with wavelet scale parameters. In order to convert the scale parameter to frequency values, a wavelet scale-to-frequency conversion is realized through Equation (3) and is illustrated in Figure 2 for the analytical Morlet wavelet function. In this study, Morlet mother wavelet is used since it exhibits a high correlation with the time domain vibration data. This mother wavelet has been used for structural damage detection [23,24]. The identified modal parameter value pairs form features f_r and ξ_r that are organized into a two-column matrix $\begin{bmatrix} f_r & \xi_r \end{bmatrix}$, where rows correspond to observations and columns correspond to features. This feature matrix is used in the next phase of the anomaly detection algorithm proposed.

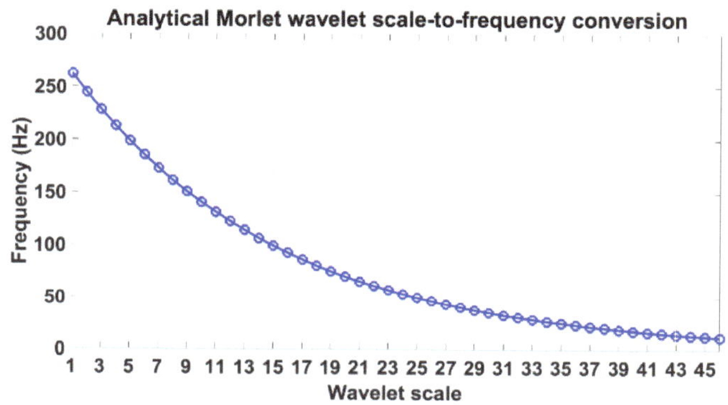

Figure 2. Wavelet scale-to-frequency conversion plot for the analytical Morlet wavelet.

3.3. Phase III—Statistical Control

Phase III is concerned with performing statistical control of the feature values. The first stage is the feature-value filtering, which is carried out by using the interquartile range (IQR) rule. The goal is to remove the outliers from the features. Unlike the threshold set to mean plus/minus two or three standard deviations, the IQR approach for outlier removal is appropriate for data that does not necessarily follow a normal distribution. The frequency values for the ridges identified at $r = 1, \ldots, R$ are filtered according to

$$Q_1(f_r) - 1.5 \times \mathrm{IQR}(f_r) < f_r < Q_3(f_r) + 1.5 \times \mathrm{IQR}(f_r), \tag{7}$$

where IQR is the interquartile range, Q_1 is the first quartile, and Q_3 is the third quartile of the filtered resonant frequencies from the previous step. The filtered resonant frequency and damping ratio values are stored as two-column vectors $\begin{bmatrix} f_r^* & \xi_r^* \end{bmatrix}$.

The next step involves exploring the underlying probability distribution of the filtered features. For this purpose, the kernel density estimate (KDE) is computed. The reason is that a kernel distribution representation of the probability density function (PDF) of the data does not make any assumptions on the underlying distribution. The KDE is defined by a smoothing function and a bandwidth value that controls the smoothness of the resulting density curve. The kernel density estimator of the data at hand (x) is given by

$$\hat{f}_h(x) = \frac{1}{nh} \sum_{i=1}^{n} K\left(\frac{x - x_i}{h}\right), \tag{8}$$

where n is the sample size, K is the kernel-smoothing function governing the shape of the curve used to generate the PDF estimate, and h is the bandwidth. In this study, the obtained vectors of filtered frequency and damping ratio values are used as the data x, and normal density is used as a kernel smoothing function since the feature values approximately follow Gaussian distribution. It is important to choose the optimum bandwidth parameter since it regulates the degree of smoothing. In this work, bandwidth optimization was carried out by the following procedure illustrated in Figure 3:

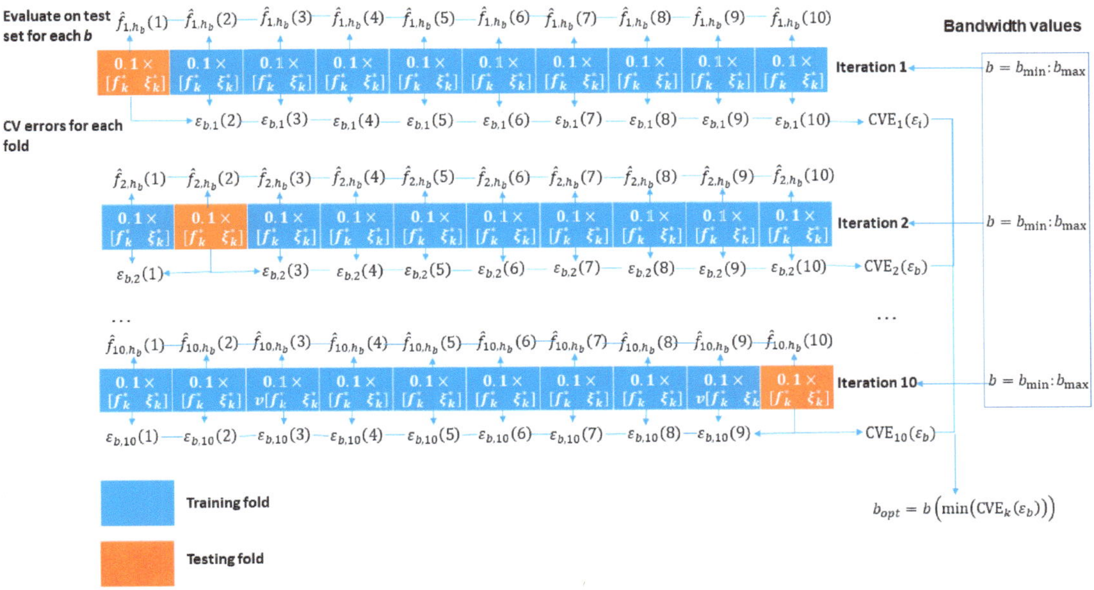

Figure 3. Optimization scheme for the kernel density bandwidth parameter.

1. Perform a cross-validation partition on the data to create 10 folds where one fold is used for testing and 9 folds are for training. Perform 10 iterations of such a partition, where a different fold is used for testing in each iteration.
2. Define a range of bandwidth parameters b to test.
3. In each training fold and the single testing fold, compute the KDE according to Equation (8) for each value of the bandwidth parameter. Then, compute an error ε between the KDEs of the testing and each training set according to

$$\varepsilon_{b,\,k}(k-1) = \frac{1}{n}\sum_i^n \log \hat{f}_{h_b}(k-1), \tag{9}$$

   where $k = 1:10$ is the number of folds.
4. Calculate the cross-validation error as a mean-squared-error of the errors in Equation (9) across all folds for each value b according to

$$\mathrm{CVE}_{b,\,k} = \frac{1}{k}\sum_k \varepsilon_{b,k}^2. \tag{10}$$

5. Find the optimum bandwidth parameter by calculating the minimum of these cross-validation errors across all bandwidth values

$$b_{opt,\,k} = b(\min(\mathrm{CVE}_{b,k},\, b)). \tag{11}$$

Afterwards, the KDEs with these optimized bandwidth parameters are calculated for reference and all monitoring cases. Then, a centroid value of the KDE for both features is calculated according to

$$C_x = \frac{\sum_i^n x_i \times \hat{f}_{h,\,i}}{\sum_i^n \hat{f}_{h,i}}, \tag{12}$$

where data x in this study are the filtered feature values. Thus, both quantities, $C_{f_r^*}$ and $C_{\xi_r^*}$, are calculated for reference and all monitoring cases. Next, the centroid values of both features are organized into a row vector for each case at hand, and the Euclidean distances between centroids at reference and each monitoring case are calculated as

$$d(C_{\text{ref}}, C_{\text{case}}) = \sqrt{\sum_i^m (C_{\text{ref},i} - C_{\text{case},i})^2}, \tag{13}$$

where $C_{\text{ref}} = \begin{bmatrix} C_{f_r^*,\text{ref}} & C_{\xi_r^*,\text{ref}} \end{bmatrix}$, $C_{\text{case}} = \begin{bmatrix} C_{f_r^*,\text{case}} & C_{\xi_r^*,\text{case}} \end{bmatrix}$, and $i = 1, 2$ since the centroid vector contains two values.

Once the Euclidean distance between the reference and all monitoring cases is calculated, a threshold value is established according to the following scheme:

6. Consider all available structures of the same type at their reference state.
7. Perform a modal parameters estimation to form feature vectors and calculate their centroid values for each structure.
8. Calculate the Euclidean distance between centroid values in all possible combinations of structure pairs.
9. Calculate the median value of these Euclidean distances and confidence bounds as

$$\text{CB} = Nq \pm z\sqrt{Nq \times (1 - q)}, \tag{14}$$

where N is the number of Euclidean distance samples, $q = 0.5$ is the quantile corresponding to median (50% of data), and z is the critical value dependent on a chosen confidence level. For confidence level 0.95, $z = 1.96$. The threshold of the Euclidean distances is set as a lower confidence bound at $T = Nq - z\sqrt{Nq \times (1 - q)}$.

4. Experimental Campaign

The algorithm proposed is validated on the modal parameters extracted from five glass-fibre-reinforced polymer composite specimens with a cylindrical shape manufactured in the scope of an SHM system prototype research project. Cylindrical structures mimic the structural components of serial production, such as a helicopter tail boom, for which the current anomaly detection algorithm is intended.

4.1. Specimens

Testing objects are the structures in the form of cylinders fabricated from a composite material with the flanges made of plywood rings. The specimens are made of 300 g/m^2 fibreglass fabric with a fibre orientation of 45° and LG 385 epoxy resin (HG 385 hardener). The weight of the specimen with the upper and lower flanges is 4.37 kg. Photo of a specimen is shown in Figure 4a. The specimen design includes (see Figure 4b):

- Item 1—composite cylinder made of fiberglass and epoxy resin;
- Items 2 and 3 —top and bottom annular flanges for cylinder fixation made of laminated plywood (30 mm thickness), respectively;
- Item 4—a network of 48 piezoelectric strain sensors;
- Item 5—wires connecting the sensors;
- Item 6—4 D-SUB type connectors at the places for connector fastening.

(a)

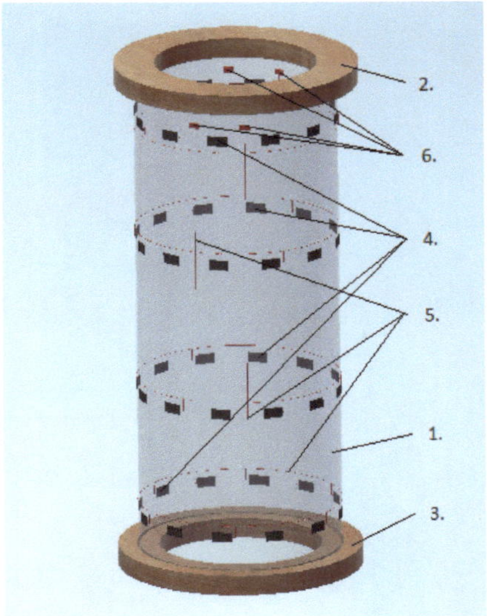

(b)

Figure 4. Specimen test object No. 1: (**a**) general view; (**b**) design.

The dimensions of the specimen are as follows: nominal diameter of 300 mm, nominal length of 710 mm (with the flanges of 773 mm) and wall thickness of 1.45 ± 0.05 mm.

4.2. Measurement Subsystem

Each specimen includes its own sensor network, which outputs the signals during testing to the measuring system, providing a registration and storage of the signals.

The sensor network of each specimen comprises 48 polyvinylidene fluoride piezo film sensors connected to four conductor terminals and wiring harnesses with connectors. These films are flexible, lightweight and have piezoelectric properties [3]. Due to these properties, a strain in the film causes a change in stress. The piezo film is located between two printed silver electrodes, forming a capacitor-like structure. The dimensions of the sensors are roughly 45 mm × 20 mm × 0.05 mm, electrical capacity of 1.3 nF, operating temperature from −40 °C to 60 °C. The view of the sensor prepared for gluing on the specimen is shown in Figure 5a. The wires used for the SHM system prototype are the small, lacquered copper wires with a diameter of 0.25 mm. These wires are glued to the sensors with a special two-component epoxy glue, after which the sensor is covered with a nonconductive insulating tape. The electrical conductors connecting the sensors with the terminals (Figure 5b) are laid in the form of bundles in the longitudinal and circumferential directions and fixed with adhesive tape. On each terminal, 12 similar (conditionally signal) conductors are assembled on 12 contacts and 12 conditionally negative conductors on one common contact. Bundles of wires are soldered to the terminal contacts. The sensors are installed on the specimens in accordance with the premade markings, as shown in Figure 5c. At this stage, the sensor network is covered with a protective composite layer, and the specimen is glued into the annular grooves of the flanges.

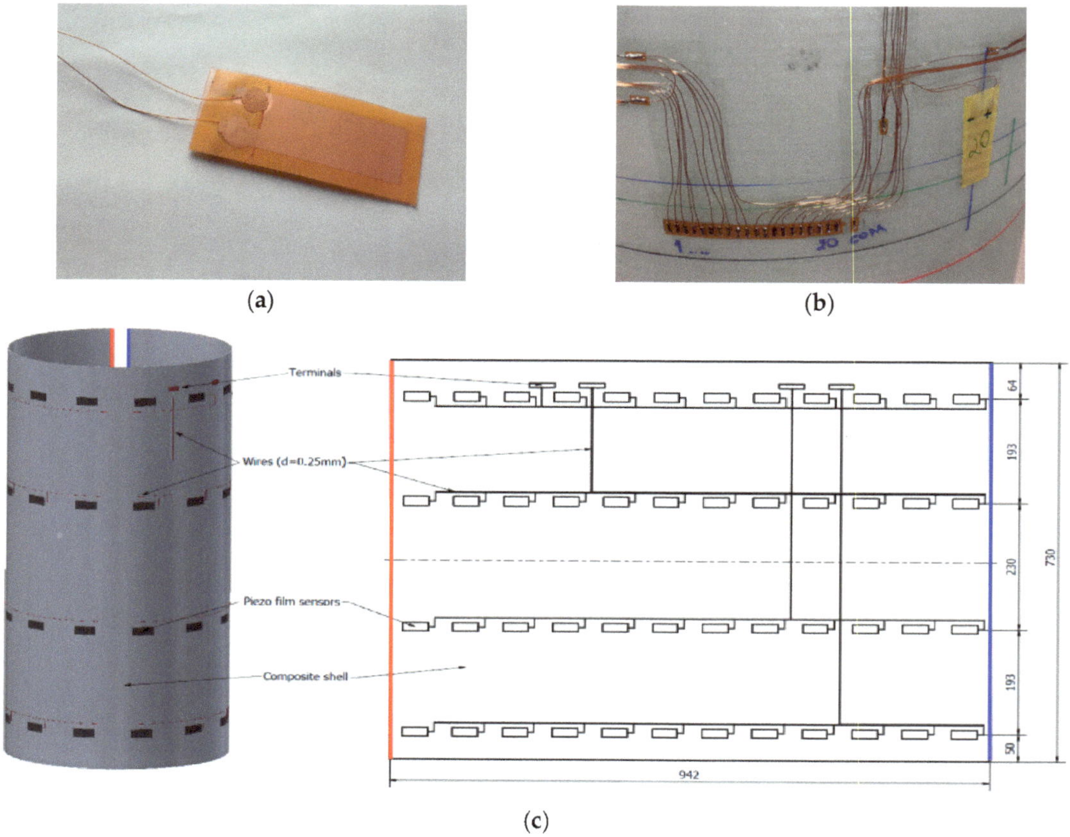

Figure 5. Installation of sensor system on the test specimens: (**a**) piezoelectric film sensor used in the measurements; (**b**) sensor terminal glued around the circumference of the cylinder; (**c**) arrangement of a sensor network on the specimen.

4.3. Modal Testing

After the instrumentation installation, each specimen is fixed on a U-shaped modal testing stand, which, in turn, is mounted on a vibration isolation base (see Figure 6). Structural excitation is performed by repeated impacts of the specimen with a plastic modal hammer in the radial and vertical directions for 120 s. This test procedure is repeated 3 times. The Brüel & Kjær (B&K, Singapore) system LAN-XI Type 3053 is used as the data-reading device. A total of four Type 3053 modules were used to measure 48 channels simultaneously. A portable computer with software from the manufacturer of measurement modules (B&K, Singapore), such as Pulse Labshop, was used for data collection, processing and management. The entire signal recording with a duration of 20 s and sampling frequency of 4096 Hz contained 4 to 6 free vibration decay responses corresponding to the 4 to 6 instances of impact excitation. The test cases realized in the current study involve a reference state and progressing damage. The damage comprises a circular hole drilled through the thickness of the specimens in the same location. The diameter of the hole is increased and has the following values: 4, 8, 16, 24, and 32 mm.

Figure 6. Modal test stand.

All test specimens manufactured were visually inspected to check for defects or damage, as well as their compliance with the specifications. Local deviations and small differences between the specimens were identified to be mainly due to the factors of hand-made technology. These factors include uneven filling of the specimens and flange joint, resin leaks on the specimen surface, air bubbles in the protective layer, undercut on the flange of specimen No. 5, slightly different location of the first hole on the flange for each specimen, resin pouring out at different locations on the inner and outer surfaces of specimens, and differences in wire layouts for sensors.

4.4. Test Cases

The test cases realized in the current study involve reference state and progressing damage, namely, a circular hole drilled through the thickness of the cylinders in the same location. Diameter of the hole is increased in five stages as follows—4, 8, 16, 24, and 32 mm. The location of the hole schematically is shown in Figure 7a, while the close-up photo view of a 4 mm hole is shown in Figure 7b.

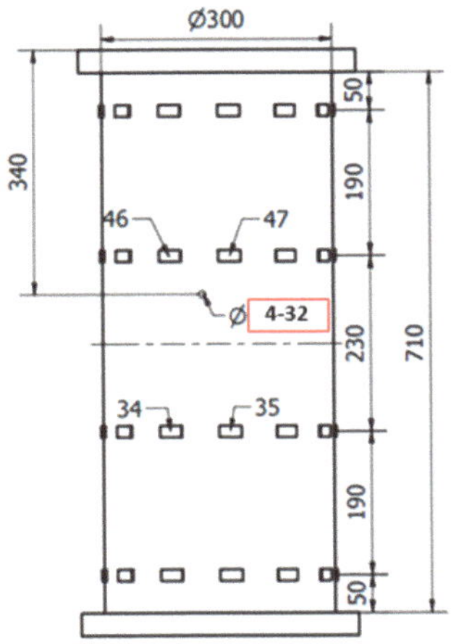

(**a**)

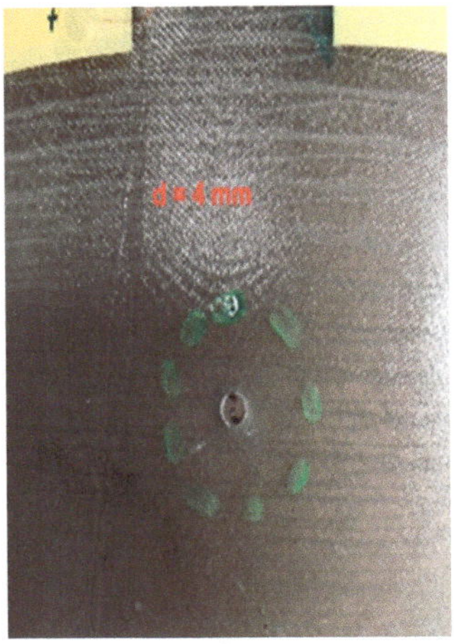

(**b**)

Figure 7. Damage in the test specimens: (**a**) schematic showing the location and size of the hole; (**b**) photo of the 4 mm hole. The hole is located close to sensors No. 46 and 47.

5. Results

5.1. Time-Frequency Analysis

An example of a response signal (specimen No. 1 from measurement channel 1) at a reference state is shown in Figure 8. The sampling frequency of the signal recording was 4096 Hz. This response signal was divided into separate free-vibration decay signals. The duration of each of these vibration signals was approximately 0.1 s. Afterwards, these responses were averaged and an averaged response for each measurement channel was obtained.

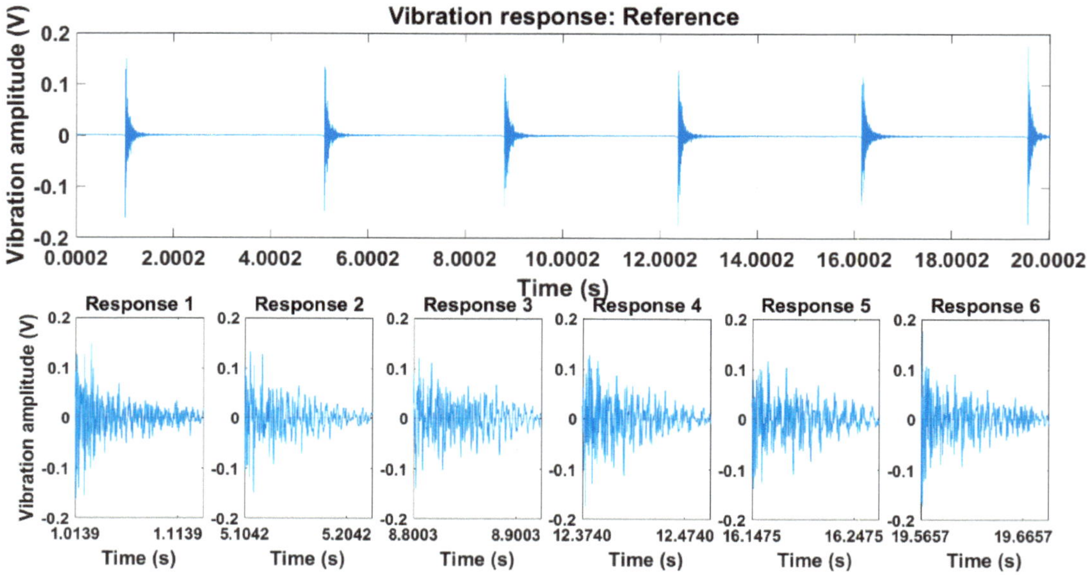

Figure 8. Recorded free vibration decay (specimen No. 1, measurement channel 1).

CWT scalograms at the reference state for the averaged time domain responses are shown in Figure 9. The scalogram is a useful analysis tool for signal analysis in joint time and frequency domains via the CWT. It is a 3D plot providing an indication of relative energy distribution across wavelet scales (related to frequencies) and time instants. Red regions mark the coordinates in time-frequency plane with high energy localization. Ridge identification using the ridge condition is shown on the bottom plots. The identified ridges and the corresponding frequencies are presented in Table 1. The frequencies correspond to the ones from the scale-to-frequency conversion from Figure 2 and do not consider variations associated with individual specimens. The vibration modes at scales 6 and 14, corresponding to 185.3 Hz and 106.4 Hz were identified for all five specimens. The vibration mode at scale 7 (172.9 Hz) was identified for all specimens, except for the second, while the vibration mode at scale 5 (198.6 Hz) was identified only for the first and fourth specimens.

Table 1. Wavelet scales and frequencies of the identified wavelet ridges of the averaged response signals.

	Specimen	**1**	**2**	**3**	**4**	**5**
Ridge #	**Scale** s **(-)**	f **(Hz)**	f **(Hz)**	f **(Hz)**	f **(Hz)**	f **(Hz)**
1	5	198.6	-	-	198.6	-
2	6	185.3	185.3	185.3	185.3	185.3
3	7	172.9	-	172.9	172.9	172.9
4	14	106.4	106.4	106.4	106.4	106.4

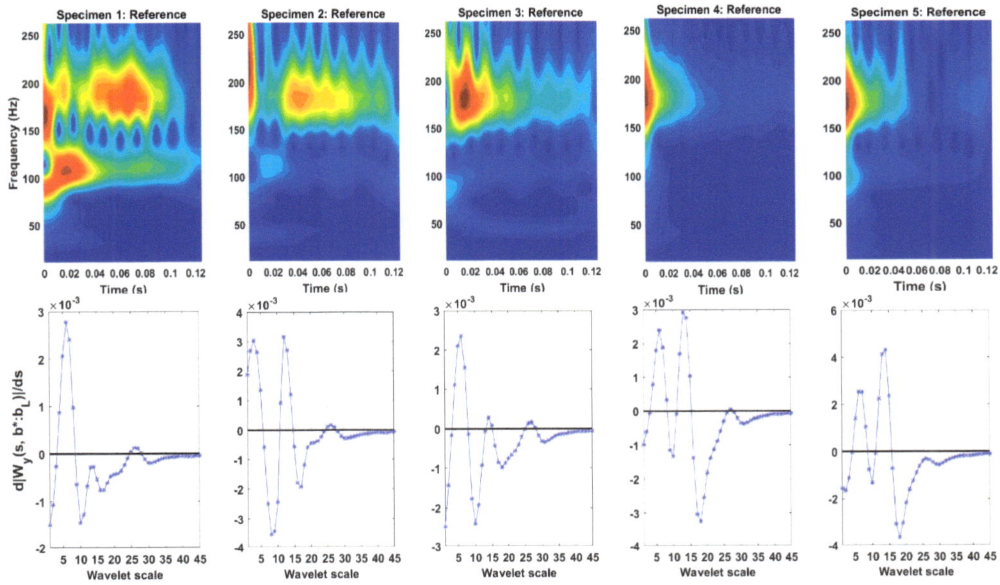

Figure 9. Resonant frequency identification with CWT. Top: CWT scalograms showing signal energy distribution at different frequencies and time instants; bottom: derivative of moduli of CWT coefficients versus wavelet scale. Scales at zero crossings (thick black line) corresponding to wavelet ridges. Note the inverse proportionality of frequencies and wavelet scales.

5.2. Modal Parameter Estimation

5.2.1. Resonant Frequencies

The estimation of resonant frequencies from wavelet phase is shown in Figure 10a. First, CWT coefficients were calculated from the averaged vibration response. Second, phase angle between the real and imaginary CWT coefficients were calculated separately for each identified wavelet ridge. Resonant frequencies were calculated as a derivative of wavelet phase according to Equation (4). The number of identified observations of resonant frequency increased with increasing the signal length and sampling frequency. Therefore, it was possible to extract large sample size of features from a long signal with high sampling frequency.

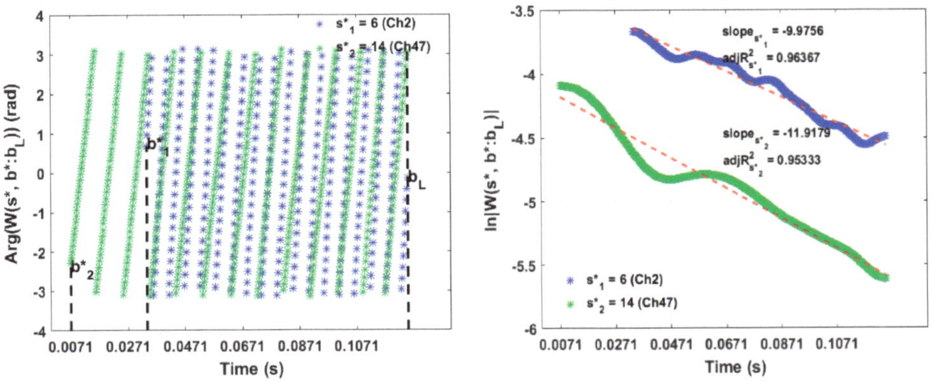

Figure 10. Estimation of modal parameters from wavelet ridges: (**a**) CWT of the signal and estimation of resonant frequencies from wavelet phase (two ridges shown); (**b**) extraction of slope of decay related to damping ratio.

5.2.2. Damping Ratio

The procedure of damping ratio estimation is shown in Figure 10b. First, the moduli of the CWT coefficients was calculated at each wavelet ridge along the time axis. Second, the natural logarithm was taken, and this result was approximated with a linear function. The slope of this fit was recorded, and the damping ratio was calculated according to Equation (6).

5.2.3. Frequency Filtering

The results of the frequency filtering for the vibration mode at 185 Hz (scale 6) at a reference state for specimen No. 2 are illustrated using histograms in Figure 11 and shown numerically in Table 2. There are significant outliers in the extracted instantaneous frequencies for all specimens. The small inset shows a histogram of the zoomed-in portion before filtering the histograms, ignoring the outlier values. The histograms on the right show the results after filtering. The bimodal character of frequency distribution can be traced. The results in Table 2 show that while the mean frequency values change only slightly, the standard deviation and, especially, frequency range reduced drastically after the filtering process.

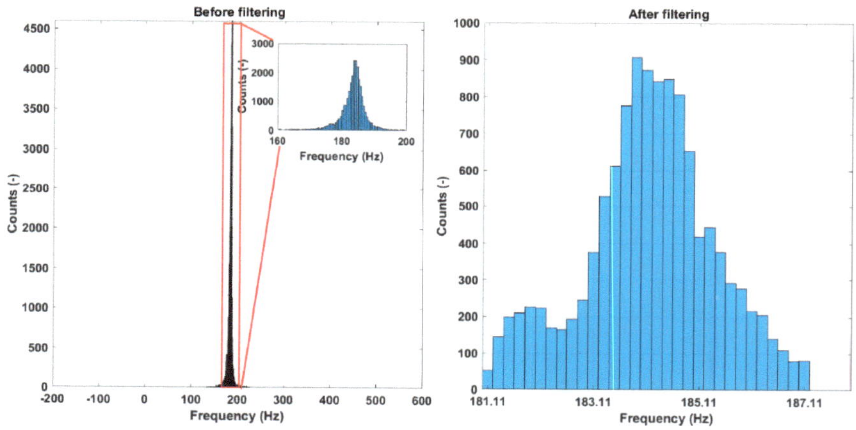

Figure 11. Resonant frequency filtering for a reference state, vibration mode at 185 Hz, specimen No. 2.

Table 2. Statistical descriptors of resonant frequency feature filtering.

Specimen No.	1	1	2	2	3	3	4	4	5	5
Filtering	No	Yes	No	Yes	No	Yes	No	Yes	No	Yes
Mean (Hz)	187.38	186.49	183.58	184.15	178.41	180.03	182.38	182.21	142.70	180.77
Variance (Hz^2)	359.10	4.97	99.40	1.54	421.48	0.94	364.81	5.29	1474.56	1.49
Range (Hz)	1512.70	11.48	701.68	6.08	2501.22	4.37	2061.54	10.86	872.45	6.02

5.3. Kernel Smoothing

After feature filtering, the next step is computation of the KDE of the filtered features. It was shown that feature values do not follow a clear normal distribution; therefore, KDE is an appropriate tool for the estimation of the underlying probability density. In the bandwidth optimization routine, bandwidth values were set from 0.01 to 1, with a total of 100 values for testing. The bandwidth parameter optimization results for specimens at the reference state are shown in Figure 12a. It can be seen that the optimum bandwidth parameters are different for different specimens even if they are designed to be equal. The range of optimum bandwidths is from 0.03 (specimen No. 2) to 0.14 (specimen No. 4). The optimized KDE of the resonant frequency distribution at scale 6 for the specimen No. 1 at reference in comparison to a KDE obtained with a default bandwidth parameter is shown in Figure 12b. The optimization procedure removes the spikes of the KDE producing a smooth curve.

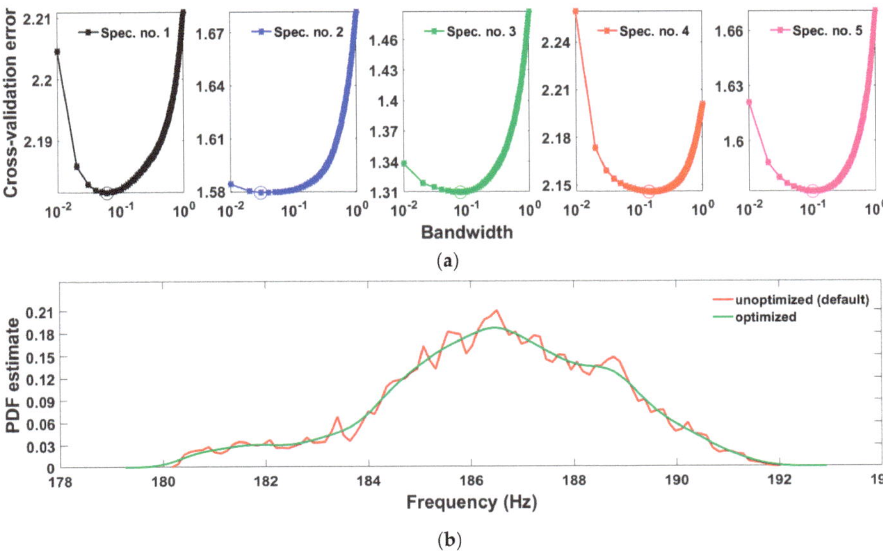

Figure 12. Bandwidth optimization. (**a**): cross-validation error for a range of bandwidth parameters where minimum value is marked with a circle; (**b**): kernel density estimate of probability density function for the optimized bandwidth parameter and default (specimen No. 1 at reference state). Optimization of bandwidth has removed the spikes of KDE smoothing out the curve.

The optimized KDEs for a specimen No. 1 are shown in Figure 13. Here, all damage scenarios are shown along with the reference for both resonant frequency and damping ratio features. Centroids are marked as filled circles of a colour corresponding to one of the associated KDE. The KDEs reveal complex multimodal distributions, indicating that the underlying probability densities are, in fact, not classical Gaussian for all damage cases.

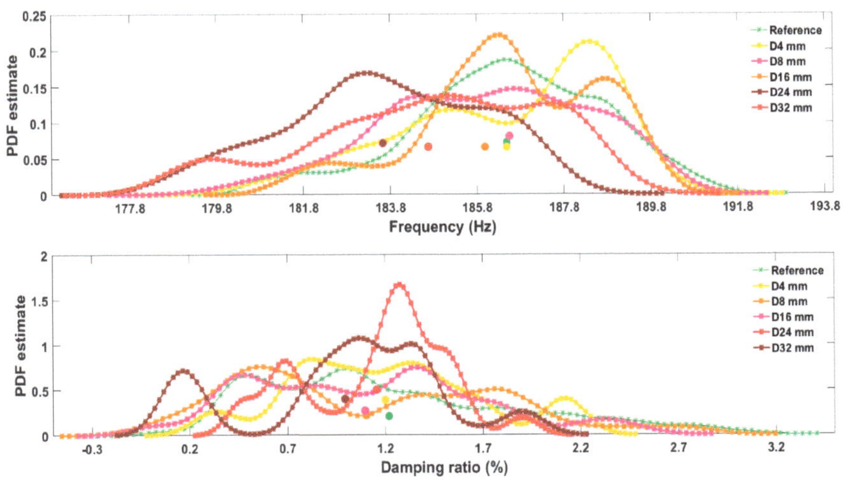

Figure 13. Kernel density estimate for specimen No. 1 at reference and damage states for resonant frequency and damping ratio modal features. Centroids are marked with filled circles. Their position with respect to the reference can be seen.

5.4. Damage Detection

5.4.1. Threshold Estimation

The threshold values for the Euclidean distances of centroids at reference are estimated from all five cylindrical specimens. Hence, the total number of combinations is 10 if that particular vibration mode is identified for all structures. The Euclidean distances calculated for three modes of vibration at 185, 176 and 106 Hz are shown in Figure 14. Only six combinations for vibration mode at 176 Hz were obtained meaning that this particular mode was identified in four out of five specimens. The values of the Euclidean distances are not uniform, indicating a relatively high standard deviation, particularly for modes at 185 and 106 Hz. Overall, the largest deviations from the reference are for the vibration mode at 185 Hz, while the smallest are the for vibration mode at 176 Hz.

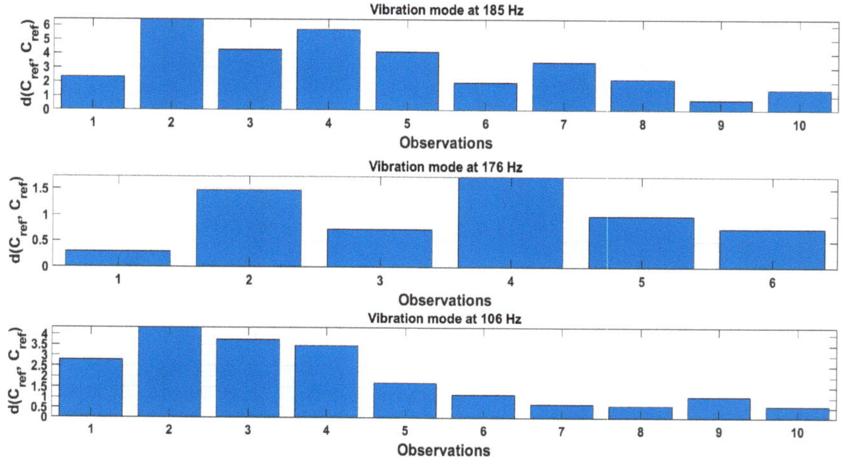

Figure 14. Euclidean distance values of centroids between reference states of all structures for the extracted vibration modes.

The median values and their confidence bounds according to Equation (14) of the calculated Euclidean distances are presented in Table 3 for all vibration modes. The threshold value is set as a lower confidence bound to ascertain that the structural change is detected.

Table 3. Threshold estimation results for each vibration mode.

Vibration Mode	Scale 14 (106 Hz)	Scale 7 (176 Hz)	Scale 6 (185 Hz)
$\text{median}\left(\text{d}\left(C_{ref}, C_{ref}\right)\right)$	1.432	0.871	2.863
CB_lower	0.614	0.289	1.455
CB_upper	3.464	2.34	4.274

5.4.2. Damage Indication

The damage detection results are displayed in Figure 15. Red horizontal line shows the threshold level. The scales for the Euclidean distances for each mode of vibration are set to the maximum value among all five specimens to compare the magnitude of deviation at the reference among different specimens. First of all, it can be seen that the values of the Euclidean distances are increasing with progression of damage, indicating that the approach proposed is sensitive to changes in damage severity. Secondly, the largest overall deviations from the reference state are observed for features from the vibration mode at 106 Hz. Furthermore, the values of Euclidean distance for all damage scenarios are above the threshold level for all specimens. From this perspective, the vibration mode at 185 Hz yields the poorest results since only the most severe damage cases are detected (above the threshold) for specimens Nos. 1, 2 and 3. No damage for this vibration mode is detected for specimens Nos. 4 and 5. Almost all cases of damage were detected for the vibration mode at 176 Hz. These observations indicate that it is crucial to consider multiple modes when detecting damage since, firstly, there might be missed detections if features were extracted only from a single mode and, secondly, the features of different vibration modes display different magnitudes of deviations from the reference.

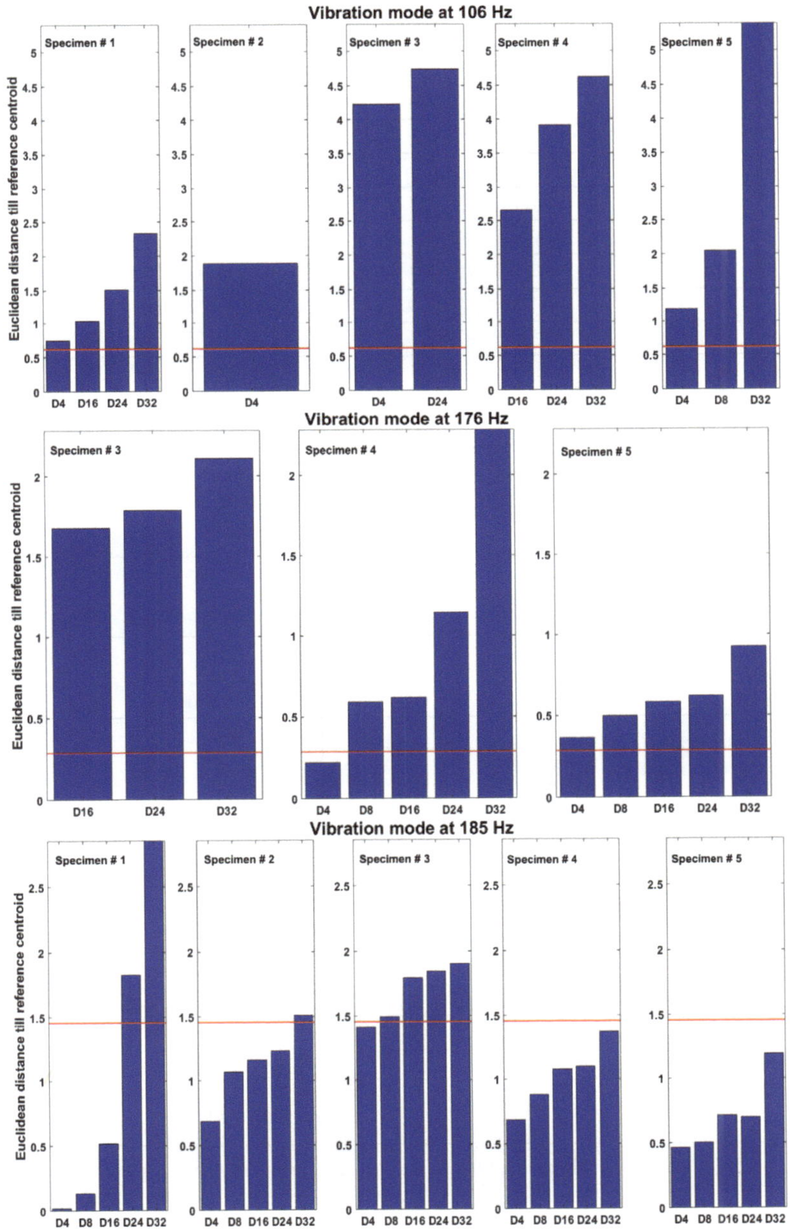

Figure 15. Damage index as Euclidean distance between feature centroids at reference state and damage states. Red horizontal line shows the threshold for each mode of vibration.

Limitations of the algorithm proposed are directly associated with the quality of the vibration signals recorded, which, in turn, strongly depend on the noise robustness of the sensors to the environmental conditions in which the structure is operating. Secondly, the algorithm does not consider the operational deflection shapes (ODSs) as an additional feature. It is anticipated that information on the ODSs could potentially enhance the discriminative power of the algorithm. It is possible to identify the ODSs with continuous wavelet transform. However, it would significantly increase the complexity of the algorithm, which was not the aim of this study. Thirdly, damage detection accuracy depends on an accurate threshold estimation. Obviously, according to statistics, a larger sample of structures would have yielded a more accurate threshold value for the Euclidean-distance damage indicator. However, there is a trade-off of accuracy and number of structures to be manufactured for such threshold calculations involving increased financial and time resources.

5.4.3. Comparison with Mahalanobis Distance

The damage identification results obtained were compared to the well-known Mahalanobis distance (MD) metric. The MD values follow a chi-squared (χ^2) distribution [25,26]. Thus, threshold for the MD values is defined as an inverse of χ^2 cumulative distribution function with v degrees of freedom and a selected probability P: $T_{MD} = \sqrt{\chi^2_{v,P}}$. The results of damage detection for specimen No. 1 are shown in Figure 16. The threshold level is selected at 95% probability and two degrees of freedom (corresponding to the two feature vectors in a feature matrix). It can be seen that while the majority of the green dots (reference MD data) lie under the threshold, the damage MD data are largely above the threshold for the vibration modes at 106 and 176 Hz. On the other hand, a poor damage detection can be seen for the vibration mode at 185 Hz, since a relatively small proportion of the damage MD data exceed the threshold.

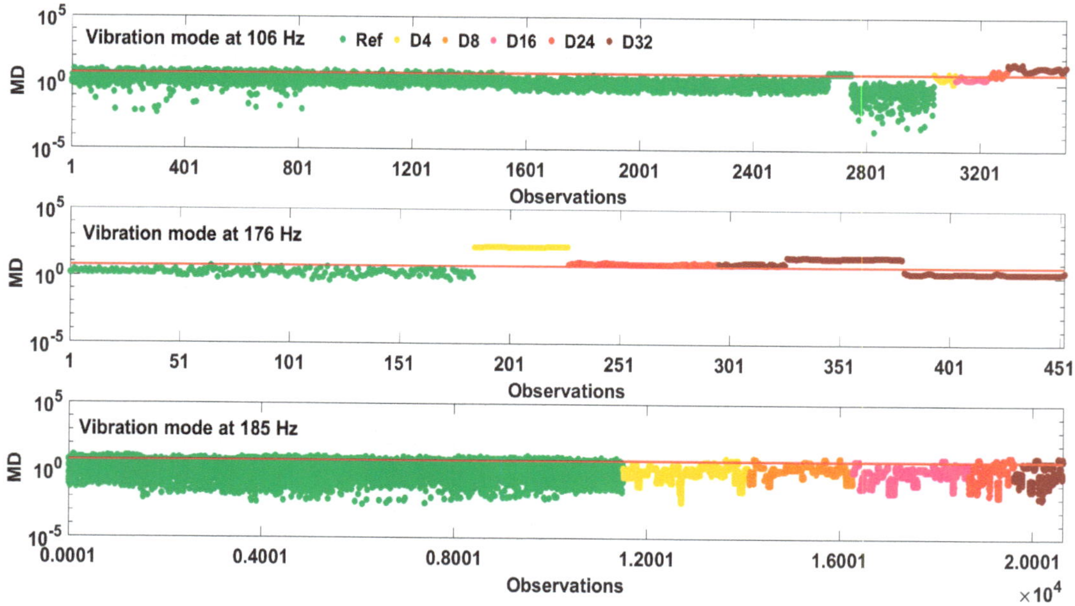

Figure 16. Mahalanobis distance (MD) metric in logarithmic scale for the identified vibration modes for specimen No. 1. Red line shows the threshold level. Different vibration modes have a different damage detection performance.

The evaluation power of both damage indicators is estimated through the false alarm rate (FAR).

$$\text{FAR} = 1 - \text{Accuracy} = 1 - \frac{\sum_{i=1}^{m} (\text{MD} > \text{T}_{\text{MD}})_i}{\sum_{i=1}^{m} \text{MD}_i}, \tag{15}$$

where m is the number of observations of the Mahalanobis distance metric for each case. The accuracy, on the other hand, is 1 − FAR and is presented in Figure 17. The colourmaps show the accuracy (in %) of damage detection for the Euclidean distance metric (on the left) and Mahalanobis distance metric (on the right). The colour corresponds to the accuracy percentage level. For the Euclidean distance, there can be either 0% or 100% accuracy (Euclidean distance either does not reach the threshold or cross it), while for the Mahalanobis distance, any intermediate accuracy can be achieved since numerous observations of this metric are obtained. Additionally, accuracy for a reference state for the Mahalanobis distance can be assessed, which is not the case for Euclidean distance. NaN (not a number) means that the vibration mode was not identified.

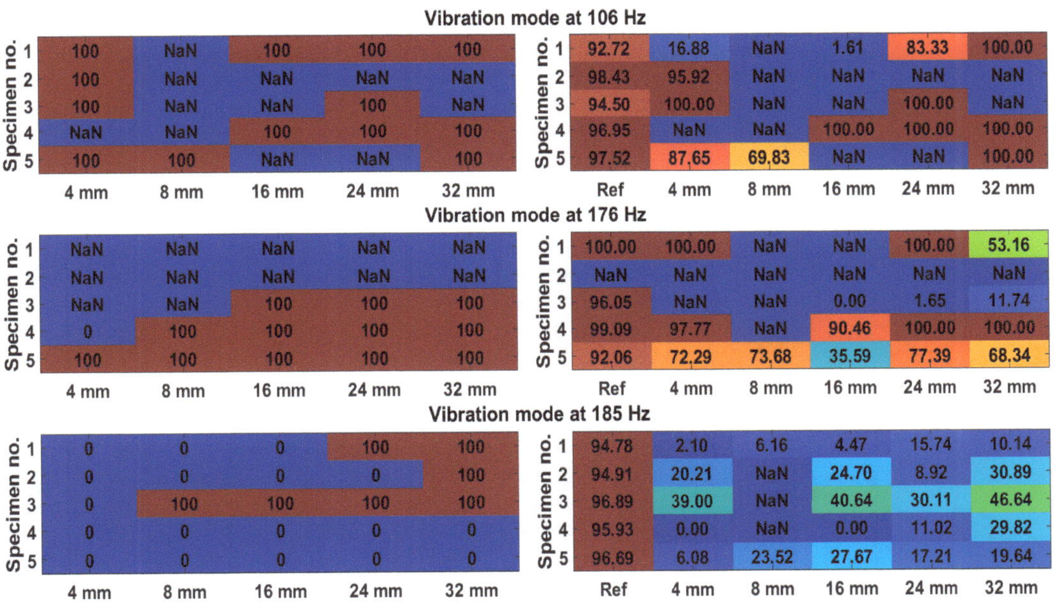

Figure 17. Colormaps of accuracy of Euclidean distance metric (**left**) and Mahalanobis distance metric (**right**) for three vibration modes. NaN means that the vibration mode was not identified for that particular case of damage.

The vibration mode at 185 Hz is the least favourable for damage detection (accuracies are low for both methods), while both other modes are roughly equal in this regard. Overall, the accuracies are comparable between both methods. In cases where there is 100% accuracy for the Euclidean distance, the corresponding accuracy for the Mahalanobis distance is usually lower, since not all observations have passed the threshold. The advantages of the method proposed are that computation for the Euclidean distance is simpler and requires less computational power since the covariance matrix does not need to be computed.

6. Conclusions

In the current study, an anomaly detection algorithm based on output-only structural vibration responses of structural components was proposed. The algorithm detects changes of modal parameters caused by structural degradation, such as the progression of damage. Phase I deals with the acquisition of vibration response signals from the sensors mounted on the structure and subsequent signal averaging and fusion. Phase II uses the modal parameters as the features identified with continuous wavelet transform routine. Phase III is carried out for decision-making regarding the state of integrity of the structure in question based on the statistical descriptors of the extracted features. Here, the feature filtering scheme based on IQR Rule is adopted to remove outlier feature values. Feature filtering revealed that the distribution for some specimens is, in fact, bimodal, while for others it is skewed and not classical Gaussian. The probability density function of the filtered features is estimated using a kernel density estimate (KDE) to avoid the assumption that the underlying distribution is normal. The damage indicator is based on the Euclidean distance between the centroid of KDE for both features at reference state and any state of damage. The following can be concluded:

1. The Euclidean distance of the centroids of the modal features KDEs between the reference and damage states can be used to detect damage.
2. The damage indicator proposed shows an upward trend for damage progression, meaning that it is effective in detecting increasing severities of damage.
3. Some vibration modes are more sensitive to damage than others. Therefore, multiple vibration modes have to be identified in order to increase the reliability of the damage detection. For example, features originating from the vibration modes at 106 and 176 Hz have significantly higher deviations from the reference than the vibration mode at 185 Hz. Therefore, these vibration modes were more effective in damage detection. On the other hand, these vibration modes could not be identified for all damage cases, while the vibration mode at 185 Hz was present in all scenarios.
4. There is a significant scatter of the feature value deviations from reference among the test samples. For the most part, this is due to inconsistencies in the sample design and instrumentation, as mentioned in Section 4.1. Specimens.
5. The damage indicator proposed was compared to the Mahalanobis distance metric for damage detection. Both methods yield comparable damage detection accuracy. Therefore, there is no reason to use a more computationally costly Mahalanobis distance approach.

Future research will be devoted to the consideration of an influence of environmental and operational factors on the structural characteristics and modal parameters for a practical SHM system. To reduce the influence of the ambient temperature, for example, some papers consider methods for suppressing this effect as interference [27], in others, methods for constructing quantitative models, that accurately predict the modal frequency that corresponds to temperature change are proposed [28]. A promising way to consider influence factors is to utilize the modal passport mentioned above, which is a method for collecting and processing all modal data of a structure while taking into account environmental and operational variances.

Author Contributions: Conceptualization, R.J.; methodology, R.J.; software, R.J.; validation, R.J.; formal analysis, R.J.; investigation, R.J.; resources, D.M.; data curation, D.M.; writing—original draft preparation, R.J.; writing—review and editing, D.M.; visualization, R.J.; supervision, D.M.; project administration, D.M.; funding acquisition, D.M. All authors have read and agreed to the published version of the manuscript.

Funding: This research was funded by the European Regional Development Fund project No. 1.1.1.1/20/A/016 "A prototype of typical structural health monitoring system of operating objects for condition based maintenance".

Institutional Review Board Statement: The study did not require ethical approval.

Acknowledgments: Authors wish to thank the staff from D un D Centrs for performing the modal tests and providing the data.

Conflicts of Interest: The authors declare no conflict of interest.

References

1. Qing, X.P.; Beard, S.J.; Kumar, A.; Ooi, T.K.; Chang, F.-K. Built-in Sensor Network for Structural Health Monitoring of Composite Structure. *J. Intell. Mater. Syst. Struct.* **2017**, *18*, 39–49. [CrossRef]
2. Liu, D.; Luo, M.; Zhang, Z.; Hu, Y.; Zhang, D. Operational modal analysis based dynamic parameters identification in milling of thin-walled workpiece. *Mech. Syst. Signal Process* **2022**, *167*, 108469. [CrossRef]
3. Janeliukstis, R.; Mironovs, D.; Safonovs, A. Statistical Structural Integrity Control of Composite Structures Based on an Automatic Operational Modal Analysis—A Review. *Polym. Mech.* **2022**, *58*, 181–208. [CrossRef]
4. Reynders, E.; De Roeck, G. Reference-based combined deterministic–stochastic subspace identification for experimental and operational modal analysis. *Mech. Syst. Signal Process* **2008**, *22*, 617–637. [CrossRef]
5. Guillaume, P.; Verboven, P.; Vanlanduit, S.; Van Der Auweraer, H.; Peeters, B. A polyreference implementation of the least-squares complex frequency domain-estimator. In Proceedings of the of the IMAC XXI, International Modal Analysis Conference, Kissimmee, FL, USA, 3–6 February 2003.
6. Zhu, Q.; Wang, Y.; Shen, G. Research and Comparison of Time-frequency Techniques for Nonstationary Signals. *J. Comput.* **2012**, *7*, 954–958. [CrossRef]
7. Hamtaei, M.R.; Anvar, S.A. Estimation of modal parameters of buildings by wavelet transform. In Proceedings of the the 14th World Conference on Earthquake Engineering 14 WCEE, Beijing, China, 12–17 October 2008.
8. Staszewski, W.J. Identification of Damping in Mdof Systems Using Time-Scale Decomposition. *J. Sound Vib.* **1997**, *203*, 283–305. [CrossRef]
9. Zhang, M.; Huang, X.; Li, Y.; Sun, H.; Zhang, J.; Huang, B. Improved Continuous Wavelet Transform for Modal Parameter Identification of Long-Span Bridges. *Shock. Vib.* **2020**, *2020*, 4360184. [CrossRef]
10. Su, W.C.; Huang, C.S.; Chen, C.H.; Liu, C.Y.; Huang, H.C.; Le, Q.T. Identifying the Modal Parameters of a Structure from Ambient Vibration Data via the Stationary Wavelet Packet. *Comput. Civ. Infrastruct. Eng.* **2014**, *29*, 738–757. [CrossRef]
11. Wei, P.; Li, Q.; Sun, M.; Huang, J. Modal identification of high-rise buildings by combined scheme of improved empirical wavelet transform and Hilbert transform techniques. *J. Build. Eng.* **2023**, *63*, 105443. [CrossRef]
12. Sun, H.; Di, S.; Du, Z.; Wang, L.; Xiang, C. Application of multisynchrosqueezing transform for structural modal parameter identification. *J. Civ. Struct. Health Monit.* **2021**, *11*, 1175–1188. [CrossRef]
13. Nagarajaiah, S.; Basu, B.; Yang, Y. Output-only modal identification and structural damage detection using time–frequency and wavelet techniques for assessing and monitoring civil infrastructures. Sensor Technologies for Civil Infrastructures. In *Volume 1: Sensing Hardware and Data Collection Methods for Performance Assessment*, 2nd ed.; Woodhead Publishing Series in Civil and Structural Engineering; Woodhead Publishing: Sawston, UK, 2022; pp. 481–529.
14. Janeliukstis, R. Continuous wavelet transform-based method for enhancing estimation of wind turbine blade natural fre-quencies and damping for machine learning purposes. *Measurement* **2021**, *172*, 108897. [CrossRef]
15. Zimek, A.; Schubert, E. Outlier Detection. In *Encyclopedia of Database Systems*; Liu, L., Özsu, M., Eds.; Springer: New York, NY, USA, 2017.
16. Helbing, G.; Ritter, M. Deep Learning for fault detection in wind turbines. *Renew. Sustain. Energy Rev.* **2018**, *98*, 189–198. [CrossRef]
17. Bangalore, P.; Patriksson, M. Analysis of SCADA data for early fault detection, with application to the maintenance management of wind turbines. *Renew. Energy* **2018**, *115*, 521–532. [CrossRef]
18. García, D.; Tcherniak, D.; Trendafilova, I. Damage assessment for wind turbine blades based on a multivariate statistical approach. *J. Phys. Conf. Ser.* **2015**, *628*, 12086. [CrossRef]
19. Movsessian, A.; Cava, D.G.; Tcherniak, D. An artificial neural network methodology for damage detection: Demonstration on an operating wind turbine blade. *Mech. Syst. Signal Process.* **2021**, *159*, 107766. [CrossRef]
20. Sarmadi, H.; Yuen, K. Early damage detection by an innovative unsupervised learning method based on kernel null space and peak-over-threshold. *Comput. Civ. Infrastruct. Eng.* **2021**, *36*, 1150–1167. [CrossRef]
21. Mironov, A.; Mironovs, D. Modal passport of dynamically loaded structures: Application to composite blades. In Proceedings of the 13th International Conference Modern Building Materials, Structures and Techniques, Vilnius, Lithuania, 16–17 May 2019. [CrossRef]
22. Mironov, A.; Doronkin, P. The Demonstrator of Structural Health Monitoring System of Helicopter Composite Blades. In Proceedings of the ICSI 2021 The 4th International Conference on Structural Integrity, Procedia Structural Integrity 37, Online, 30 August–2 September 2022.
23. Yang, G.; Yang, Z.-B.; Zhu, M.-F.; Tian, S.-H.; Chen, X.-F. Directional wavelet modal curvature method for damage detection in plates. *J. Phys. Conf. Ser.* **2022**, *2184*, 12027. [CrossRef]
24. Dziedziech, K.; Staszewski, W.J.; Mendrok, K.; Basu, B. Wavelet-Based Transmissibility for Structural Damage Detection. *Materials* **2022**, *15*, 2722. [CrossRef]
25. Colone, L.; Hovgaard, M.; Glavind, L.; Brincker, R. Mass detection, localization and estimation for wind turbine blades based on statistical pattern recognition. *Mech. Syst. Signal Process.* **2018**, *107*, 266–277. [CrossRef]
26. Yang, C.; Zhang, S.; Liu, Y.; Yu, K. Damage detection of bridges under changing environmental temperature using the characteristics of the narrow dimension (CND) of damage features. *Measurement* **2022**, *189*, 110640. [CrossRef]

27. Luo, J.; Huang, M.; Lei, Y. Temperature Effect on Vibration Properties and Vibration-Based Damage Identification of Bridge Structures: A Literature Review. *Buildings* **2022**, *12*, 1209. [CrossRef]
28. Shan, W.; Wang, X.; Jiao, Y. Modeling of Temperature Effect on Modal Frequency of Concrete Beam Based on Field Monitoring Data. *Shock. Vib.* **2018**, *2018*, 8072843. [CrossRef]

sensors

Article

Random Traffic Flow Simulation of Heavy Vehicles Based on R-Vine Copula Model and Improved Latin Hypercube Sampling Method

Hailin Lu [1,2], Dongchen Sun [1] and Jing Hao [1,*]

[1] School of Civil Engineering and Architecture, Wuhan Institute of Technology, Wuhan 430074, China
[2] Hubei Provincial Engineering Research Center for Green Civil Engineering Materials and Structures, Wuhan 430073, China
* Correspondence: jing_hao@stu.wit.edu.cn

Abstract: The rationality of heavy vehicle models is crucial to the structural safety assessment of bridges. To establish a realistic heavy vehicle traffic flow model, this study proposes a heavy vehicle random traffic flow simulation method that fully considers the vehicle weight correlation based on the measured weigh-in-motion data. First, a probability model of the key parameters in the actual traffic flow is established. Then, a random traffic flow simulation of heavy vehicles is realized using the R-vine Copula model and improved Latin hypercube sampling (LHS) method. Finally, the load effect is calculated using a calculation example to explore the necessity of considering the vehicle weight correlation. The results indicate that the vehicle weight of each model is significantly correlated. Compared to the Monte Carlo method, the improved LHS method better considers the correlation between high-dimensional variables. Furthermore, considering the vehicle weight correlation using the R-vine Copula model, the random traffic flow generated by the Monte Carlo sampling method ignores the correlation between parameters, leading to a weaker load effect. Therefore, the improved LHS method is preferred.

Keywords: weigh-in-motion; random traffic flow; correlation; R-vine Copula; Latin hypercube sampling

Citation: Lu, H.; Sun, D.; Hao, J. Random Traffic Flow Simulation of Heavy Vehicles Based on R-Vine Copula Model and Improved Latin Hypercube Sampling Method. *Sensors* **2023**, *23*, 2795. https://doi.org/10.3390/s23052795

Academic Editors: Yongbo Li, Bing Li and Khandaker Noman

Received: 26 January 2023
Revised: 24 February 2023
Accepted: 1 March 2023
Published: 3 March 2023

1. Introduction

Transportation structures such as roads and bridges are designed to carry moving traffic loads. However, with the rapid economic development, the load capacity and occupancy of heavy vehicles are increasing [1,2], generating greater safety hazards to in-service highway bridges and even leading to serious accidents [3]. The heavy vehicle weight parameters indicate strong randomness and significant correlation between parameters; therefore, it is of great importance to fully study the randomness and correlation of heavy vehicle weights and propose a more realistic simulation method for heavy vehicle flow to evaluate the safety of bridge structures.

Several scholars have implemented random traffic flow simulations considering several parameters, such as vehicle type, vehicle weight, axle weight, and vehicle speed, based on data measured utilizing a dynamic weighing system, weigh-in-motion (WIM). For example, Zhouhong et al. [4], Yang et al. [5], and Liang and Xiong [6] developed random traffic flow models applicable to specific regions using Monte Carlo simulation methods. Notably, mass parameters, such as vehicle weight and axle weight, are important for the load effect, and there is a significant correlation between the mass parameters of each traffic model. To build a model closer to a real traffic flow, numerous scholars have used the Copula theory to describe the nonlinear correlation of random parameters. For example, Li et al. [7] analyzed the correlation between vehicle axle weights using the t-Copula function and established a random traffic flow model based on a Monte Carlo simulation. Li [8] analyzed the axle weight correlation according to the Copula distribution function, established a two-dimensional compound Poisson process for vehicle speed and weight using the Levy Copula function, and finally established a random traffic flow model utilizing Monte Carlo simulations. Torres-Alves et al. [9] used the vine Copula model to analyze the axle weight, wheelbase, and vehicle distance to establish a random traffic flow model that considered the correlations through random sampling. Sorianoa et al. [10] used the binary Copula function to construct a joint distribution model for overweight trucks with regard to occupancy and average daily traffic flow. In general, the accuracy of random traffic flow simulations can be improved by considering the correlation between the mass parameters of each vehicle. However, existing simulation methods have the following two shortcomings: First, the parameters of the C-vine and D-vine Copula models used in random traffic flow simulations are all based on fixed-type subjective assumptions, whereas the correlation structure between the variables of each dimension in actual engineering is complex and variable. Furthermore, accurately describing the parameters using a fixed structure is difficult. Therefore, accurately constructing a high-dimensional variable correlation model using the vine Copula model still has certain limitations [11,12]. Second, random traffic flow simulations are mainly conducted by considering the correlation between parameters through the Copula theory and Monte Carlo sampling. The correlation between parameters is difficult to determine using solely the Monte Carlo sampling method because it leads to inaccurate sampled parameters when correlating them. Therefore, a more rational sampling method is urgently needed.

A literature review found the following theories to resolve the above two problems. Morales-Nápoles et al. [13] proposed an R-vine Copula model for topology optimization based on the data-driven nonparametric estimation of the decomposed Copula function, which has better flexibility and practicality. Latin hypercube sampling (LHS), proposed by McKay et al. [14], can achieve stratified sampling to avoid the sampling aggregation phenomenon induced in Monte Carlo sampling while achieving improved accuracy and efficiency. Iman and Conover [15] proposed a simulation method independent of the data distribution. It derives the expected rank correlation matrix using multi-parameter input random variables through matrix transformation to fully preserve the data correlation characteristics. This method can be applied to any type of distribution sampling.

In light of this, 2020 WIM data was collected from the 49,010 Census Station of Interstate 80 in the U.S. to analyze the statistical characteristics of daily traffic flow, vehicle type, vehicle weight, vehicle speed, and other heavy vehicle parameters. Based on this, a scholastic traffic flow model for heavy vehicles was established using the R-vine Copula model with an improved LHS method. The applicability and superiority of the method were verified. This method provides a reference for vehicle load modeling and load design limit optimization.

2. Statistical Characterization of Heavy Vehicle Load Parameters

A WIM system equipped with dual loop sensors was installed on the 49,010 Census Station of Interstate 80, the second largest freeway in Vacaville, California, U.S., as shown in Figure 1. Monitoring data, which includes parameters such as vehicle type, axle weight, vehicle weight, daily traffic volume, vehicle speed, and lane location, was collected, a total of 181,800 data for the year 2020.

Figure 1. Sensor layout at the mainline traffic survey station.

2.1. Model Classification and Lane Occupancy

According to the classification standard published by the Federal Highway Administration (FHWA) [16], vehicle classification is based on the number of vehicle axles (Axle) and the truck towing method. The towing method is subdivided into single unit (SU), single trailer (ST), and multiple trailer (MT). The vehicle occupancy of each lane at the assessment site is listed in Table 1, and the last six vehicle types in the table are defined by the FHWA as heavy vehicles.

Table 1. Lane occupancy by vehicle type.

Vehicle Type	Lane 0/‱	Lane 1/‱	Lane 2/‱	Lane 3/‱	Total/‱
2 Axle, 4T SU	1790.3000	1788.9550	2333.1040	965.8129	2985.4270
Bus	108.3499	107.5772	56.7047	104.5149	169.5703
2 Axle, 6T SU	3311.0860	3310.1770	3487.1130	2365.9790	5127.4970
3 Axle SU	302.7385	302.6917	418.3685	334.2686	119.7357
4+ Axle SU	33.4853	32.5274	91.9235	18.3512	7.454761
<4 Axle ST	422.6222	421.4048	370.9372	582.9789	140.6703
5 Axle ST	3178.3480	3178.6290	2240.7240	4686.4910	941.3933
6+ Axle ST	19.2045	18.9485	12.2632	29.1005	6.1783
<5 Axle MT	185.4344	185.2683	174.9283	257.5469	44.1158
6 Axle MT	86.8883	86.2973	36.8910	139.6438	27.8277
7+ Axle MT	3.1873	3.1622	3.0405	4.4789	0.6127

2.2. Vehicle Weight Statistics

Parametric and nonparametric methods are commonly used for estimating probability density functions. The parametric method assumes that the random variation of variables conforms to a known distribution and performs parameter estimations based on the monitoring data, whereas the nonparametric method can estimate probability density functions without assuming the type of probability distribution. The parametric method has the following limitations: First, the process of assuming the distribution type is often subjective. Second, the random behavior and dispersion of the monitoring data make it difficult to determine the best assumption of the distribution type. Therefore, based on the monitoring data, the nonparametric kernel density estimation method was used to fit the vehicle weight distribution of heavy vehicles.

Estimating the nonparametric kernel density of the variable x as $\hat{f}(x)$, the expression becomes as follows:

$$\hat{f}(x) = \frac{1}{nh}\sum_{i=1}^{n} K(\frac{x - x_i}{h}), \tag{1}$$

where h is the smooth parameter, n is the sample capacity, and K is the kernel function. The kernel functions typically used in such applications are the Gaussian, Box, trigonometric, and Epanechnikov.

The nonparametric fit and R^2 goodness-of-fit tests revealed that the probability density and distribution functions of the weight of the six types of heavy vehicles are well-described by the different kernel functions, as indicated in Figure 2. Among them, the values of the R^2 goodness-of-fit of the Gaussian kernel density estimation for the weight distribution of the six heavy vehicles were all greater than 0.98. These values were slightly higher than the nonparametric fitting results of the other three kernel functions. In this study, the Gaussian kernel function was adopted, whose expression is as follows:

$$K_{gausssian} = \frac{1}{2\pi} e^{(\frac{-u^2}{2})}. \tag{2}$$

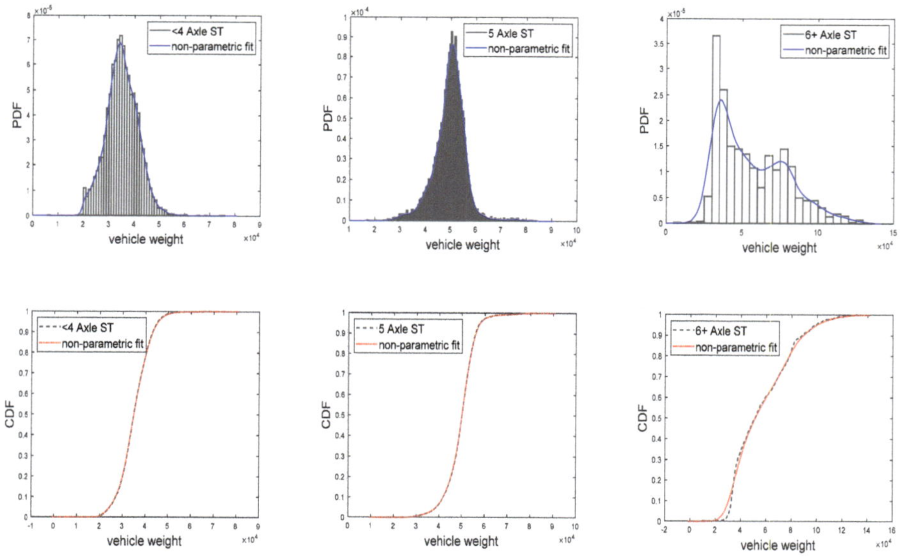

Figure 2. *Cont.*

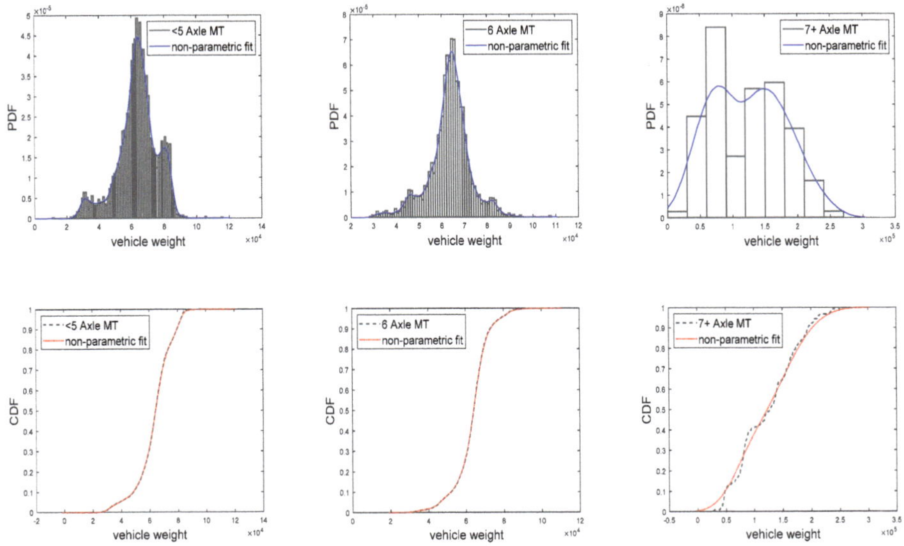

Figure 2. Weight of heavy vehicles.

Vehicle weight is the most critical parameter influencing the vehicle load effect. Overloaded heavy vehicles are especially hazardous to the safety of highway bridges. The vehicle weight of each vehicle type in a heavy traffic flow is random, and there is a correlation between the weights of different types of vehicles. To realize an accurate simulation of the heavy vehicle traffic flow, the correlation between the weights of different types of vehicles must be analyzed in addition to considering the random behavior of the weight of each vehicle type. The Pearson correlation coefficient is mainly used to describe a linear correlation applicable to a single-peaked normal distribution. The Kendall rank correlation coefficient can accurately measure the consistency of variation trends and the degree of variation between variables, and is applicable to various distributions. Considering the nonlinear correlation between the vehicle weights of each heavy vehicle type, the Kendall rank correlation coefficient was used as the index to evaluate the correlation. The calculation is expressed as follows:

$$\tau = \sum_{i<k} (\text{sign}(x[j] - x[i]) \times \text{sign}(y[j] - y[i])). \tag{3}$$

The Kendall rank correlation coefficients of heavy vehicle weights are listed in Table 2. These coefficients indicate that the correlation of the vehicle weights of certain types of heavy vehicles is more significant; thus, the correlation needs to be considered when modeling a heavy vehicle traffic flow.

Table 2. Correlation coefficients of heavy vehicle weight.

	<4 Axle ST	5 Axle ST	6+ Axle ST	<5 Axle MT	6 Axle MT	7+ Axle MT
<4 Axle ST	1	0.3799	0.1725	0.4082	0.4032	0.0886
5 Axle ST	0.3799	1	0.2246	0.3463	0.3101	0.1076
6+ Axle ST	0.1725	0.2246	1	0.27	0.1613	0.194
<5 Axle MT	0.4082	0.3463	0.2700	1	0.4011	0.1313
6 Axle MT	0.4032	0.3101	0.1613	0.4011	1	0.0874
7+ Axle MT	0.0886	0.1076	0.1940	0.1313	0.0874	1

2.3. Daily Traffic Statistics

The cumulative probability density of the daily traffic flow is shown in Figure 3. The daily traffic flow was found to be mainly concentrated from 2500 to 6000 and from 7500 to 11,000 vehicles. When the daily traffic flow is less than 6000 vehicles, it is called the general operation state, and when it exceeds 6000 vehicles, it is called the intensive operation state. The average daily traffic volume is approximately 4900 vehicles/day for the general operation state, 9600 vehicles/day for the intensive operation state, and 7800 vehicles/day for the annual average daily traffic volume, not considering the operation state.

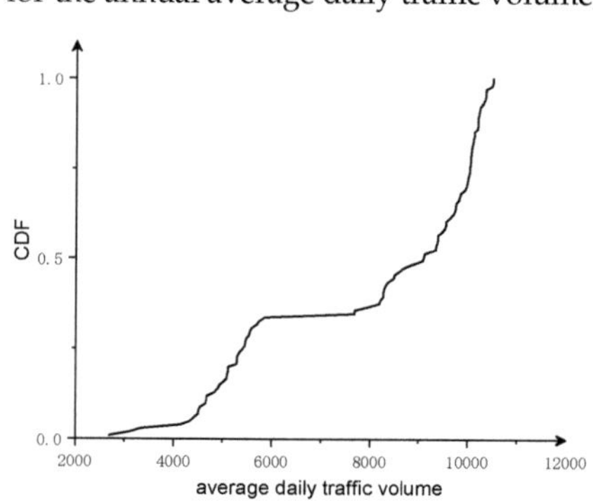

Figure 3. Cumulative distribution of average daily traffic volume.

2.4. Vehicle Speed Statistics

A statistical analysis of the measured speed for each vehicle in the traffic flow was conducted. The measured speed data of each vehicle was fitted by Gaussian and multi-peaked Gaussian distributions. The goodness-of-fit R^2 results for each vehicle speed was greater than 0.96. The results indicated that the speed of each vehicle in the traffic flow conformed to Gaussian and multi-peaked Gaussian distributions. The fitting formula is shown in Equation (4), and the fitting parameters of Gaussian and multi-peak Gaussian distributions are listed in Table 3.

$$f(x) = \sum_{1}^{i} a_i x e^{-\frac{(x-b_i)^2}{ci}}. \tag{4}$$

Table 3. Vehicle speed fitting parameters for each vehicle type.

Vehicle Type		Parameters	
<4 Axle ST	a = 230.2	b = 58.78	c = 4.506
5 Axle ST	a1 = 1588	b1 = 60.68	c = 2.288
	a2 = 432	b2 = 62.06	c = 5.881
6+ Axle ST	a = 838.8	b = 61.44	c = 3.828
<5 Axle MT	a = 1046	b = 61.2	c = 3.003
6 Axle MT	a1 = 306.4	b1 = 61.44	c1 = 1.296
	a2 = 273.2	b2 = 60.57	c2 = 4.082
7+ Axle MT	a = 188.9	b = 60.58	c = 5.524

3. Six-Dimensional Joint Distribution Model for Heavy Vehicle Weight

3.1. Six-Dimensional Joint Distribution Model for Vehicle Weight Based on R-Vine Copula

The Copula theory enables the modeling of joint distributions of multidimensional random variables. Sklar's theorem [17] provides the relationship between the joint distribution function $F(x_1, x_2, \cdots, x_n)$ and the Copula distribution function $C(u_1, u_2, \cdots, u_n)$,

$$F(x_1, x_2, \cdots, x_n) = C(u_1, u_2, \cdots, u_n). \tag{5}$$

By deriving Equation (5), the corresponding probability density function is obtained as follows:

$$f(x_1, x_2, \cdots, x_n) = c(u_1, u_2, \cdots, u_N)\prod_{n=1}^{N} f_n(x_n). \tag{6}$$

where $c(u_1, u_2, \cdots, u_N) = \frac{\partial C(u_1, u_2, \cdots, u_N)}{\partial u_1 \partial u_2 \cdots \partial u_N}$ is the Copula density function, and N is the density function of the marginal probability density function $f_n(x_n)$ (n = 1, 2, $\cdots$, N).

Although the above Copula model can effectively describe the correlation between random variables, its application to high-dimensional random variables elicits the problems of dimensional disaster and insufficient model accuracy. To resolve these problems, the R-vine Copula model can decompose the high-dimensional Copula function into the product of several two-dimensional Copula functions [18,19]. An n-dimensional R-vine Copula model consists of $n-1$ layer trees $T_1, T_2, \ldots\ T_{n-1}$, where each edge in the tree corresponds to a two-dimensional Copula distribution function, and the set of nodes in the tree is denoted as N = {$N1$, $N2$, $N3$, ... , Nn}. An n-dimensional R-vine structure is subject to the following conditions:

(1) A tree T_1 containing n vertices and $n-1$ edges.
(2) The tree T_i contains n − i + 1 vertices and n − i edges.
(3) If an edge of the tree T_i connects two nodes, the two edges in the $Ti-1$ tree corresponding to these two nodes share the same node.

The symbol e represents an edge in the tree and the set of edges E in E = (E_1, E_2, $\cdots$, E_n−1); the edge e = $a(e)$, $b(e)$ | $D(e)$ of Ei represents $D(e)$ as a condition of a(e), b(e), and a subset consisting of conditional variables. Each edge e = {a, b} ∈ Ei consists of two nodes connected by an edge e. The density function corresponding to edge e is denoted as $Ca(e)$, $b(e)$ | $D(e)$. The n random variables are $X1$, $X2$, ... , Xn, and the subvectors denoted by $XD(e)$ are determined by the condition set $D(e)$. The i random variables of the marginal probability density function are fi. Based on this, the final joint density function f is shown in Equation (7).

$$f(x_1, x_2, \cdots x_n) = \prod_{k-1}^{n} f_k(x_k) \prod_{i=1}^{n-1} \prod_{e \in Ei} c_{a(e),b(e)|D(e)}\left(F(x_{a(e)|D(e)}), F(x_{b(e)|D(e)})\right). \tag{7}$$

Equation (7) shows that once the marginal probability density function $f_k(x_k)$ of the six heavy vehicle weights and the two-dimensional Copula distribution function $c_{a(e),b(e)|D(e)}$ corresponding to each edge in the tree are determined, the joint probability density function $f(x_1, x_2, \cdots x_n)$ can be determined, and the initial construction of the six-dimensional joint distribution model of R-vine Copula can also be realized.

3.2. Optimization of the Joint Distribution Model of R-Vine Copula

For the six-dimensional joint distribution model of the above six-dimensional R-vine Copula, there exist $(6!/2) \times 2^{(6-2)!/[2(6-4)!]}$ possible topologies, and the correlation between the random variables varies with topology. Therefore, determining the best correlation between the random variables becomes a critical problem to resolve for the correlation between high-dimensional random variables. Thus, the joint distribution models of the six heavy vehicle weights were optimized in terms of the connection structure of each layer of the tree, and the joint distribution model as follows: (1) Maximum spanning tree optimization was conducted on the connection structure of each group of trees in the R-vine structure according to the edge weight coefficients. The empirical Kendall weights $\hat{\tau}_{ij}$ were used as the evaluation index, and the optimization formula for its structure is as follows:

$$\max \sum_{\text{edges } e=\{i,j\} \text{ in spanning tree}} \left|\hat{\tau}_{ij}\right|. \tag{8}$$

(2) To ensure the goodness-of-fit of each marginal distribution model with the final generated joint distribution model, the optimal Copula distribution function was selected using two criteria, namely, the Akaike information criterion (*AIC*) and Bayesian information criterion (*BIC*), to optimize each joint distribution model. The *AIC* and *BIC* were calculated as follows:

$$AIC = 2k - 2\ln\{RVine\}(\theta|u). \tag{9}$$

$$BIC = \ln n \times k - 2\ln\{RVine\}(\theta|u|r). \tag{10}$$

where k is the number of parameters, and $\{RVine\}(\theta \mid u \mid r)$ denotes the set of parameters as θ, u, and r.

In addition, the BIC can solve the problrm that the sample size n result to the complex model possesing of large amount of calculation. Moreover, smaller values of AIC and BIC indicate a more accurate description of the correlation between random variables.

The vine structure of the joint distribution model of the six heavy vehicle weights is shown in Figure 4. The marginal distribution models of each layer of the tree, the Copula distribution function, Copula distribution function coefficients (par1 and par2), and AIC and BIC results of each marginal distribution model are listed in Table 4. The AIC and BIC of the joint distribution models of the six heavy vehicles after optimization were −969.2181 and −921.9249, respectively.

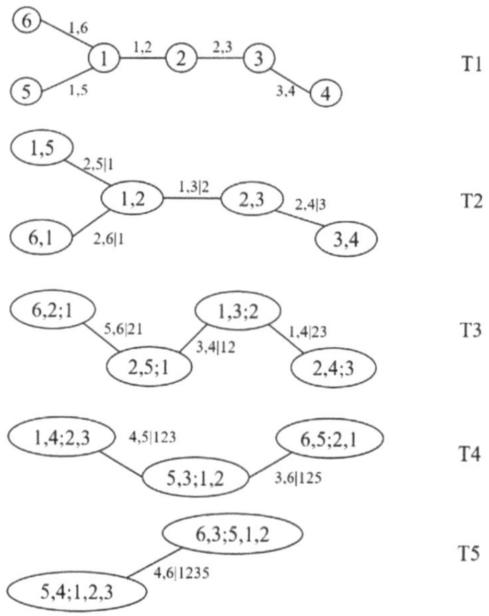

Figure 4. R-vine structure of weight.

Table 4. Parameters of R-vine Copula model.

Tree	Edge	Copula	Par1	Par2	AIC	BIC
1	1,6	Frank	−0.7304		0.0157	2.9716
	1,5	Frank	2.1041		−13.9431	−10.9872
	1,2	Clayton	0.3003		−4.8657	−1.9099
	2,3	Rotated Clayton	1.3532		−25.6906	−22.7348
	3,4	Gumbel	1.7000		−900.3170	−897.3612
2	2,5 ∣ 1	Frank	−0.5347		1.2936	4.2492
	2,6 ∣ 1	Student	−0.0234		−39.9868	−34.0751
	1,3 ∣ 2	Frank	−0.0837		1.9577	4.9134
	2,4 ∣ 3	Rotated Joe	1.0329		1.9902	4.9460
3	5,6 ∣ 21	Gaussian	0.0137		1.9094	4.8652
	3,4 ∣ 12	Gumbel	1.0623		2.0451	5.0009
	1,4 ∣ 23	Frank	0.2530		2.0387	4.9946
4	4,5 ∣ 123	Frank	−0.6243		0.4744	3.4302
	3,6 ∣ 125	Frank	−0.2696	5.5349	2.0123	4.9682
5	4,6 ∣ 1235	Clayton	0.0544		1.8483	4.8041

4. Application of Improved Latin Hypercube Sampling

The Monte Carlo method is often used for sampling in existing random traffic simulations because of its advantages of simplicity and ease of implementation. However, when the number of simulations is small, this method exhibits an aggregation phenomenon, resulting in the neglect of small probability events. Furthermore, this method tends to destroy the correlation between parameters when sampling multidimensional random variables. The LHS method avoids data aggregation by stratifying the probability distribution and is suitable for multidimensional variable sampling with high accuracy and efficiency. Diagrams of the two sampling methods are shown in Figures 5 and 6.

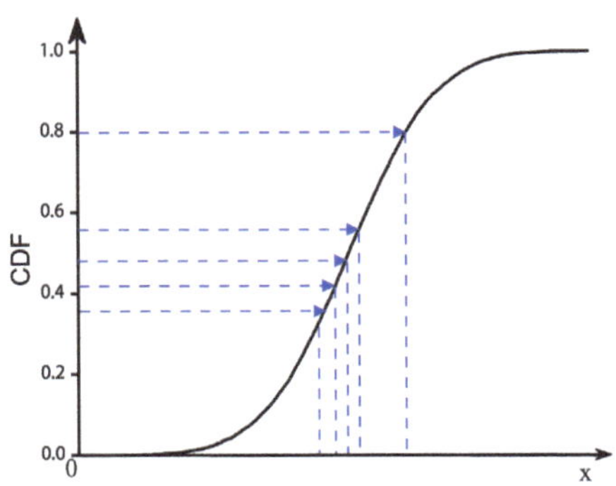

Figure 5. Monte Carlo sampling.

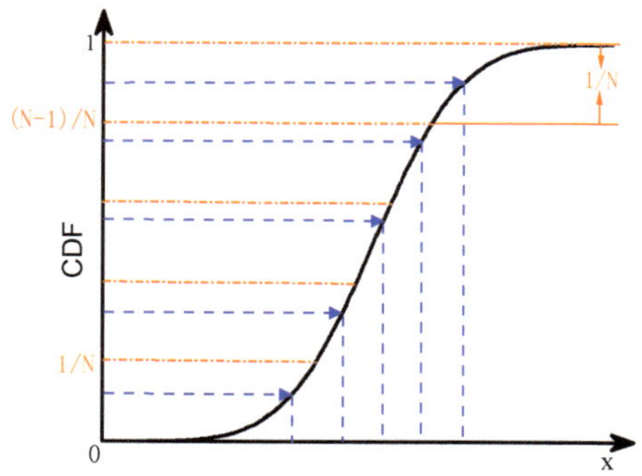

Figure 6. Latin hypercube sampling (LHS).

Fundamentals of Improved Latin Hypercube Sampling

The improved random simulation method proposed by Iman and Conover [15] preserves the correlation between random variables and is applicable to any distribution type. This method is based on the following principle.

If the elements of the random vector x are uncorrelated and there is a correlation matrix I, C is the expected correlation matrix generated by transforming *x*. C is positive, definite, and symmetric and is equal to the target correlation coefficient matrix C*. According to the Cholesky determinant used by Scheuer and Stoller [20], a lower triangular matrix *P* can be obtained such that *PP′* = C. The desired correlation matrix C is obtained by transforming the vector *XP′*. The Cholesky determinant used is as follows:

$$p_{i,i} = (c_{i,i} - \sum_{k=1}^{i-1} p_{i,k}^2)^{\frac{1}{2}}, \tag{11}$$

$$p_{i,j} = (c_{i,j} - \sum_{k=j}^{i-1} p_{i,k}p_{j,k}) \div p_{j,j}, \tag{12}$$

where $c_{i,\,i}$ and $p_{i,\,i}$ are the diagonal elements in the matrix; *i* and *j* represent the rows and columns in the matrix, respectively; and $c_{i,\,k}$ represents the elements of the *i*-th row and *k*-th column in matrix *C*.

5. Simulation of Random Traffic Flow of Heavy Vehicles and Analysis of Load Effect

Based on the results of the analysis of statistical characteristics of heavy vehicle load parameters, the six-dimensional joint distribution model of vehicle weight, and the improved LHS method mentioned above, the simulation flow chart of the random traffic flow of heavy vehicles is shown in Figure 7. based on the idea of this figure, the simulation program for the random traffic flow of heavy vehicles was prepareand, and this random traffic flow contains 300 vehicles during one hour, which considering the vehicle weight correlation. Notably, the wheelbase-to-axle weight distribution ratios for the six types of heavy vehicles were calculated according to the standard vehicle model provided by the FHWA [21]. Due to sensor performance limitations, the WIM device was not able to collect the following distance during system acquisition, so the authors used the average distance in this article.

The traffic flow samples generated were used with the R-vine Copula model and improved LHS method, called working condition I. To further verify the superiority of this method, working condition II (R-vine Copula model and Monte Carlo sampling method) and working condition III (Monte Carlo sampling method) was also set up, and its samples were calculated separately. The comparison results between the samples generated by the two working conditions and actual model occupancy are listed in Table 5. Working condition I was found to be closer to the monitoring data than working conditions II and III, indicating that the method proposed in this study is superior.

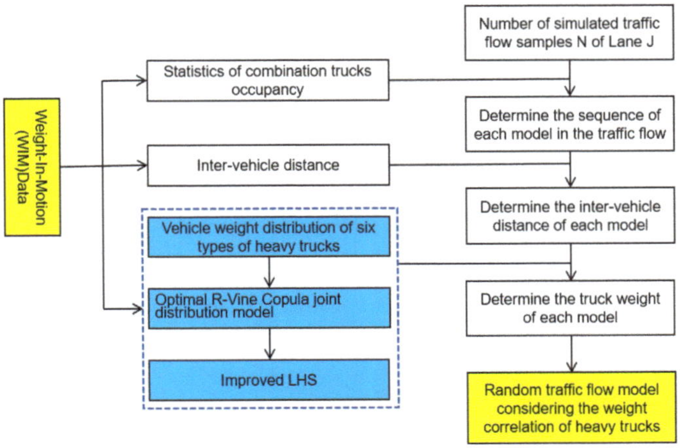

Figure 7. Flow chart of the heavy vehicle random traffic flow model.

Table 5. Total weight of each vehicle in traffic.

	<4 Axle ST	5 Axle ST	6+ Axle ST	<5 Axle MT	6 Axle MT	7+ Axle MT
Monitoring data	0.0881	0.1329	0.1463	0.1637	0.1614	0.3075
Working condition 1	0.0883	0.1330	0.1463	0.1637	0.1614	0.3074
Working condition 2	0.0905	0.1343	0.1427	0.1682	0.1592	0.3055
Working condition 3	0.0891	0.1348	0.1460	0.1645	0.1609	0.3047

In order to further verify the necessity of considering the correlation of heavy vehicle weight parameters, the load effects of three one-spans simply-supported beams under three working conditions were calculated separately. Firstly, ANSYS, a finite element analysis software, was used to build one-spans of 10 m, 20 m, and 30 m, respectively. The vehicle load samples under the three working conditions were input into the structure to obtain the maximum bending moment of the span section in turn, and the results are shown in Table 6. When the correlation is not considered, the bending moment is the smallest; when the correlation is considered by the R-vine Copula and the Monte Carlo sampling is used, the bending moment is the second largest; when the traffic load sample is obtained by the method of this paper, the bending moment is the largest. This shows that the load effect is conservative if the correlation of the heavy vehicle weight is not fully considered. This is since even if the R-Vine Copula theory is used to consider the vehicle weight correlation, the sampling method still uses Monte Carlo, which leads to the concentration of the sample on the lighter vehicle weight models and, thus, leads to the small load effect results, which is noteworthy.

Table 6. The total weight share of each model in the traffic flow.

	Working Condition 1/kN·m	Working Condition 2/kN·m	Working Condition 3/kN·m
10 m	3725	3348	3288
20 m	7511	7062	6491
30 m	12,255	11,253	10,776

6. Conclusions

In this study, based on the monitored traffic data, a random traffic sample of heavy vehicles considering vehicle weight correlations was generated using the optimal R-vine Copula model and improved LHS method. The following conclusions were obtained.

(1) The nonparametric kernel density estimation can effectively estimate the probability distribution function of the vehicle weight, and there is a correlation between the weight of each type of heavy vehicle. The vehicle speed conforms to the Gaussian and multi-peaked Gaussian distributions.

(2) Various Copula distribution functions of the R-vine Copula model can be selected to connect the marginal distribution functions of each dimension flexibly. Using the maximum spanning tree to choose the optimal topology, the AIC and BIC selected the R-vine Copula model to achieve an accurate description of the joint distribution of the vehicle weight of each vehicle model in the heavy vehicle traffic flow.

(3) The Monte Carlo sampling method destroys the correlation between multidimensional variables, whereas the improved LHS method adequately preserves the data correlation.

(4) The random traffic samples of heavy vehicles generated by considering the vehicle weight correlation based on the optimal R-vine Copula model and improved LHS method are more in line with actual scenarios than other methods. Moreover, the calculated load effect will be smaller if the vehicle weight correlation is not considered.

The authors concluded that the correlation between heavy vehicle weights may be correlated with the industrial distribution and industrialization of each region. Subsequent in-depth exploration of the statistical laws of heavy vehicle weight correlation needs to be investigated based on a large amount of WIM data, using the improved method proposed in this paper, and in conjunction with stochastic process theory.

Author Contributions: Conceptualization, methodology, resources, validation, H.L., D.S. and J.H.; funding acquisition, supervision, writing—review and editing, H.L., D.S. and J.H.; project administration, visualization, writing—original draft, D.S. and J.H.; data curation, formal analysis, investigation, software, D.S. All authors have read and agreed to the published version of the manuscript.

Funding: This research was funded by the Science and Technology Plan of Wuhan Urban and Rural Construction Commission, grant number 201831.

Institutional Review Board Statement: Not applicable.

Informed Consent Statement: Not applicable.

Data Availability Statement: The data presented in this study are available on request from the corresponding author. The data are not publicly available due to privacy.

Acknowledgments: Support from the Science and Technology Plan of Wuhan Urban and Rural Construction Commission (No. 201831) is acknowledged.

Conflicts of Interest: The authors declare no conflict of interest.

References

1. Keogh, D.L.; O'Brien, E. *Bridge Deck Analysis*; E & FN Spon; CRC Press: London, UK, 1999; pp. 1–34.
2. Caprani, C.C. Probabilistic Analysis of Highway Bridge Loading Events. Ph.D. Thesis, Dublin Institute of Technology, Dublin, Ireland, 2005; pp. 29–45.
3. Zhang, G.; Liu, Y.; Liu, J.; Lan, S.; Yang, J. Causes and statistical characteristics of bridge failures: A review. *Sci. Direct.* **2022**, *9*, 288–406. [CrossRef]
4. Zhou, Z.; Xue, C.; Yang, Z.; Yuan, W.; Xia, Z. Vehicle Load Model for Highway Bridges in Jiangsu Province Based on WIM. *J. Southeast Univ.* **2020**, *50*, 143–152.
5. Yang, D.-H.; Guan, Z.-X.; Yi, T.-H.; Li, H.-N.; Ni, Y.-S. Fatigue Evaluation of Bridges Based on Strain Influence Line Loaded by Elaborate Stochastic Traffic Flow. *J. Bridge Eng.* **2022**, *27*, 04022082. [CrossRef]
6. Liang, Y.; Xiong, F. Measurement-based bearing capacity evaluation for small and medium span bridges. *Measurement* **2020**, *149*, 106938. [CrossRef]
7. Li, M.; Liu, Y.; Yang, X. Random Vehicle Flow Load Effect Considering Axle Load. *J. Zhejiang Univ.* **2019**, *53*, 78–88.

8. Liu, Y.; Zhang, H.; Deng, Y.; Li, D. Fatigue Reliability Assessment for Orthotropic Steel Deck Details Using Copulas: Application to Nan-Xi YangtzeRiver Bridge. *J. Bridge Eng. (ASCE)* **2017**, *23*, 04017123. [CrossRef]
9. Torres-Alves, G.A.; Morales-Nápoles, O.; Jonkman, S.N. Structural reliability analysis of a submerged floating tunnel under copula-based traffic load simulations. *Eng. Struct.* **2022**, *269*, 114752. [CrossRef]
10. Soriano, M.; Casas, J.R.; Ghosn, M. Simplified probabilistic model for maximum traffic load from weigh-in-motion data. *Struct. Infrastruct. Eng.* **2017**, *13*, 454–467. [CrossRef]
11. Bedford, T.; Cooke, R.M. Probability Density Decomposition for Conditionally Dependent Random Variables Modeled by Vines. *Ann. Math. Artif. Intell.* **2001**, *32*, 245–268. [CrossRef]
12. Bedford, T.; Cooke, R. Vines—A new graphical model for dependent random variables. *Ann. Stat.* **2002**, *30*, 1031–1068. [CrossRef]
13. Morales Napoles, O. *About the Number of Vines and Regular Vines on N Nodes*; TU Delft Library: Delft, The Netherlands, 2016.
14. McKay, M.D.; Beckman, R.J.; Conover, W.J. Comparison of Three Methods for Selecting Values of Input Variables in the Analysis of Output from a Computer Code. *Technometrics* **1979**, *21*, 239–245.
15. Iman, R.; Conover, W. A Distribution-Free Approach to Inducing Rank Correlation among Input Variates. *Commun. Stat.-Simul. Comput.* **1982**, *11*, 311–334. [CrossRef]
16. Hallenbeck, M.E.; Selezneva, O.I.; Quinley, R. *Verification, Refinement, and Applicability of Long-Term Pavement Performance Vehicle Classification Rules*; FHWA: Washington, DC, USA, 2014.
17. Sklar, A. Fonctions de Repartition an Dimensions et Leurs Marges. *Publ. De L'institut De Stat. De L'université De Paris* **1959**, *8*, 229–231.
18. Zhao, Y.; Guo, X.; Su, B.; Sun, Y.; Zhu, Y. Multi-Lane Traffic Load Clustering Model for Long-Span Bridge Based on Parameter Correlation. *Mathematics* **2023**, *11*, 274. [CrossRef]
19. Mu, H.; Liu, H.; Shen, J. Copula-Based Uncertainty Quantification (Copula-UQ) for Multi-Sensor Data in Structural Health Monitoring. *Sensors* **2020**, *20*, 5692. [CrossRef] [PubMed]
20. Scheuer, E.M.; Stoller, D.S. On the Generation of Normal Random Vectors. *Technometrics* **1962**, *4*, 278–281. [CrossRef]
21. Federal Highway Administration. *Highway Safety and Truck Crash Comparative Analysis Technical Report*; FHWA: Washington, DC, USA, 2015; pp. 56–57.

sensors

MDPI

Article

Crack Monitoring in Rotating Shaft Using Rotational Speed Sensor-Based Torsional Stiffness Estimation with Adaptive Extended Kalman Filters

Young-Hun Park, Hee-Beom Lee and Gi-Woo Kim *

Department of Mechanical Engineering, Inha University, Incheon 22212, Republic of Korea
* Correspondence: gwkim@inha.ac.kr; Tel./Fax: +82-32-860-7313

Abstract: In this study, we present an alternative solution for detecting crack damages in rotating shafts under torque fluctuation by directly estimating the reduction in torsional shaft stiffness using the adaptive extended Kalman filter (AEKF) algorithm. A dynamic system model of a rotating shaft for designing AEKF was derived and implemented. An AEKF with a forgetting factor (λ) update was then designed to effectively estimate the time-varying parameter (torsional shaft stiffness) owing to cracks. Both simulation and experimental results demonstrated that the proposed estimation method could not only estimate the decrease in stiffness caused by a crack, but also quantitatively evaluate the fatigue crack growth by directly estimating the shaft torsional stiffness. Another advantage of the proposed approach is that it uses only two cost-effective rotational speed sensors and can be readily implemented in structural health monitoring systems of rotating machinery.

Keywords: crack monitoring; rotating shaft; torsional stiffness estimation; rotational speed sensors; adaptive extended Kalman filter; forgetting factor update

Citation: Park, Y.-H.; Lee, H.-B.; Kim, G.-W. Crack Monitoring in Rotating Shaft Using Rotational Speed Sensor-Based Torsional Stiffness Estimation with Adaptive Extended Kalman Filters. *Sensors* **2023**, *23*, 2437. https://doi.org/10.3390/s23052437

Academic Editors: Yongbo Li and Bing Li

Received: 4 February 2023
Revised: 19 February 2023
Accepted: 21 February 2023
Published: 22 February 2023

1. Introduction

Rotating machinery (or turbomachinery) has steadily been in the field of interest for industrial applications in internal combustion engines, power generators, turbines, and high-speed machining [1]. Rotating machinery generally consists of a rotor and a non-rotating part (stator), with torque transmitted through a rotating shaft. Cracks in rotary shafts are among the most dangerous and significant defects. The crack occurs in rotating shafts because of various mechanisms such as high and low cycle fatigue, stress corrosion, or unbalanced force caused by the rotor offset [2]. The shafts of the above-mentioned machines are typically subjected to harsh working conditions, such as loading and temperature variations. Thus, successive failures can lead to enormous economic and human resource losses. If a crack propagates continuously and is not detected in advance, an abrupt failure may occur, leading to catastrophic consequences. Thus, real-time monitoring of crack damage in the rotating shaft is essential.

Generally, contact sensors, which provide high data accuracy and convenience, can be used to detect such cracks in a rotating shaft. However, rotating and internal parts are generally difficult to measure directly. Thus, it is challenging to monitor shaft cracks using a contact sensor, such as a strain gauge. Therefore, in recent years, fault diagnosis studies on rotating machinery have focused on indirect detection methods through vibration response characteristic analysis of components, such as bearings and gears. As a result, numerous vibration-based crack detection techniques have been developed over the last decades [3]. These techniques include experimental signal-based and model-based methods. Several model-based crack detection methods, such as wavelet transform [4], have been developed to enhance fault diagnosis. Experimental signal-based methods using nonlinear vibration responses have also been widely used for damage detection in structures [5,6].

For more precise, reliable, and effective detection, various non-destructive techniques (NDT), such as radiography, magnetic particle inspection, and ultrasonic methods, are attempted to diagnose and monitor the behavior of rotating machines, although these techniques consume more time and are expensive [7]. However, high-frequency amplitudes are too small to detect cracks, and responses can be generated by assembly tolerances, manufacturing state noise, and other defects.

These disadvantages of the current technology necessitate developing non-traditional technology for detecting structural surface damage, such as cracks in the rotating shaft [8–12]. Non-model-based crack detection has also been attempted as a statistical-based data analysis method, using trained models from artificial neural networks [13] and genetic algorithms [14]. Recently, machine learning has been studied as a solution for detecting defects effectively without human experts [15]. However, this method is data-inefficient because we cannot acquire sufficient experimental data on actual crack sequences for large systems. Over the last decades, some research on structural health monitoring has been conducted based on the adaptive extended Kalman filter algorithm (AEKF) [16–20]. For example, the Kalman filter with the forgetting factor method had been applied to several systems, such as a lithium-ion battery, to consider the variation of system model parameters [21]. However, it is still necessary to study a new detection method, although previous studies have shown promising results in detecting cracks in rotating shafts.

Therefore, this study primarily aims to provide an alternative solution for detecting crack damages in rotating shafts by directly estimating the change in stiffness using the adaptive extended Kalman filter algorithm (AEKF) with a forgetting factor update. To the best of our knowledge, we report for the first time that it is possible to achieve a new means of detecting the torsional crack in a rotating shaft using AEKF with a forgetting factor update algorithm. Cracks of varying geometry are caused by different types of stress-field directions and are classified according to their orientation with respect to the shaft axis, as shown in Figure 1a. The direction of the stress field depends on the type of stress (such as bending or torsion) and geometric factors. When high cyclic stress is repeated, the crack propagates such that the crack plane is perpendicular to the direction of the tensile stress field. When bending stress is applied to the shaft, a stress field forms along the axis, and the crack propagates into the shaft section, creating a transverse crack, which is frequently called a breathing crack [22]. Torsional stress forms a tensile stress field in the direction of 45° to the shaft axis. In this study, we focused on torsional slant cracks of shafts and attempted to use Kalman-filter-based torsional stiffness estimation. When fatigue cracks occur in rotating shaft systems under alternating torque excitation, the cracks gradually grow larger over time as they are repeatedly opened and closed. As the cross-sectional area decreases, the torsional stiffness of the shaft suddenly decreases, as shown in Figure 1b. The AEKF-based estimator of shaft torsional stiffness using a dynamic model of rotating machinery is described in Section 2. Simulation results using the proposed algorithm under sinusoidal torque input are presented in Section 3. The simulation results were experimentally validated, as described in Section 4.

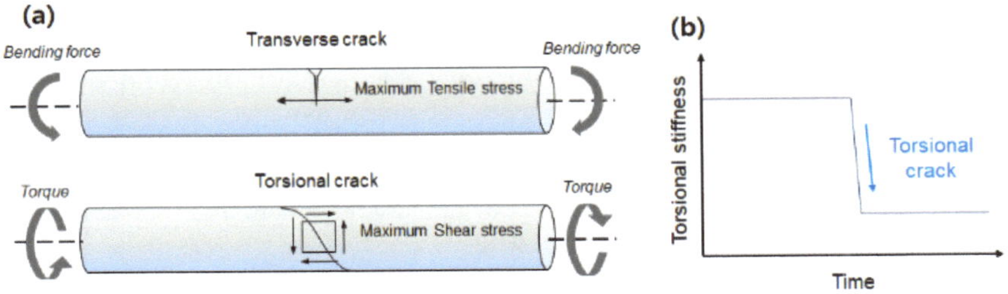

Figure 1. Characterization of fatigue cracks: (**a**) two types of crack propagation: transverse and torsion crack, (**b**) sudden torsional stiffness reduction.

2. Design of Adaptive Extended Kalman Filters

2.1. Dynamic Modeling of Rotating Shaft

Although there are various connecting structures, such as bearings and shafts, in a real rotating shaft system, the system model was formulated based on a dynamic circular shaft, to which torque and rotational speed were applied together. It is assumed to be a lumped-parameter model with two primary masses (i.e., a semi-definite system with a single natural frequency) because it is unlikely to excite the higher modes in our simple test bed system (single frequency excitation). The system elements simulated the driving-load motor dynamo for experimental verification of the proposed algorithm. The stiffness of the bellows coupling connecting the shaft and the damping effect of the bearing stand were neglected. As shown in Figure 2, the rotating shaft model comprised four components. The governing equation for the shaft rotation is given as follows:

$$J_m\ddot{\theta}_m + c_m\dot{\theta}_m + k_s(\theta_m - \theta_l) = T_m, \tag{1}$$

$$k_s(\theta_m - \theta_l) - J_l\ddot{\theta}_l = 0, \tag{2}$$

where J_m is the moment of inertia (driving motor), J_l is the moment of inertia (load motor), k_s is the torsional stiffness of the shaft, and c_m is the damping coefficient of the viscous friction of the driving motor. When torque is applied by the load motor, the angular velocity difference between the two sides is caused by the stiffness of the shaft connecting the two motors. It is necessary to express the dynamic model into a state-space model to implement the Kalman filter algorithm.

$$\dot{x} = Ax + Bu, \tag{3}$$

$$y = Cx + Du, \tag{4}$$

where x is the state vector, u is the input vector, and $\dot{x}$ is the time derivative of the state vector. In Equation (3), A is a state matrix, and B is an input matrix. In Equation (4), y is a measurement variable, C is a measurement matrix, and D is a feed-forward matrix.

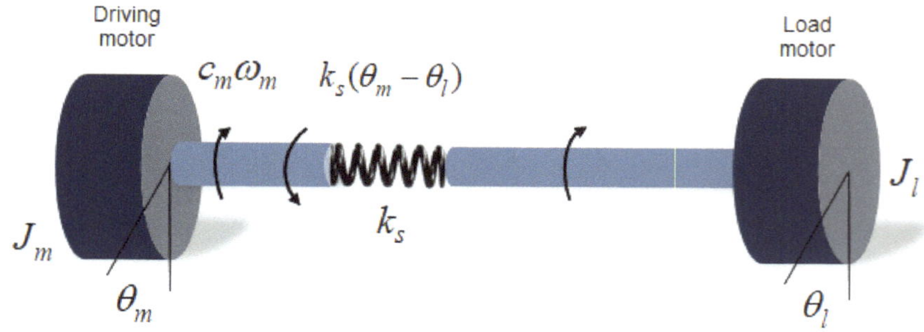

Figure 2. Schematic of the rotating shaft model with shaft torsional stiffness.

The four state variables for stiffness estimation are selected as shown in Equation (5). In this study, the time-varying stiffness is treated as a state variable, and it is assumed to be linearly proportional to the crack sizes for designing Kalman filters, although it can be changed by the nonlinear dynamics of the rotating shaft system [23]. $\theta_m - \theta_l$ is the difference in angular displacement on both sides and ω_l is the angular velocity of the load motor. The fourth state variable is the angular velocity of the driving motor. The driving motor torque is an input for the system. From Equations (1) and (2), the state space equation was derived as follows:

$$x = [x_1 x_2 x_3 x_4]^T = [(\theta_m - \theta_l)\omega_l k_s \omega_m]^T, \tag{5}$$

$$u = T_m, \tag{6}$$

$$\dot{x} = f(x,u) = \begin{bmatrix} \dot{x}_1 \\ \dot{x}_2 \\ \dot{x}_3 \\ \dot{x}_4 \end{bmatrix} = \begin{bmatrix} x_4 - x_2 \\ \frac{x_3 x_1}{J_l} \\ 0 \\ \frac{-c_m x_4 - x_3 x_1 + u}{J_m} \end{bmatrix}, \tag{7}$$

$$y = h(x) = \begin{bmatrix} x_2 \\ x_4 \end{bmatrix} = \begin{bmatrix} 0 & 1 & 0 & 0 \\ 0 & 0 & 0 & 1 \end{bmatrix} \begin{bmatrix} \theta_m - \theta_l \\ \omega_l \\ k_s \\ \omega_m \end{bmatrix}. \tag{8}$$

The reformulated system model $f(x,u)$ is a nonlinear model, and the measurement model $h(x)$ is a linear model with an actual measurable angular velocity value as the output. For the estimation of torsional stiffness, an extended Kalman filter (EKF) that linearizes a nonlinear model is required, and an adaptive EKF (AEKF) with a P-adaptive loop is proposed to improve estimation performance. As the proposed AEKF algorithm is based on the discrete-time domain, the continuous equation was discretized using the Euler method, as shown in Equation (9).

$$\dot{x} = \frac{x(k) - x(k-1)}{\Delta t} \rightarrow x(k) = \dot{x}(k)\Delta t + x(k-1), \tag{9}$$

where Δt is the time step, and k and $k - 1$ represent the time instant at $t = k\Delta t$ and $t = (k - 1)\Delta t$, respectively. Substituting Equation (7) into Equation (9), Equations (10) and (11) are defined as follows:

$$\begin{cases} x_k = f_{k-1}(x_{k-1}, u_{k-1}) \\ y_k = h_k(x_k) \end{cases}. \tag{10}$$

2.2. Adaptive Extended Kalman Filters

Kalman filtering is a state-estimation technique developed by Rudolf Kalman in 1960. It features a recursive structure and optimally estimates the state of a linear dynamic system based on measurements contaminated by noises. Kalman filters are used in many industrial fields, such as computer vision, robotics, and vehicular electronics [24,25]. The general linear discrete-time system model required to design the KF is given as

$$\begin{matrix} x_{k+1} = Ax_k + Bu_k + w_k \\ y_k = Hx_k + v_k \end{matrix}, \tag{11}$$

where w_k is a multivariate Gaussian distribution system noise variable with a covariance matrix, and v_k is a multivariate Gaussian distribution measurement noise variable with a covariance matrix. In this study, there was no input in the measurement model, and the application of the EKF was based on the nonlinear model. The general discrete-time equation is as follows:

$$\begin{cases} x_k = f_{k-1}(x_{k-1}, u_{k-1}) + w_{k-1} \\ y_k = h_k(x_k) + v_k \end{cases}. \tag{12}$$

The extended Kalman filter assumes differentiability of the state-change function instead of linearity of the model. The nonlinear system model was linearized using the Jacobian, and the Jacobian matrix was calculated based on the previous estimate.

$$A_{k-1} = \left.\frac{\partial f_{k-1}}{\partial x}\right|_{\hat{x}_{k-1}} B_{k-1} = \left.\frac{\partial f_{k-1}}{\partial u}\right|_{\hat{x}_{k-1}}, \tag{13}$$

$$H(k) = \left.\frac{\partial h_k}{\partial x}\right|_{\hat{x}_{k|k-1}}. \tag{14}$$

Matrices A and B of the rotating shaft system model linearized using Equations (13) and (14) are as follows:

$$A^* = \frac{\partial f(x,u)}{\partial x} = \begin{bmatrix} 0 & -1 & 0 & 1 \\ \frac{x_3}{J_L} & 0 & \frac{x_1}{J_L} & 0 \\ 0 & 0 & 0 & 0 \\ -\frac{x_3}{J_m} & 0 & -\frac{x_1}{J_m} & \frac{C_m}{J_m} \end{bmatrix}, \tag{15}$$

$$B^* = \frac{\partial f(x,u)}{\partial u} = \begin{bmatrix} 0 & 0 & 0 & \frac{1}{J_m} \end{bmatrix}^T, \tag{16}$$

$$H = \begin{bmatrix} 0 & 1 & 0 & 0 \\ 0 & 0 & 0 & 1 \end{bmatrix}, \tag{17}$$

where (*) indicates the system model matrix linearized using the Jacobian.

The discrete-time EKF algorithm has the following form:

- Initial estimation stage at $k = 0$

$$\begin{cases} \hat{x}_0 = E[x_0] \\ P_0 = E[(x_0 - \hat{x}_0)(x_0 - \hat{x}_0)^T] \end{cases}. \tag{18}$$

where E represents the expected value of the random variable.

- Prediction stage

$$\begin{array}{l} \hat{x}(k|k-1) = f_{k-1}(\hat{x}_{k-1}, u_{k-1}, 0) \\ P(k|k-1) = A(k,k-1)P(k-1)A^T(k,k-1) + Q(k-1) \end{array}. \tag{19}$$

A matrix was differentiated using the Jacobian in Equation (13). In the prediction stage, the variables predicted are a priori state variable and an error covariance matrix.

- Correction stage

$$K(k) = P(k|k-1)H^T(k)\left(H(k)P(k\middle|k-1)H^T(k) + R(k)\right)^{-1}, \tag{20}$$

$$\begin{array}{l} \hat{x}(k) = \hat{x}(k\middle|k-1) + K(k)[y(k) - h_k(\hat{x}_{k|k-1}, u_k, 0)] \\ P(k) = [I - K(k)H(k)]P(k|k-1) \end{array}. \tag{21}$$

In general, it is difficult to estimate the time-varying parameter (shaft stiffness) using the EKF because filter estimation relies on past data, and state estimation can diverge when past data are not adequate for recursive estimation methods. In this study, an AEKF with a forgetting factor (λ) was used to resolve this technical limitation [26]. The updated forgetting factor corrects the error covariance matrix, and the Kalman gain matrix is increased by the inverse of the forgetting factor. In general, the forgetting factor is considered a constant tuning parameter. However, convergence decreases when the uncertainty is large, such as in a nonlinear model. In this study, an adaptive loop was employed for more weighting to recent data using the residual between the measured and estimated values [27,28]. The AEKF equation is identical to the EKF in Equation (19), except for the forgetting factor in the error covariance equation.

$$P(k+1\middle|k) = \lambda(k+1)A(k+1,k)P(k)A^T(k+1,k) + Q(k), \tag{22}$$

with $\lambda(k) \geq 1$. Thus, divergence is prevented by considering the influence of the most recently measured data on the state and parameter. The performance of the AEKF is the most important factor because it completely depends on the forgetting factor. The residual $z(k)$ is defined as the difference between the measured and predicted values of the measurement. The residual is a white noise sequence when the optimal filtering gain is used.

$$z(k) = y(k) - H(k)\hat{x}(k|k-1). \quad (23)$$

For any gain, the covariance of the residuals is expressed as:

$$C_0(k) = E[z(k)z^T(k)] = H(k)P(k\big|k-1)H^T(k) + R(k). \quad (24)$$

The auto covariance of the residual is

$$\begin{aligned} C_j(k) &= E[z(k+j)z^T(k)] \\ &= H(k+j)A(k+j,k+j-1) \\ &\times[I-K(k+j-1)H(k+j-1)]\cdots A(k+2,k+1) \\ &\times[I-K(k+1)H(k+1)]A(k+1,k) \\ &\times[P(k|k-1)H^T(k)-K(k)C_0(k)] \\ \forall j &= 1,2,3,\cdots \end{aligned} . \quad (25)$$

In general, $C_j(k)$ in Equation (25) is equal to zero when Equations (20) and (24) are substituted into Equation (25), implying that the residual sequences are uncorrelated when the optimal gain is applied. However, the actual covariance of the residual $C_0(k)$ is different from the theoretical covariance, owing to errors in the system model parameters and noise covariance. Therefore, $C_j(k)$ may not be equal to zero. In Equation (25), we can choose a forgetting factor such that the last term of $C_j(k)$ for all is zero.

$$P(k\big|k-1)H^T(k) - K(k)C_0(k) = 0. \quad (26)$$

In the optimal condition, $S(k)$ and $g(\lambda,k)$ are as follows:

$$S(k) = P(k\big|k-1)H^T(k) - K(k)C_0(k), \quad (27)$$

$$g(\lambda,k) = \frac{1}{2}\sum_{i=1}^{n}\sum_{j=1}^{m} S_{ij}^2(k). \quad (28)$$

The optimality of the Kalman filter can be determined through Equation (27), which is a scalar function, and $S_{ij}(k)$ is (i,j) th element of $S(k)$. As the smaller $g(k)$ yields more optimal filter, the forgetting factor $\lambda(k)$ should be selected to minimize $g(k)$.

Various studies have been conducted based on the least-squares estimation (LSE) approach to better track time-varying parameters of dynamic systems. In this study, a recursive estimation method with a forgetting factor update was introduced to track time-varying parameters. The constant forgetting factor was optimally updated based on the following Equation (i.e., the gradient descent method):

$$\lambda^{l+1}(k) = \lambda^l(k) + \varphi\frac{\partial g^l(\lambda,k)}{\partial\lambda^l(k)}\forall l = 0,1,2,\ldots, \quad (29)$$

with initial conditions

$$\lambda^0(1) = 1, \lambda^0(k) = \lambda(k-1), \quad (30)$$

where k is the time series and l is the iteration time of the time instant. φ is the step length (i.e., learning rate, $0 < \varphi < 1$). If Equation (31) is satisfied in the p-th iteration (i.e., converges), the iteration is stopped, and the optimal forgetting factor is determined using Equation (32).

$$\left|\lambda^{p}(k)-\lambda^{p-1}(k)\right|<\varepsilon. \tag{31}$$

$$\lambda(k)=\max\{1,\lambda^{p}(k)\}. \tag{32}$$

However, the iterative numerical method does not guarantee real-time processing. Finally, a one-step AEFK algorithm was used to resolve this computational burden. In the given system, state Equations (12), (18), and (19) have the following assumption:

Assumption 1. $Q(k), R(k)$ and $P(0)$ *are positive definite.*

Assumption 2. *The measurement matrix* $H(k)$ *is fully ranked, and the optimal forgetting factor can be calculated as*

$$\lambda(k)=\max\{1,trace[N(k)]/trace[M(k)]\}, \tag{33}$$

where

$$M(k)=H(k)A(k,k-1)P(k-1)A^{T}(k,k-1)H^{T}(k), \tag{34}$$

$$N(k)=C_{0}(k)-H(k)Q(k-1)H^{T}(k)-R(k). \tag{35}$$

The $C_0(k)$ value was estimated using the recursive equation as follows:

$$C_{0}(k)=G_{1}(k)/G_{2}(k), \tag{36}$$

$$G_{1}(k)=G_{1}(k-1)/\lambda(k-1)+z(k)z^{T}(k), \tag{37}$$

$$G_{2}(k)=G_{2}(k-1)/\lambda(k-1)+1 \tag{38}$$

with initial conditions $G_1(0)=0$ and $G_2(0)=0$. The proofs of Equations (33)–(35) was derived by substituting Equation (20), which derives the Kalman gain value into Equation (26).

$$\begin{aligned}&P(k|k-1)H^{T}(k)\\&\times\left\{I-\left[H(k)P(k|k-1)H^{T}(k)+R(k)\right]^{-1}C_{0}(k)\right\}=0\end{aligned} \tag{39}$$

$$H(k)P(k|k-1)H^{T}(k)=C_{0}(k)-R(k). \tag{40}$$

Equation (40) implies that, with Assumptions 1 and 2, the optimality condition described in Equation (26) is equivalent to Equation (24). Substituting Equation (14) into Equation (40), and then reconstructing it yields the following:

$$\begin{aligned}&\lambda(k)H(k)A(k,k-1)P(k-1)A^{T}(k,k-1)H^{T}(k)\\&=C_{0}(k)-H(k)Q(k-1)H^{T}(k)-R(k)\end{aligned}. \tag{41}$$

The overall estimation process using the AEKF algorithm with a forgetting factor update is shown in Figure 3.

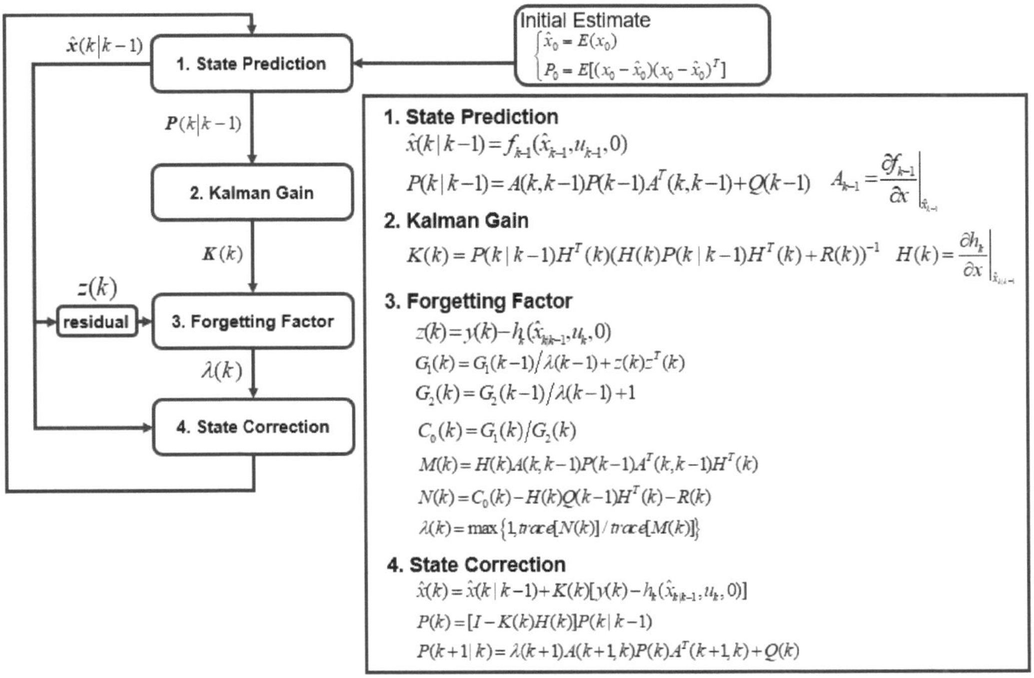

Figure 3. Overall flow chart of the AEKF algorithm for estimating the time-varying shaft torsional stiffness.

3. Estimation of Shaft Torsional Stiffness

3.1. Simulation Scheme

Based on the proposed algorithm, a situation in which cracks occur owing to shaft damage was simulated using MATLAB®. The parameter values of the system model are listed in Table 1. In this study, to mimic crack fatigue due to persistent cyclic excitation, a sinusoidal torque input was applied (frequency of 1 Hz, $T_m(t) = 10{,}000\sin(2\pi t)\ Nmm$). The angular velocity measurement data from the simulation model was set to be contaminated by the white Gaussian random noise $v(k) = N(0, (10^{-3})^2)$.

Table 1. Parameters for the estimation of torsional stiffness.

Parameters (Unit)	Value
Inertia moment of load motor $J_l(Nmm^2)$	580
Inertia moment of driving motor $J_m(Nmm^2)$	180
Damping constant c_m (Nmm·s/rad)	1000
Shaft torsional stiffness k_s (Nmm/rad)	735,000

To evaluate the response time of the proposed estimator, step response to a sudden downward step input is used for the crack initiation scenario, which corresponds to large cracks in an experiment. Then, it was assumed that the torsional stiffness suddenly decreased from 735,000 to 345,000 Nmm/rad in 10 s, as shown in Figure 4.

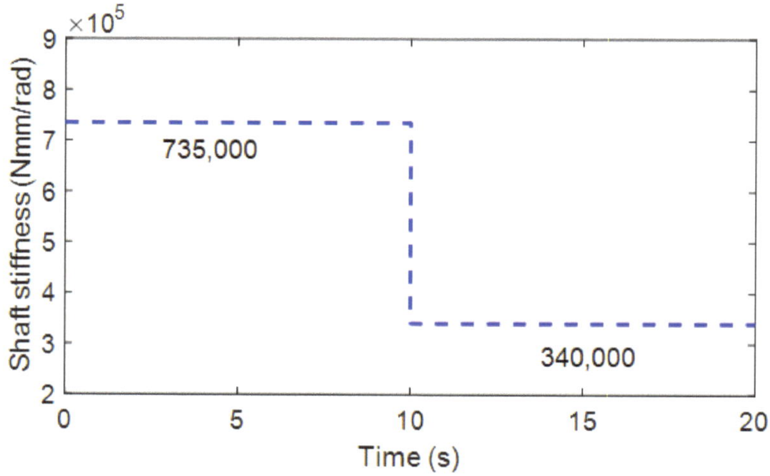

Figure 4. Crack scenario for sudden shaft torsional stiffness drop (735,000 → 340,000 Nmm/rad).

The initial state value and error covariance for estimation are as follows:

$$\begin{array}{l} x_0 = \begin{bmatrix} 0 & 0 & 735,000 & 0 \end{bmatrix} \\ P_0 = diag\ (\begin{bmatrix} 0.01 & 1 & 800,000 & 1 \end{bmatrix}) \end{array}. \tag{42}$$

The system noise covariance matrix Q and the measurement noise covariance R for Equations (21) and (22) were tuned in various cases as follows, and the optimal estimates were derived:

$$Q = \begin{bmatrix} 10^{-8} & & & \\ & 10^{-7} & & \\ & & 10^{-7} & \\ & & & 10^{-7} \end{bmatrix}, R = \begin{bmatrix} 10^{-3} & \\ & 10^{-3} \end{bmatrix}. \tag{43}$$

To evaluate the basic estimation performance of the AEKF, the root-mean-squared error (RMSE) at the kth time instant was calculated for a more rigorous analysis.

$$RMSE(k) = \sqrt{\frac{1}{k}\sum_{i=1}^{k} (p(i) - \hat{p}(i))^2}, \tag{44}$$

where k is the time instant at $t = k\Delta t$, $p(i)$ and $\hat{p}(i)$ are the true (i.e., Figure 4) and estimated values, respectively. The steady-state mean of the RMSE (MRMSE) was then calculated to exclude the effect of transient behavior. The basic estimation results for the sudden torsional stiffness drop are shown in Figure 5. The AEKF accurately estimated the sudden torsional stiffness change. In contrast, the EKF did not track the time-varying shaft stiffness change. The forgetting factor was appropriately changed by the P-adaptive loop when the stiffness rapidly decreased in 10 s. Additional scenarios with different reduction rates are applied to the simulation model to investigate the effectiveness of the proposed algorithm. These different scenarios allow for the evaluation of the tracking performance of the proposed algorithm under the same conditions, such as process and measurement noise covariance matrices. As shown in Figure 6, a gradual reduction from 1.5×10^4 Nmm/rad starts at approximately 5 s, drops to 0.4×10^4 Nmm/rad at 35 s (simulating a situation where the crack is propagating). When a crack growth is propagating and a gradual torsional stiffness drop occurs, the AEKF can deal appropriately.

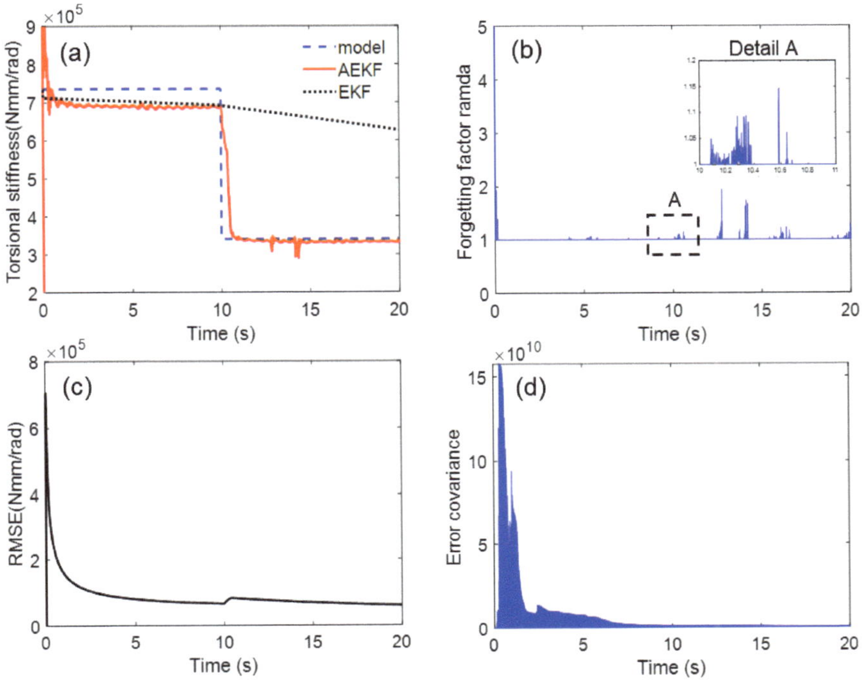

Figure 5. Simulation result for tracking sudden torsional stiffness drop: (**a**) time response of torsional stiffness, (**b**) corresponding time history of forgetting factor, and (**c**) RMSE, (**d**) convergence history of covariance.

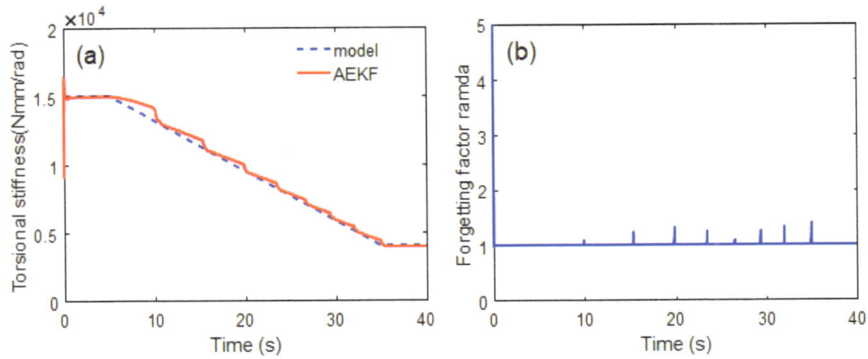

Figure 6. Simulation result for tracking gradual torsional stiffness drop for 30 s: (**a**) time response of torsional stiffness, (**b**) corresponding time history of forgetting factor.

3.2. Robustness Analysis

The robustness of the proposed estimation model under noise and parametric model uncertainty was analyzed by introducing perturbations to sensor noise and main parameters. To evaluate the robustness under noise and parametric uncertainties, the relative error to the nominal value (i.e., normalized performance measure) was quantitatively calculated.

$$\text{Relative Error} = \frac{|MRMSE_{\text{Perturbed}} - MRMSE_{\text{Nominal}}|}{MRMSE_{\text{Nominal}}}. \quad (45)$$

As sensor information is inherently contaminated by electrical noise, the effect of electrical noise on the estimated performance was examined. The sensor data were contaminated by adding a white Gaussian random noise. A probability density function is shown in Figure 7a,b as an example. Considering the random noise (error) distribution can be fitted to a normal Gaussian distribution with variance ($\sigma^2 = 0.00197$, Case 1; $\sigma^2 = 0.00298$, Case 2), it was confirmed by white Gaussian random noise. The proposed AEKF appeared to be robust against the Gaussian random noise extracted from the sensor data because the estimation results appeared to be similar to the original data, with no significant discrepancy, as shown in Figure 7c,d.

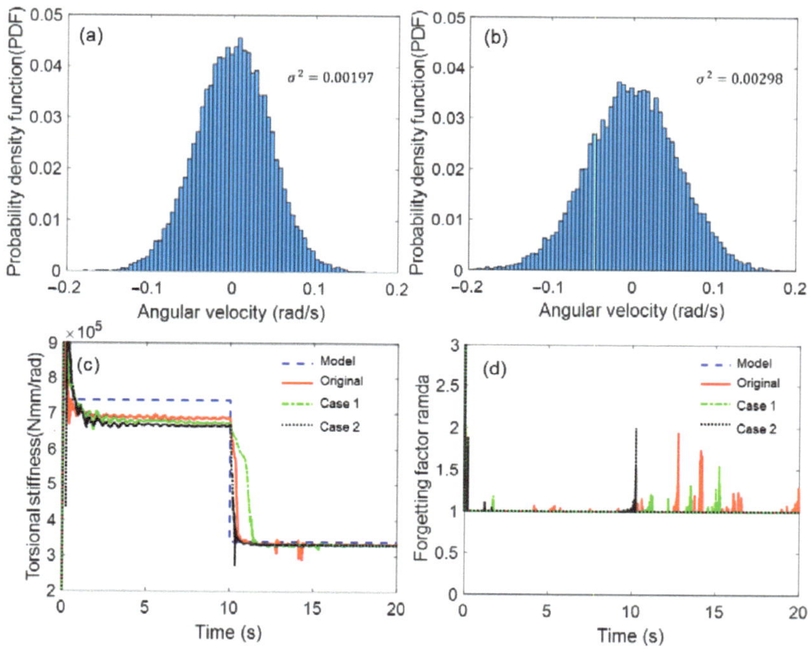

Figure 7. Simulation results of shaft stiffness estimation under noise uncertainty. Gaussian random distribution: (**a**) Case 1, and (**b**) Case 2, (**c**) torsional shaft stiffness, (**d**) forgetting factor.

The estimation performance of the proposed AEKF model was evaluated under parametric uncertainty, such as the moment of inertia. The moment of inertia on both sides is an important model uncertainty because it depends on the size, weight, and connection structure of the coupling. The nominal value for the moment of inertia of the load motor (580 Nmm^2) was perturbed by -20% (464 Nmm^2) and +20% (696 Nmm^2), and the nominal inertia moment of the driving motor (180 Nmm^2) was also perturbed by -20% (144 Nmm^2) and +20% (216 Nmm^2). The damping coefficient varied under normal operating conditions (800~1200 Nmm·s/rad), depending on the bearing lubrication condition. As the estimation results were similar to the nominal values within a reasonable range under various parametric uncertainties, the robustness of the proposed model was demonstrated, as shown in Figure 8.

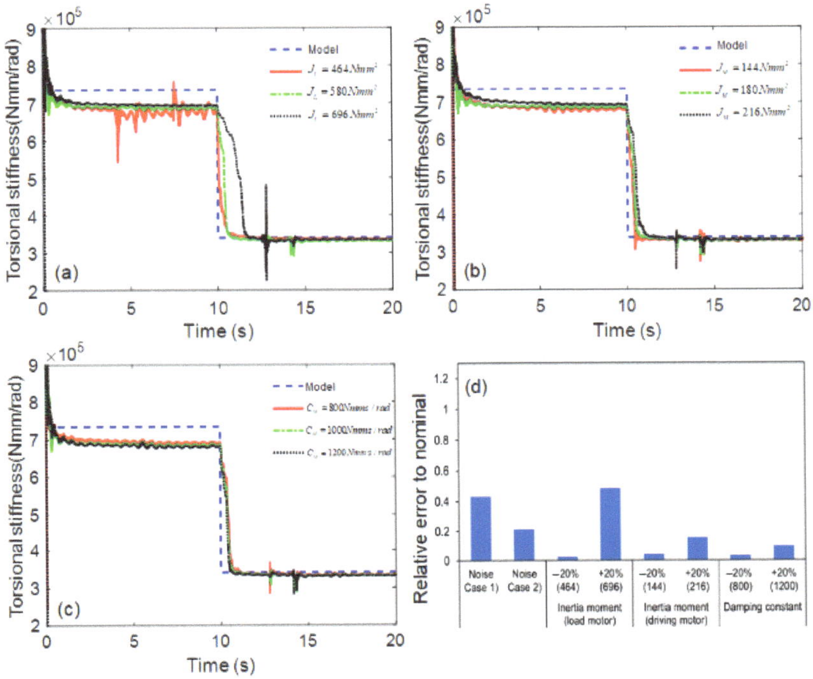

Figure 8. Simulation results of shaft stiffness estimation under parametric uncertainty: (**a**) inertia moment of load motor, (**b**) inertia moment of driving motor, (**c**) damping constant, and (**d**) relative errors.

4. Experimental Validation

4.1. Experimental Set-Up

The proposed shaft health-monitoring method was experimentally validated using a torque dynamo. An aluminum hollow-rod specimen (D_o: 20 mm, D_i: 18.2 mm, L: 550 mm) was used for the rotating shaft, as shown in Figure 9. The torque dynamo comprises a driving motor and torque-controlled load motor (Mitsubishi HG-SR152, 10 Hz bandwidth). Sinusoidal torque ($T_m(t) = 10{,}000 \sin(2\pi t)$ Nmm) was applied at a rotating speed of 5.23 rad/s (50 RPM). For the shaft crack scenario, the shaft was exchanged in turn from a normal shaft without cracks to a cracked shaft in the 45° direction (Figure 9b). The crack depth was set to 5 mm to ensure that the shaft stiffness could suddenly drop from the original value. To examine the possibility of applying a non-contact angular velocity sensor (tachometer), the measurement model considered the angular velocity values on both sides of the rotating shaft. The angular velocities of both sides were measured using a photoelectric detector-type rotational velocity sensor (ONO SOKKI, model: LG-930), which calculates the rotation speed by counting the light reflected on the gear per rotation as a pulse. The real-time monitoring performance was evaluated using a dSPACE® system (DS1104).

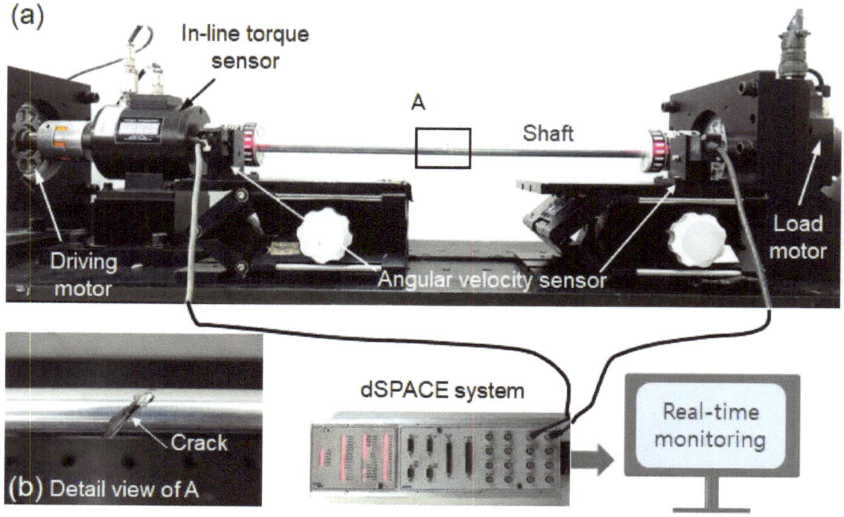

Figure 9. Schematic of experimental set-up: (**a**) overall photograph, and (**b**) details of cracked shaft.

In this study, the recursive least square estimator (RLSE) was used to identify the unknown model parameters. The rotating shaft model expressed in Equations (1) and (2) were reformulated in the matrix form as follows:

$$y_k = h_k^T \theta_k + v_k \tag{46}$$

where

$$y_k = \begin{bmatrix} T_m \\ 0 \end{bmatrix},\ h_k^T = \begin{bmatrix} \ddot{\theta}_m & 0 & \dot{\theta}_m & (\theta_m - \theta_l) \\ 0 & -\ddot{\theta}_l & 0 & (\theta_m - \theta_l) \end{bmatrix},\ \theta_k = [J_m,\ J_l,\ c_m,\ k_s]^T \tag{47}$$

In addition to the measured data from the sensors, other information was required for the two matrices y_k and h_k^T. First, the input torque (T_m) in matrix y_k was measured using an in-line torque sensor (model: YDR-2K), as shown in Figure 9. The angular displacement ($\theta_m - \theta_l$) and two angular accelerations ($\ddot{\theta}_m, \ddot{\theta}_l$) for the matrix h_k^T was obtained by directly differentiating and integrating using the low-pass filtering of the angular velocity signal. The RLSE was then designed as follows:

- Initial estimates

$$\hat{\theta}_0 = E[\theta] \tag{48}$$

$$P_0 = E\left[(\theta - \hat{\theta}_0)(\theta - \hat{\theta}_0)^T\right] \tag{49}$$

- Kalman gain calculation

$$K_{k+1} = P_k h_{k+1}\left(h_{k+1}^T P_k h_{k+1} + w_{k+1}^{-1}\right)^{-1} \tag{50}$$

- Parameter update

$$\hat{\theta}_{k+1} = \hat{\theta}_k + K_{k+1}\left(y_{k+1} - h_{k+1}^T \hat{\theta}_k\right) \tag{51}$$

- Covariance update

$$P_{k+1} = \left(I - K_{k+1}\boldsymbol{h}_{k+1}^{T}\right)P_k \tag{52}$$

All parameters of the rotating system are successfully estimated because they converge to a steady-state final positive value after 5000 iterations, as shown in Figure 10. The identified system parameters of the rotating shaft model are listed in Table 2.

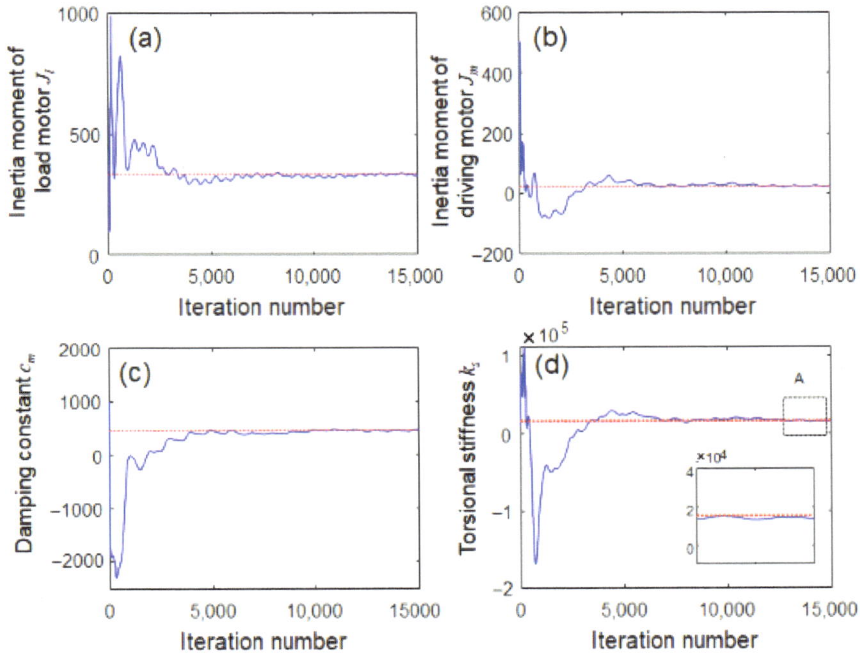

Figure 10. Convergence histories in system identification: (**a**) inertia moment of load motor, (**b**) inertia moment of driving motor, (**c**) damping constant, and (**d**) shaft torsional stiffness (inset: zoomed view of A).

Table 2. Identified system parameters of the rotating shaft model.

Parameters (Unit)	Value
Inertia moment of load motor J_l(Nmm^2)	595
Inertia moment of driving motor J_m(Nmm^2)	20
Damping constant c_m (Nmm·s/rad)	280
Shaft torsional stiffness k_s (Nmm/rad)	15,000

4.2. Results and Discussion

For the AEKF estimation model, the initial states were set, and the two noise covariance matrices (Q and R) were tuned by trial and error, as listed in Table 3. The shaft stiffness estimated using the proposed algorithm was compared in Figure 11. The estimated stiffness became steady-state and converged after 15 s in both cases. In the case of the normal state (no crack), the convergence value was identical to the system identification value (i.e., 15,000 Nmm/rad). When the stiffness changes owing to the sudden drop of crack (crack depth 5 mm) from 15,000 Nmm/rad (normal) to a certain value (abnormal crack, in this case approximately 7500), the proposed algorithm can detect this sudden drop. However, it was difficult to conform to the shift in shaft stiffness by naked eyes from the two angular velocity inputs.

Table 3. Tuning parameters for the AEKF estimation model.

P_0	$diag[0.1\ 1\ 650{,}000\ 1]$
Q	$diag[1\ 2.1\ 2\ 2.1] \times 10^{-5}$
R	$diag[9\ 9] \times 10^{-8}$

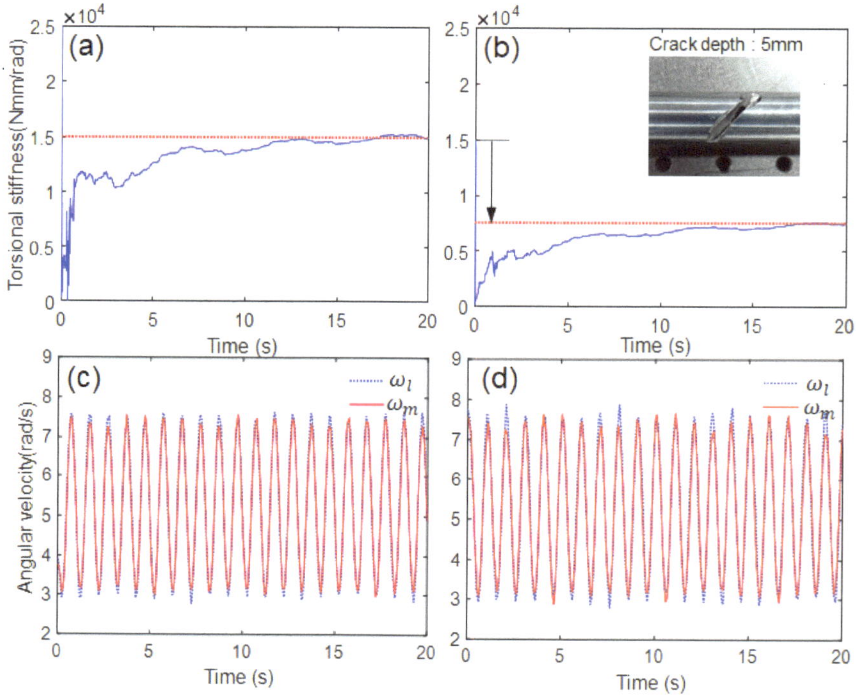

Figure 11. Experimental results: (**a**,**b**) estimated responses of torsional stiffness, (**c**,**d**) angular velocity inputs, (**a**,**c**) without crack, (**b**,**d**) with crack (crack depth 5 mm).

A different crack scenario was established to further investigate the effectiveness of the proposed algorithm. The crack depth was further increased (11 mm) to suddenly drop from 15,000 Nmm/rad to below 7500 Nmm/rad owing to the reduction in cross-sectional area (the ratio of the crack segment area to the original cross-sectional area was 65%) [29,30]. Similar to Figure 11, the proposed algorithm can track this stiffness drop due to the heavy crack, as shown in Figure 12. The proposed estimation model could not only estimate the decrease in stiffness caused by a crack, but also quantitatively evaluate the fatigue crack growth by directly estimating the shaft torsional stiffness. The robustness of the proposed estimation model under noise uncertainty was evaluated by introducing the perturbations in sensor noise. The original sensor signal was filtered by a digital moving average filter (no phase delay). Two corrupted signals were generated; the low-pass filtering is small (Case 1, less contaminated) and off (Case 2, more contaminated, i.e., raw data). The probability density distribution of sensor noise extracted from the original sensor signal was similar to Gaussian distribution, as shown in Figure 13c. The proposed estimation model seemed to be robust against the Gaussian random noise in all sensor data because the estimation results appeared to be similar regardless of the degree of contamination, as shown in Figure 13a,b. In the case of heavy crack (crack depth 11 mm), the proposed estimation model turned out to be more robust against the Gaussian random noise, as shown in Figure 13d. The robustness of the proposed estimation model under model uncertainty was not investigated in the experiment because it was unlikely to significantly change two main model parameters (inertia moment of load and driving motor).

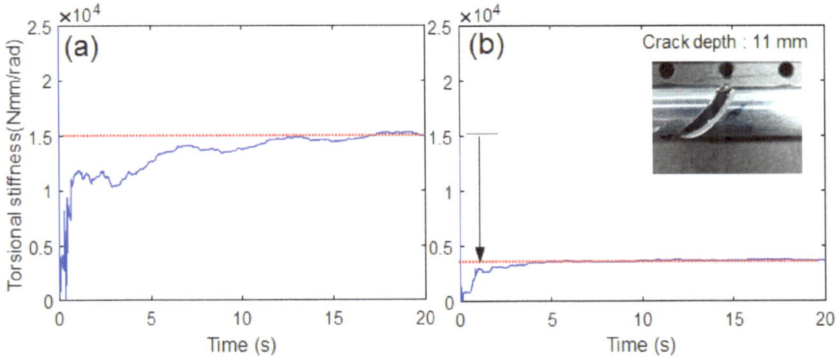

Figure 12. Experimental results for different crack scenario: (**a**) without crack, (**b**) with crack (crack depth: 11 mm) [Supplementary Materials].

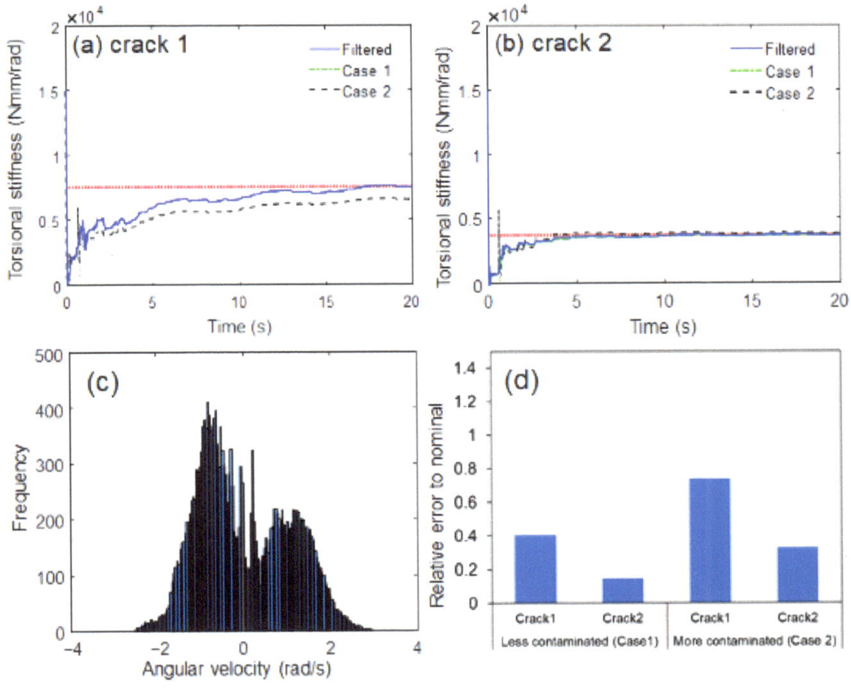

Figure 13. Estimated torsional stiffness responses under electrical sensor noise uncertainty: (**a**) crack depth 5 mm, (**b**) crack depth 11 mm, (**c**) probability density distribution of sensor noise extract from original sensor signal, (**d**) relative error to nominal.

5. Conclusions

In this study, the torsional crack in the rotating shaft was successfully detected in real-time by estimating the reduction of torsional stiffness in the rotating shaft using the AEKF approach with forgetting factor update. The main contributions of this study are summarized as follows:

- We concluded that the proposed approach is a promising alternative means for detecting torsional cracks in rotating shafts despite the difficulty in tuning the Q and R matrices of the AEKF.
- The proposed estimation model could not only estimate the decrease in stiffness caused by a crack but also quantitatively evaluate the fatigue crack growth by directly estimating the shaft torsional stiffness.
- Another advantage of the proposed approach is that it uses only two cost-effective rotational speed sensors; therefore, it does not require noncontact-type torque sensors, which are typically expensive and suffer from durability limitations.

With these advantages, the proposed approach can be readily implemented in structural health monitoring systems of rotating machinery. In future research, we will continue to address some of the ongoing issues. In particular, the localization of cracks in rotating shafts should be studied further. In addition, if the input variables cannot be measured, an advanced algorithm should be applied to simultaneously estimate unknown input and state variables.

Supplementary Materials: The following are available online at https://www.mdpi.com/article/10.3390/s23052437/s1, Video S1: Crack Monitoring in Rotating Shaft Using Torsional Stiffness Estimation with Adaptive Extended Kalman Filters.

Author Contributions: G.-W.K. (the corresponding author) as the principal investigator takes the primary responsibility for this research. H.-B.L. and Y.-H.P. performed the experiments and analyzed the results. All authors drafted and reviewed this manuscript. All authors have read and agreed to the published version of the manuscript.

Funding: This work was supported by an INHA UNIVERSITY Research Grant.

Institutional Review Board Statement: Not applicable.

Informed Consent Statement: Not applicable.

Data Availability Statement: The datasets used and/or analyzed during the current study are available from the corresponding author upon reasonable request.

Acknowledgments: The authors are thankful to Hyundai Doosan Infracore for their assistance with the measurement instruments.

Conflicts of Interest: The authors declare no conflict of interest.

References

1. Sabnavis, G.; Kirk, R.G.; Kasarda, M.; Quinn, D. Cracked shaft detection and diagnostics: A literature review. *Shock Vib. Dig.* **2004**, *36*, 287. [CrossRef]
2. Liu, J.; Tang, C.; Pan, G. Dynamic modeling and simulation of a flexible-rotor ball bearing system. *J. Vib. Control* **2022**, *28*, 3495–3509. [CrossRef]
3. Feng, K.; Ji, J.; Ni, Q.; Beer, M. A review of vibration-based gear wear monitoring and prediction techniques. *Mech. Syst. Signal Process.* **2023**, *182*, 109605. [CrossRef]
4. Pricop, M.; Pazara, T.; Pricop, C.; Novac, G. Crack detection in rotating shafts using combined wavelet analysis. *J. Physics: Conf. Ser.* **2019**, *1297*, 012031. [CrossRef]
5. Gradzki, R.; Kulesza, Z.; Bartoszewicz, B. Method of shaft crack detection based on squared gain of vibration amplitude. *Nonlinear Dyn.* **2019**, *98*, 671–690. [CrossRef]
6. Kim, G.-W.; Johnson, D.R.; Semperlotti, F.; Wang, K.-W. Localization of breathing cracks using combination tone nonlinear response. *Smart Mater. Struct.* **2011**, *20*, 055014. [CrossRef]
7. Bansode, V.M.; Billore, M. Crack detection in a rotary shaft analytical and experimental analyses: A review. *Mater. Today Proc.* **2021**, *47*, 6301–6305. [CrossRef]
8. Rathna Prasad, S.; Sekhar, A. Detection and localization of fatigue-induced transverse crack in a rotor shaft using principal component analysis. *Struct. Health Monit.* **2021**, *20*, 513–531. [CrossRef]
9. Liang, H.; Zhao, C.; Chen, Y.; Liu, Y.; Zhao, Y. The Improved WNOFRFs Feature Extraction Method and Its Application to Quantitative Diagnosis for Cracked Rotor Systems. *Sensors* **2022**, *22*, 1936. [CrossRef]
10. Sathujoda, P. Detection of a slant crack in a rotor bearing system during shut-down. *Mech. Based Des. Struct. Mach.* **2020**, *48*, 266–276. [CrossRef]

11. Hidle, E.L.; Hestmo, R.H.; Adsen, O.S.; Lange, H.; Vinogradov, A. Early Detection of Subsurface Fatigue Cracks in Rolling Element Bearings by the Knowledge-Based Analysis of Acoustic Emission. *Sensors* **2022**, *22*, 5187. [CrossRef]
12. Reitz, T.; Fritzen, C.-P. A novel baseline-free approach for acousto-ultrasonic crack monitoring of rotating axles. *Struct. Health Monit.* **2021**, *20*, 990–1003. [CrossRef]
13. Wang, C.; Zheng, Z.; Guo, D.; Liu, T.; Xie, Y.; Zhang, D. An Experimental Setup to Detect the Crack Fault of Asymmetric Rotors Based on a Deep Learning Method. *Appl. Sci.* **2023**, *13*, 1327. [CrossRef]
14. Nath, A.G.; Udmale, S.S.; Singh, S.K. Role of artificial intelligence in rotor fault diagnosis: A comprehensive review. *Artif. Intell. Rev.* **2021**, *54*, 2609–2668. [CrossRef]
15. Kim, Y.; Yi, S.; Ahn, H.; Hong, C.-H. Accurate Crack Detection Based on Distributed Deep Learning for IoT Environment. *Sensors* **2023**, *23*, 858. [CrossRef]
16. Yang, J.; Pan, S.; Huang, H. An adaptive extended Kalman filter for structural damage identifications II: Unknown inputs. *Struct. Control Health Monit.* **2007**, *14*, 497–521. [CrossRef]
17. Yang, J.N.; Lin, S.; Huang, H.; Zhou, L. An adaptive extended Kalman filter for structural damage identification. *Struct. Control Health Monit.* **2006**, *13*, 849–867. [CrossRef]
18. Zhou, L.; Wu, S.; Yang, J.N. Experimental study of an adaptive extended Kalman filter for structural damage identification. *J. Infrastruct. Syst.* **2008**, *14*, 42–51. [CrossRef]
19. Wang, Y.; He, M.; Sun, L.; Wu, D.; Wang, Y.; Qing, X. Weighted adaptive Kalman filtering-based diverse information fusion for hole edge crack monitoring. *Mech. Syst. Signal Process.* **2022**, *167*, 108534. [CrossRef]
20. Wang, Y.; He, M.; Sun, L.; Wu, D.; Wang, Y.; Zou, L. Improved Kalman filtering-based information fusion for crack monitoring using piezoelectric-fiber hybrid sensor network. *Front. Mater.* **2020**, *7*, 300. [CrossRef]
21. Shrivastava, P.; Soon, T.K.; Idris, M.Y.I.B.; Mekhilef, S.; Adnan, S.B.R.S. Combined state of charge and state of energy estimation of lithium-ion battery using dual forgetting factor-based adaptive extended Kalman filter for electric vehicle applications. *IEEE Trans. Veh. Technol.* **2021**, *70*, 1200–1215. [CrossRef]
22. Al-hababi, T.; Alkayem, N.F.; Zhu, H.; Cui, L.; Zhang, S.; Cao, M. Effective identification and localization of single and multiple breathing cracks in beams under gaussian excitation using time-domain analysis. *Mathematics* **2022**, *10*, 1853. [CrossRef]
23. Lin, Y.; Chu, F. The dynamic behavior of a rotor system with a slant crack on the shaft. *Mech. Syst. Signal Process.* **2010**, *24*, 522–545. [CrossRef]
24. Bishop, G.; Welch, G. An introduction to the Kalman filter. *Proc. SIGGRAPH Course* **2001**, *8*, 41.
25. Chen, B.-C.; Wu, Y.-Y.; Hsieh, F.-C. Estimation of engine rotational dynamics using Kalman filter based on a kinematic model. *IEEE Trans. Veh. Technol.* **2010**, *59*, 3728–3735. [CrossRef]
26. Xia, Q.; Rao, M.; Ying, Y.; Shen, X. Adaptive fading Kalman filter with an application. *Automatica* **1994**, *30*, 1333–1338. [CrossRef]
27. Akhlaghi, S.; Zhou, N.; Huang, Z. In Adaptive adjustment of noise covariance in Kalman filter for dynamic state estimation. In Proceedings of the IEEE Power & Energy Society General Meeting, Chicago, IL, USA, 16–20 July 2017; pp. 1–5.
28. Lee, D.-H.; Yoon, D.-S.; Kim, G.-W. New indirect tire pressure monitoring system enabled by adaptive extended Kalman filtering of vehicle suspension systems. *Electronics* **2021**, *10*, 1359. [CrossRef]
29. Bhalerao, G.N.; Patil, A.A.; Waghulde, K.B.; Desai, S. Dynamic analysis of rotor system with slant cracked shaft. *Mater. Today Proc.* **2021**, *44*, 4268–4281. [CrossRef]
30. Muñoz-Abella, B.; Montero, L.; Rubio, P.; Rubio, L. Determination of the Critical Speed of a Cracked Shaft from Experimental Data. *Sensors* **2022**, *22*, 9777. [CrossRef]

sensors

MDPI

Article

Single-Sensor Engine Multi-Type Fault Detection

Daijie Tang [1], Fengrong Bi [1], Jiangang Cheng [1], Xiao Yang [1], Pengfei Shen [1] and Xiaoyang Bi [2,*]

[1] State Key Laboratory of Engines, Tianjin University, Tianjin 300350, China
[2] State Key Laboratory of Reliability and Intelligence Electrical Equipment, Hebei University of Technology, Tianjin 300130, China
* Correspondence: xy_bi@hebut.edu.cn

Abstract: Engine fault detection is conducive to improving equipment reliability and reducing maintenance costs. In practical scenarios, high-quality data is difficult to obtain. Usually, only single-sensor data is available. This paper proposes a fault detection method combining Variational Mode Decomposition (VMD) and Random Forest (RF). At first, the spectral energy distribution is obtained by decomposing and statistic the engine data of multiple working conditions. Based on the spectral energy distribution, the overall optimal mode number was identified, and the quadratic penalty term was optimized using SNR. The improved VMD (IVMD) improves mode aliasing and iterative efficiency and unifies feature dimensions. Decomposition of real signals demonstrates the effectiveness. The paper designs a feature vector composed of seven types of attributes, including unit bandwidth energy, center frequency, maximum singular value and so on. The feature vector is then fed to RF for classification. Features are selected in order of importance to classification to improve the training efficiency. By comparing with various algorithms, the proposed method has higher accuracy and faster training efficiency in single-speed, multi-speed and cross-speed single-sensor data diagnosis. The results show that the method has application prospects with little training data and low hardware requirements.

Keywords: fault detection; single-sensor data; variational mode decomposition; vibration; random forest

Citation: Tang, D.; Bi, F.; Cheng, J.; Yang, X.; Shen, P.; Bi, X. Single-Sensor Engine Multi-Type Fault Detection. *Sensors* **2023**, *23*, 1642. https://doi.org/10.3390/s23031642

Academic Editors: Yongbo Li and Bing Li

Received: 14 December 2022
Revised: 22 January 2023
Accepted: 28 January 2023
Published: 2 February 2023

1. Introduction

As one of the critical power sources, the reliability of engines has received more attention in recent years. In time, engine fault detection can detect weak faults, which is conducive to fault prevention and repair. Data-driven approaches usually require large amounts of high-quality data for training. However, engine labelled-data is challenging to obtain and mostly comes from a single sensor due to cost constraints. Research on single-sensor engine fault detection based on small data amounts and low hardware requirements is necessary [1].

Vibration acceleration signals are widely used in fault detection research because of their rich component condition information and ease of measurement [2,3]. Ma et al. proposed a multi-channel Lanczos quaternion singular spectrum analysis to extract fault characteristic frequencies from multiple vibration sensor signals [4]. Ribeiro et al. proposed a multi-head one-dimensional convolutional neural network to diagnose six motor faults using vibration signals from two directions [5]. However, the engine vibration signal has a wide frequency band (up to about 12,000 Hz), and sensors with high sampling accuracy and a wide frequency band with good stability are usually costly. Moreover, the hardware conditions of the engine control system are ordinary, so the research on single-sensor fault detection under low hardware requirements is gradually gaining attention. Basuraj et al. proposed a single-sensor online filtering method for recursive singular spectrum analysis based on the concept of first-order feature perturbation, which proved its effectiveness in several data sets. [6]. Ayati et al. used KNN for single-sensor fault classification after extracting features using fast Fourier transform and wavelet packet transform [7]. In general, it is challenging to diagnose faults in different cylinders of an engine separately using a single sensor.

Engine fault detection methods can be roughly divided into three categories: knowledge-driven, model-driven, and data-driven. Wang et al. proposed an aero-engine dynamic threshold fault detection based on the isolated forest method, which requires only normal data for training to achieve high accuracy [8]. Ellefsen et al. proposed an online diagnosis method for marine diesel engine degradation based on variational autoencoder and expert knowledge [9]. Knowledge-driven methods usually diagnose a single type of fault and require solid expert knowledge. Liu et al. proposed a model-based aero-engine soft fault detection method, which achieved fault diagnosis by comparing smooth residuals and preset thresholds [10]. Wang et al. established a mapping model between the shaft radial vibration average and the misalignment value based on shaft shape characteristics. A new monitoring scheme has been designed and the accuracy of detecting misalignment is greater than 90% [11]. Model-driven method research can help explore the failure mechanism, but it is usually challenging to achieve. Data-driven methods are widely used due to their ease of implementation and high accuracy [12,13]. Deep learning methods have been widely used in engine fault detection in recent years due to their powerful data mining capabilities [14,15]. These methods require less expert knowledge and more high-quality training data. However, high-dimensional and huge data processing capability leads to higher hardware requirements for deep learning methods. In practical scenarios, high-quality training data is difficult to obtain because of the dangers of engine failure simulation experiments. Under the constraints of low hardware conditions and lack of data, the combination of signal processing methods and simple pattern classification methods still has potential to be explored [16,17].

Variational Mode Decomposition (VMD) is an advanced signal processing method capable of decomposing a signal into several intrinsic mode functions (IMFs) [18]. Compared with empirical mode decomposition (EMD), VMD effectively suppresses mode aliasing and improves the quality of decomposition [19]. However, the mode number K and quadratic penalty term α, predefined in VMD, strongly influence the decomposition and are difficult to determine [20,21]. For these reasons, scholars have proposed many optimization ideas for adaptively selecting K and α [21,22]. The adaptive VMD method leads to a varying number of IMFs, so component screening is usually performed after decomposition [23,24]. The process of screening IMFs requires expert knowledge and is time-consuming and labor-intensive. In addition, many scholars optimize (K, α) through swarm intelligence optimization algorithms [25,26]. This method ignores the problem that the VMD efficiency drops sharply as K increases (as shown in Section 4).

The unsupervised clustering method is ineffective in diagnosing engine faults because there are many types of failure, complex operating conditions, and large signal noise [27,28]. Supervised pattern classification methods such as deep neural networks (DNN) are more suitable due to their powerful learning capabilities. Shahid et al. used a one-dimensional convolutional neural network (1DCNN) to identify the crankshaft angle degree of the engine and successfully diagnosed the misfire fault [29]. Zhang et al. proposed a long short-term memory recurrent neural network (LSTM-RNN) for evaluating bearing degradation and proposed waveform entropy to improve the accuracy effectively [30]. Lee et al. compared the performance of multilayer perception (MLP), residual network (ResNet), LSTM, and ResNet-LSTM in diagnosing production failure cases and found that ResNet-LSTM works best [31]. However, the effectiveness of DNN is built on sufficient high-quality labeled data. Due to the complex calculation of DNN, the training time is long, and it is challenging to optimize and retrain the model [32]. Li et al. first used a simplified DNN to extract the fault features of rotating machinery and then combined random forest (RF) for fault classification, which has higher efficiency and accuracy than advanced DNN methods [32]. RF has faster training and classification speed than DNN and may be suitable for engine fault detection.

The paper aims to propose a single-sensor, cross-speed fault detection method that is applicable to low hardware requirements and small data amounts. The work has resulted in the following contributions.

(1) A new overall K and α optimization method based on spectral energy distribution and SNR is proposed for VMD, avoiding IMF screening and unifying the feature dimension to prepare for quick diagnosis.
(2) The center frequencies are preset based on spectral energy distribution, which reduces the number of VMD iterations and mode aliasing.
(3) A feature set was designed for IVMD-RF to achieve single-sensor fault diagnosis. Further filtering of features by feature importance ranking improves efficiency. Different single-sensor datasets demonstrate the effectiveness of the method.

The rest of the article is organized as follows. Section 2 introduces the basic principles of the methods used in the paper. In Section 3, the fault data collection experiment of the diesel engine is presented. Section 4 introduces the optimization of the VMD method and the verification of its decomposition effect. In Section 5, IVMD-RF is presented and compared with various DNN methods on two diagnostic cases.

2. Theories

2.1. Variational Mode Decomposition

The purpose of VMD is to decompose an actual signal into several ideal narrowband signals while satisfying the constraint that the sum of their bandwidths is the smallest. Assume that each IMF closely surrounds its center frequency in the frequency domain. Therefore, the objective can be summarized as the following constrained variational problem:

$$\begin{cases} \min\limits_{\{u_k\},\{\omega_k\}} \{\sum\limits_k \left\| \partial_t[(\delta(t)+\frac{j}{\pi t}) * u_k(t)]e^{-j\omega_k t} \right\|_2^2\} \\ s.t. \sum\limits_k u_k = f \end{cases}, \tag{1}$$

where $\{u_k(t)\} = \{u_1(t), u_1(t), \ldots, u_k(t)\}$ and $\{\omega_k\} = \{\omega_1, \omega_2, \ldots, \omega_k\}$ represent the decomposed IMFs and the corresponding center frequencies, respectively. $\delta(t)$ is the shock function.

The reconstruction constraint can be addressed by introducing a quadratic penalty α and Lagrange multipliers λ. The constrained variational problem of (1) is transformed into an unconstrained one by introducing these two parameters. The obtained augmented Lagrangian is shown in (2):

$$L(\{u_k\},\{\omega_k\},\lambda) := \alpha \sum_k \left\| \partial_t[(\delta(t)+\frac{j}{\pi t}) * u_k(t)]e^{-j\omega_k t} \right\|_2^2 + \left\| f(t) - \sum_k u_k(t) \right\|_2^2 + \left\langle \lambda(t), f(t) - \sum_k u_k(t) \right\rangle, \tag{2}$$

This problem can be solved by Parseval/Plancherel Fourier isometry under the norm. The expressions of $\widehat{u}_k^{n+1}(\omega)$ and ω_k^{n+1} are shown in (3) and (4).

$$\widehat{u}_k^{n+1}(\omega) = \frac{\widehat{f}(\omega) - \sum_{i \neq k} \widehat{u}_i(\omega) + \frac{\widehat{\lambda}(\omega)}{2}}{1 + 2\alpha(\omega - \omega_k)^2}, \tag{3}$$

$$\omega_k^{n+1} = \frac{\int_0^\infty \omega \left| \widehat{u}_k(\omega) \right|^2 d\omega}{\int_0^\infty \left| \widehat{u}_k(\omega) \right|^2 d\omega}, \tag{4}$$

where (3) is equivalent to the Wiener filter of the current residual $\widehat{f}(\omega) - \sum_{i<K} \widehat{u}_i(\omega)$. IMF can be obtained by inverse Fourier transform of $\widehat{u}_k^{n+1}(\omega)$.The flow of VMD is shown in Algorithm 1. The default ε value is 1×10^{-7}.

Algorithm 1: VMD

Input: A signal f, mode number K and quadratic penalty α.
Output: A set of IMFs
Initialize $\{\widehat{u}_k^1\}, \{\widehat{\omega}_k^1\}, \{\widehat{\lambda}^1\}$, n ← 0
repeat
  for k ← 1 to K **do**
Update $\widehat{u}_k$ for all $\omega \geq 0$ by (3)
Update ω_k by (4)
  end for
  Dual ascent for all $\omega \geq 0$:
  $\widehat{\lambda}^{n+1}(\omega) \leftarrow \widehat{\lambda}^{n}(\omega) + \tau\left[\widehat{f}(\omega) - \sum_k \widehat{u}_k^{n+1}(\omega)\right]$
until convergence: $\sum_k \left\|\widehat{u}_k^{n+1} - \widehat{u}_k^{n}\right\|_2^2 / \left\|\widehat{u}_k^{n}\right\|_2^2 < \varepsilon$.

2.2. Random Forests

The random forest algorithm was proposed by Breiman [33], which is suitable for solving data prediction and classification. A random forest is a combination of decision tree classifiers. Each tree depends on the value of an independently sampled random vector and has the same distribution for all trees in the forest.

(1) Suppose the original sample is $X = \{(x_1, y_1), (x_2, y_2), \ldots, (x_n, y_n)\}$, where x_i and y_i represent feature values and labels, respectively. T training samples $X_1, X_2, \ldots, X_T$ are extracted from the original dataset X by bootstrap sampling with return, and $X_i(i = 1, 2, \ldots T)$ and X have the same number of samples.
(2) Build a decision tree $hi(X_i, \Theta_k)$ for each training sample $X_i(i = 1, 2, \ldots T)$, where i =1, 2, … T, k =1, 2, …. The decision tree model used in the paper is shown in (5) and (6).

$$d(x_1, x_2, \ldots, x_n, h_t) = \begin{cases} label(h_t) & h_t \text{ is the leaf node} \\ d(x_1, x_2, \ldots, x_n, h_t) & h_t \text{ is the inner node} \end{cases} \quad (5)$$

$$hi(X_i, \Theta_k) = d(x_1, x_2, \ldots, x_n, root(h_t)) \quad (6)$$

where *root*(h_t) is the root node of the decision tree. $d(x_1, x_2, \ldots, x_n, h_t)$ is the division criterion of the decision tree. The segmentation criterion consists of segmentation variables and predictions measured by the impurity function.

The Gini coefficient is proportional to the impurity level. The optimal split is to find the largest split of the Gini coefficient as follows:

$$Gini(t) = 1 - \sum_{j=1}^{J} \{p(j|t)\}^2 \quad (7)$$

where $p(j|t)$ is the probability of the jth category in node t, that is, the ratio of the jth category to the total number of sample labels J.

Before selecting attributes for each non-leaf node, randomly select m attributes from M attributes as the set of categorical attributes for the current node. Take $m = \text{int}(\sqrt{M})$, where int is the rounding function. The nodes are divided according to the optimal division method of m attributes, and a complete decision tree is established. The growth of each decision tree is not pruned until the leaf node grows.

A random forest generated from T decision trees is used to classify the test samples. Each tree has voting power to decide the classification result. Summarize the output categories of the decision tree, and the category with the most votes is the final classification result. The classification decision model *H(x)* is shown in (8).

$$H(x) = \underset{\gamma}{\operatorname{argmax}} \sum_{i=1}^{T} I(hi(X_i, \Theta_k) = \gamma) \tag{8}$$

where γ is the label variable of the output and I is the indicator function.

3. Diesel Engine Faults Simulation Experiment

To verify the effectiveness of the proposed method, our team conducted a fault simulation experiment on an in-line 6-cylinder diesel engine. The specific parameters of the engine are shown in Table 1. The experiment was performed on a bench base supported by an air spring. The engine and the dynamic dynamometer adopt a flexible connection. The photoelectric pulse speed sensor is placed at the position of the vertical connecting shaft to measure the engine speed. The vibration acceleration sensors are arranged on the cylinder head and block as shown in Figure 1. The data used in this paper are vibration acceleration signals in the Y-direction in Figure 1. The signal is input to the computer for processing and recording after passing through the acquisition front end. The models of the instruments used in the experiment are shown in Table 2.

Table 1. Parameters of diesel engine.

Items	Parameters
Displacement	7.14 L
Rated power/Rated speed	220 kW/2300 rpm
Maximum torque/Speed range	1250 Nm/1200–1600 rpm
Intake/Exhaust valve clearance	0.30 m/0.50 m

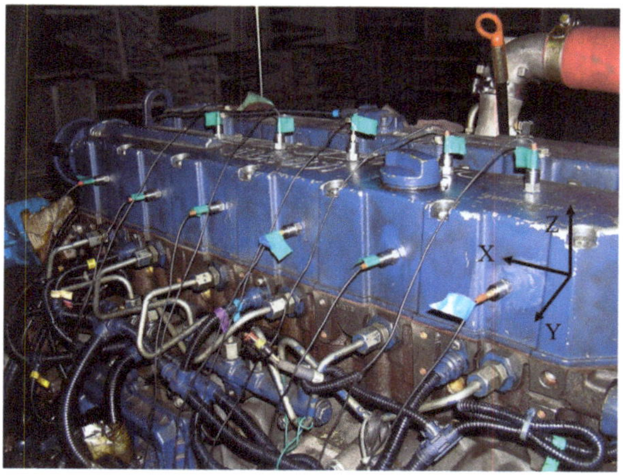

Figure 1. Sensor positions and coordinate direction.

Table 2. Experimental instrument parameters.

Instruments	Parameters
Dynamic dynamometer	CAC380, Xiangyi Power
Vibration acceleration sensor	621B40, PCB
Photoelectric pulse speed sensor	SPSR-115/230, Monarch
Data acquisition front end	SCADAS05, LMS

Components with the highest failure probability are fuel injection and oil supply equipment (25.1%), water leakage (13.1%), and valves and sealing (17.4%) [34]. Generally, leak failure is easy to detect by water temperature sensors. The paper focuses on two other types of failures. The paper simulates three faults for the fuel supply equipment: abnormal common rail pressure, abnormal fuel supply, and abnormal injection advance angle. Abnormal rail pressure is set to simulate the fault of the common rail system, and the insufficient fuel supply is to simulate injector failure. The weak power combustion abnormality is simulated by slightly changing the injection advance angle. In addition, the abnormal valve clearance is simulated by adjusting the opening of intake and exhaust valves with a plug gauge. The abnormal valve clearance conditions all occurred on the first cylinder only. Experiments were performed at the following rotational speeds: 700 rpm, 1300 rpm, 1600 rpm, 2000 rpm, and 2300 rpm. The parameters of normal working conditions under each speed condition are shown in Table 3. The fault settings at rated speed (2300 rpm) are shown in Table 4, where the Roman numerals represent different fault conditions. The fault conditions of other speeds are also adjusted to the same extent as those in Table 4 on the basis of the normal parameters in Table 3. The abnormal advance angle failure simulation is not carried out under 700 rpm idling conditions. The load range of the engine includes 100% and 50%.

Table 3. Normal working conditions under different speeds.

Speed (rpm)	Valve Clearance-Intake, Exhaust (mm)	Fuel Supply (mg/cyc)	Rail Pressure (bar)	Injection Advance Angle (°CA)
700	(0.30, 0.50)	60.0	405	-
1300	(0.30, 0.50)	117.0	1250	9.49
1600	(0.30, 0.50)	117.0	1350	12.98
2000	(0.30, 0.50)	117.0	1500	15.00
2300	(0.30, 0.50)	112.5	1550	18.45

Table 4. Fault type and degree parameter setting (2300 rpm).

Mark	Valve Clearance (Intake, Exhaust)/mm	Fuel Supply	Rail Pressure/bar	Injection Advance Angle/°CA
I	(0.30, 0.50)	100%	1550	18.45
II	(0.20, 0.40)	100%	1550	18.45
III	(0.35, 0.55)	100%	1550	18.45
IV	(0.40, 0.60)	100%	1550	18.45
V	(0.30, 0.50)	75%	1550	18.45
VI	(0.30, 0.50)	25%	1550	18.45
VII	(0.30, 0.50)	100%	1350	18.45
VIII	(0.30, 0.50)	100%	1150	18.45
IX	(0.30, 0.50)	100%	1550	17.45
X	(0.30, 0.50)	100%	1550	16.45
XI	(0.30, 0.50)	100%	1550	19.45
XII	(0.30, 0.50)	100%	1550	20.45

Note: The shaded green marks the location of the faulty parameter.

4. Optimization of Variational Mode Decomposition

VMD's denoising ability is better than EMD [35], and the decomposed IMFs have a better signal-to-noise ratio (SNR). However, the decomposition effect of VMD is greatly affected by parameter settings, especially the mode number K and the quadratic penalty term α. Improper K value setting will lead to over-decomposition or under-decomposition. In addition, as K increases, the efficiency of the original VMD decreases drastically. Figure 2 shows the effect of different K values on the decomposition time of each IMF. The results show that the efficiency of VMD is much higher when $K \leq 3$. From Figure 2, traversing K to find the optimal value and using various swarm intelligence optimization algorithms are both inefficient. Therefore, Ref. [36] proposes an adaptive recursive variational mode decomposition (ARVMD) that dynamically selects the K in recursive loops. ARVMD effectively improves efficiency and reduces recursive mode aliasing. The process of ARVMD is shown in Algorithm 2.

Algorithm 2: ARVMD

Input: A signal f_0, Sampling frequency Fs and quadratic penalty α.
Output: A set of IMFs.
$f = f_0$; IMFs = []; E_u = []; i = 0;
while $E_{ui} > E_{th}$ do
  $i = i + 1$;
  $P_f \leftarrow$ Power spectral density (f);
  $(P_{max}, F_{max}) \leftarrow$ Maximum, corresponding frequency (P_f);
  $N_{peak} \leftarrow$ Numbers of maxima points in $[F_{max} \pm 0.027 \times Fs]$;
  $\{F_1, F_2, \ldots, F_n\} \leftarrow$ Corresponding frequencies of maxima points;
  $$K_i = \begin{cases} 1, N_{peak} < 2 \\ 2, N_{peak} = 2 \\ 3, N_{peak} > 2 \end{cases}$$
  $\{u_1, u_2, \ldots, u_{Ki}\} \leftarrow$ VMD (f, K_i, α, $\{F_1, F_2, \ldots, F_{Ki}\}$);
  $\{E_{u1}, E_{u2}, \ldots, E_{uKi}\} \leftarrow$ Unit bandwidth energy ($\{u_1, u_2, \ldots, u_{Ki}\}$);
  IMFs $\leftarrow$ IMFs $\cup \{u_1, u_2, \ldots, u_{Ki}\}$;
  $E_u \leftarrow E_u \cup \{E_{u1}, E_{u2}, \ldots, E_{uKi}\}$;
  $f = f - \sum_1^{Ki} u_{Ki}(t)$;
end while
IMFs $\leftarrow$ Selection by $E_{ui} > E_{th}$ (IMFs)
return IMFs

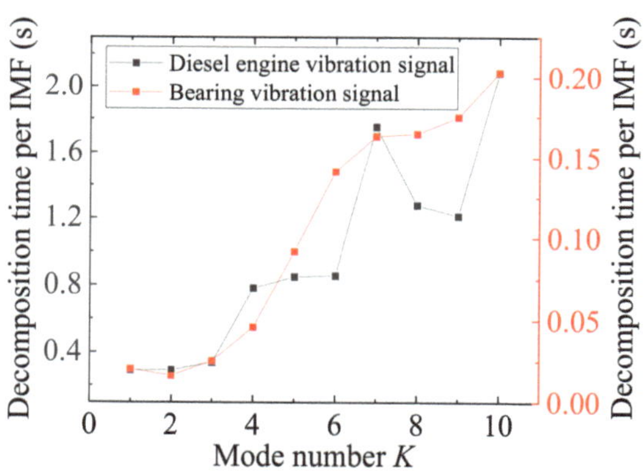

Figure 2. Decomposition time per IMF of VMD. The curve data come from the average of normal signals of various speeds. The engine data come from the experiment of Section 3. The bearing data come from the bearing dataset of Case Western Reserve University [37].

Complex types and working conditions characterize engine faults. However, the component number obtained by ARVMD is variable, resulting in inconsistent feature vector dimensions, which is not conducive to diagnosing multi-speed engine vibration data. A K-value optimization method based on the energy distribution in the frequency domain is proposed to unify the feature dimension. First, ARVMD decomposes the signals of various engine working conditions and obtains many IMFs. These conditions contain data for different speeds and faults (I to XII, as shown in Table 4). Then, the unit bandwidth energy [36] of each IMFs is calculated, and the center frequency of the IMFs is recorded. The unit bandwidth energy is shown in (9):

$$E_u = \frac{e_{IMFi}}{B_{IMFi}} \tag{9}$$

where e_{IMFi} is the energy of IMF and B_{IMFi} is the bandwidth of IMF. The bandwidth is the width of the spectrum when the amplitude of the power spectral density is reduced by 99%. The frequency band [0 Hz, 12,800 Hz] is divided into 128 segments, and the width of each segment is 100 Hz. The unit bandwidth energy of the components located in each segment is counted and averaged. Figure 3 shows the unit bandwidth energy spectrum displayed on divided frequency bands. The results show that there are six prominent energy frequencies: 150 Hz, 1450 Hz, 1950 Hz, 2450 Hz, 4950 Hz, and 7050 Hz. Here, each frequency segment uses the frequency in the middle as the value of the abscissa. Therefore, the engine data will be uniformly decomposed using K equal to 6. This approach can improve the consistency of data processing and help reduce the randomness caused by adaptive decomposition. It also ensures that the dimension of the feature vectors at different speeds is uniform.

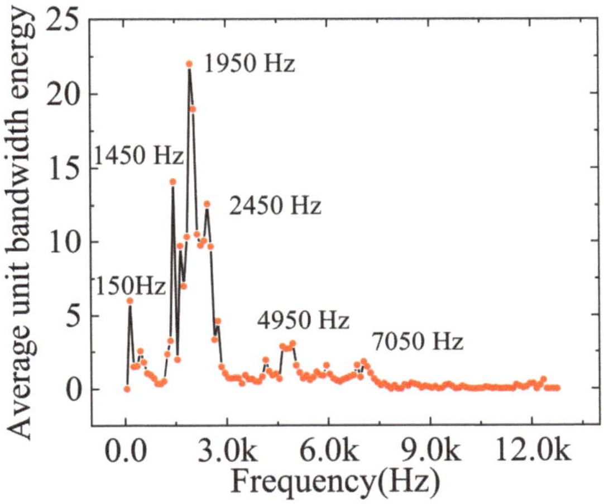

Figure 3. Frequency domain distribution of unit bandwidth energy of engine data.

Furthermore, the iterations of the center frequency of the original VMD are zero-based. It is beneficial for decomposing low-frequency components, but the decomposition time for high-frequency components is longer. Presetting suitable initial center frequencies can significantly improve the efficiency of the VMD [18]. Therefore, the six significant frequencies in Figure 3 are used as the initial center frequencies to iterate.

The quadratic penalty term α is a parameter introduced to improve the convergence when solving the variational model. The role of α in the decomposition is reflected in the noise reduction of the signal. The SNR is the best criterion for choosing a suitable α. However, it is difficult to obtain the SNR of the actual signal after decomposition. Therefore, a set of simulated signals is constructed according to the spectral energy distribution of Figure 3. The expression of the simulated signal is as (10). $\{s_1, s_2, \ldots, s_6\}$ are single-frequency components, which restore the amplitude ratio and frequency of each component in Figure 3. The amplitude of s_3 is set to 100, and the other components are reduced proportionally. s_7 is the noise component with a power of 25 dbW. S_1 is decomposed using VMD, where K is six, and the initial center frequency is preset. Set the variation range of α to [1000, 20,000], and the step size is 100. Calculate the SNR between IMFs and $\{s_1, s_2, \ldots, s_6\}$, and the results are shown in Figure 4. With the increase of α, the SNR has a trend of increasing first and then decreasing. Summing the SNR of each component, it is found that the total SNR does not change much when α is 6000 to 8000. The value of α used in the paper is 6800, and the inset of Figure 4 shows that the SNR reaches the maximum at this value. After the optimized α is obtained, it is used in the mode number optimization for reverse verification, and the results show that it does not affect the results in Figure 3.

$$\begin{cases} s_1(t) = 27\sin(2\pi * 150t), 0 \le t \le 0.053 \\ s_2(t) = 64\sin(2\pi * 1450t), 0 \le t \le 0.053 \\ s_3(t) = 100\sin(2\pi * 1950t), 0 \le t \le 0.053 \\ s_4(t) = 57\sin(2\pi * 2450t), 0 \le t \le 0.053 \\ s_5(t) = 13\sin(2\pi * 4950t), 0 \le t \le 0.053 \\ s_6(t) = 8\sin(2\pi * 7050t), 0 \le t \le 0.053 \\ s_7(t) = \eta \\ S_1 = s_1 + s_2 + s_3 + s_4 + s_5 + s_6 + s_7 \end{cases} \quad (10)$$

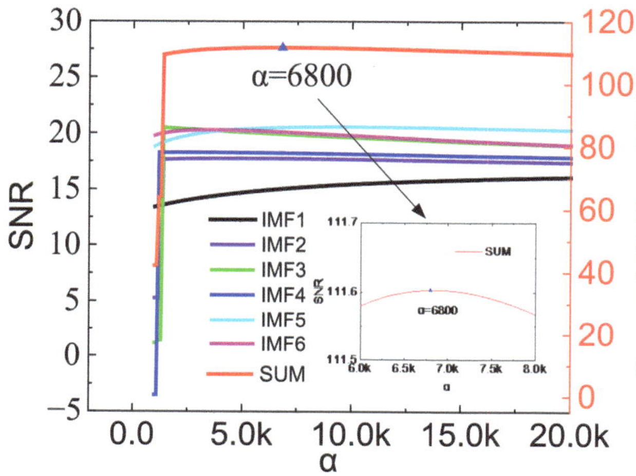

Figure 4. The effect of α on the decomposition SNR.

The optimization of K, α and the iterative optimization of the center frequency have been completed. Next, decompose an actual signal using the improved VMD (IVMD) to verify the effect. A signal of valve clearance increase at 1600 rpm (Condition III in Table 4) was randomly selected for decomposition. The signal's time and frequency domain are shown in Figure 5a,b. Decompose this signal using VMD and IVMD. Figure 6 shows the frequency domain image of the decomposed IMFs. VMD decomposes four components in the [2000 Hz, 3000 Hz] while IVMD decomposes three. The results show that using the same K, VMD focuses on decomposing low-frequency components, while IVMD is more balanced. The average bandwidth aliasing ratio R_{ABA} is introduced to measure the effect of suppressing mode aliasing [36]. The expression of R_{ABA} is shown in (11):

$$R_{ABA} = \sum_{i=1}^{K} \frac{1}{K} \frac{B_A}{B_{IMFi}}, i = 1, 2, \ldots, K. \quad (11)$$

where K is the mode number, B_A is the aliasing bandwidth of the IMFi and other components, and B_{IMFi} is the bandwidth of IMFi. The smaller the R_{ABA}, the better the effect of suppressing mode aliasing. The R_{ABA} for VMD and IVMD results is 0.13 and 0.05, respectively. IVMD suppresses mode aliasing better than VMD. In addition, the center frequency iteration curves of IMFs are shown in Figure 7. Figure 7 shows that IVMD performs 87 iterations, less than VMD's 194 iterations, effectively improving efficiency. The results show that the presetting center frequency can significantly improve the iteration efficiency.

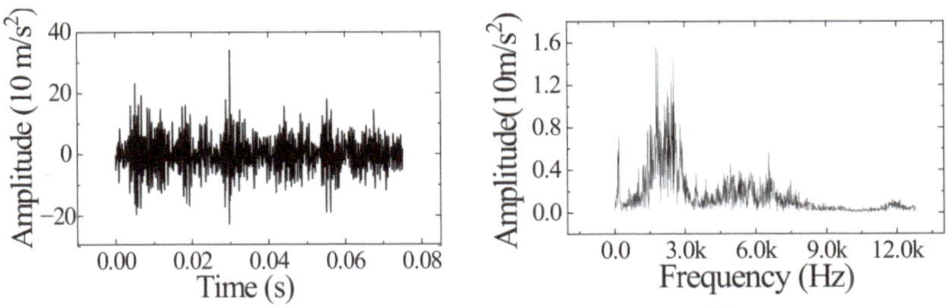

(a) (b)

Figure 5. A signal of valve clearance increase at 1600 rpm. (**a**) The signal in time domain. (**b**) The signal in frequency domain.

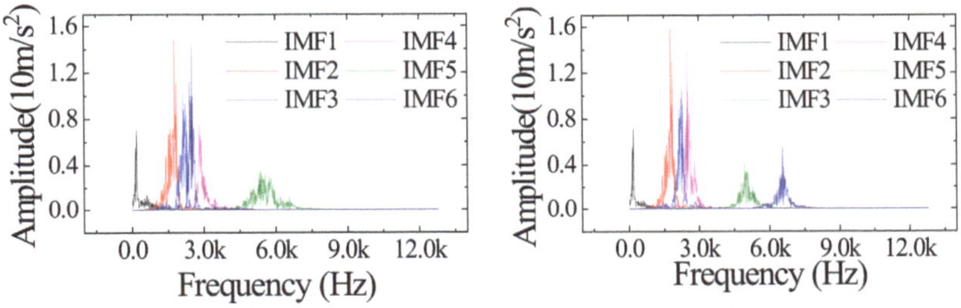

(a) (b)

Figure 6. Frequency domain image of the decomposed IMFs. (**a**) Result of VMD. (**b**) Result of IVMD.

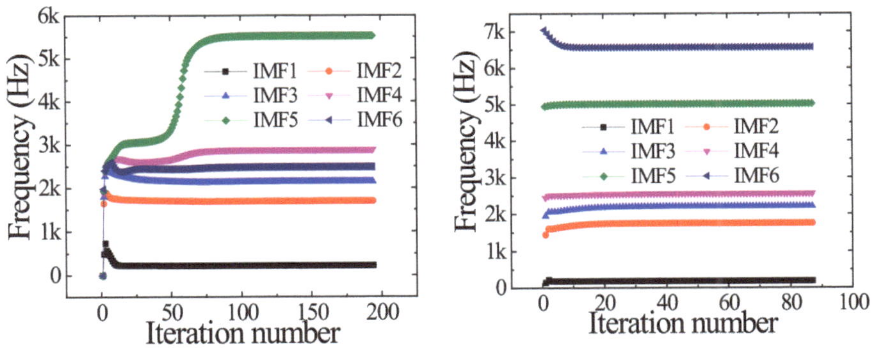

(a) (b)

Figure 7. Center frequency iterative curves for IMFs. (**a**) Iterative curves of VMD. (**b**) Iterative curves of IVMD.

5. VMD-RF Fault Detection Method

After the IVMD decomposition of the engine signal, calculating proper features is beneficial to improve the diagnostic accuracy. The RF method can automatically select a subset of features for classification by bootstrap sampling with return. Instead of considering the feature dimension, features are required to describe the data information as comprehensively as possible. Therefore, the used features include overall features and local features. Finally, seven types of local features are selected. Namely, maximum singular value, energy, unit bandwidth energy, kurtosis, variance, root mean square value (RMS), and center frequency. The seven types of features are calculated for the six IMFs obtained by IVMD. In addition, maximum singular value, energy, RMS, and variance are calculated for the original signal. Feature names and symbols are shown in Table 5.

Table 5. Attribute names and symbols.

Attribute Name	Symbols
Maximum singular value	S_1, S_2, S_3, S_4, S_5, S_6
Energy	E_1, E_2, E_3, E_4, E_5, E_6
Unit bandwidth energy	Eu_1, Eu_2, Eu_3, Eu_4, Eu_5, Eu_6
Kurtosis	K_1, K_2, K_3, K_4, K_5, K_6
Variance	V_1, V_2, V_3, V_4, V_5, V_6
Root mean square	R_1, R_2, R_3, R_4, R_5, R_6
Center frequency	C_1, C_2, C_3, C_4, C_5, C_6
Original signal attributes	S, E, R, V

Note: The features calculated for IMF1 to IMF6 are denoted by the symbols with suffixes 1 to 6, respectively. The symbols without suffixes indicate the features calculated for the original signal.

5.1. Case 1: Diesel Engine Fault Diagnosis

Once the complete feature set is obtained, the feature set can be fed into the RF for classification. The whole flow of fault diagnosis is shown in Figure 8. The data of the first cylinder head (1H), the third cylinder head (3H), and the first cylinder block (1B) at 2300 rpm and the data of the first cylinder head at 2000 rpm were selected for preliminary verification of the algorithm's validity. Each dataset contains four types of faults in Table 4, with a total of 12 fault conditions. Each fault condition includes 200 samples, and the training/test ratio is 4:1. The number of decision trees in RF is 100. The depth of the decision tree is not limited. Then, the training samples are used to generate a random forest. The results of the diagnostic accuracy are shown in Table 6. The proposed method is compared with sequential minimum optimization for support vector machines (SMO-SVM), Multilayer Perceptron (MLP) [31], one-dimensional convolutional neural networks (1DCNN) [29], long and short term memory recurrent neural networks (LSTM-RNN) [30], and residual neural networks (ResNet) [31]. The 1DCNN and LSTM-RNN ran for 200 epochs, while ResNet ran for 30 epochs. The parameters of each algorithm are as follows:

(1) SVM: The RBF kernel is chosen, and the penalty term C is set to 1. The inverse of the radius of influence of the support vector gamma is set to 0.1.
(2) MLP: Two hidden layers are used, both with 30 neurons. The momentum is 0.2, and the learning rate is 0.3.
(3) 1DCNN: The network consists of two convolutional layers (kernel size = 5), two maximum pooling layers (kernel size = 2), and a linear layer. The activation function is ReLU, and the optimizer is Adam.
(4) LSTM-RNN: The network contains two LSTM layers with 64 nodes in each layer.
(5) ResNet: The network uses the 18-layer ResNet model, as described in Ref. [38].

SMO-SVM, MLP, 1DCNN, and IVMD-RF achieved high accuracy from the results of single-speed data. The LSTM-RNN had the lowest accuracy, which shows its poor classification ability for non-time series. Compared to the 1H data set, the diagnostic accuracy of the 1B and 3H datasets decreased significantly due to the increased distance of the sensor location from the combustion chamber and valve. The vibration signal may be distorted or coupled with other disturbances when it is transmitted.

The next step is to use these methods to diagnose multiple speed conditions. All types of failure data for the first cylinder head (1H) at 700 rpm, 1300 rpm, 1600 rpm, 2000 rpm, and 2300 rpm were made into one dataset. Since there are no abnormal injection advance angle faults in the 700 rpm data, a total of 56 labeled categories of data are included. The diagnostic results are shown in Table 6. Compared to the single-speed dataset for the first cylinder head (1H), the accuracy of each method decreases to varying degrees as the number of failure types increases. The accuracy of SMO-SVM dropped the most. SMO-SVM method is suitable for single-speed data classification but not as effective as other methods for multi-speed and multi-class data. The proposed method still maintains high accuracy. The results show that IVMD-RF has advantages for multi-speed and multi-type fault diagnosis scenarios.

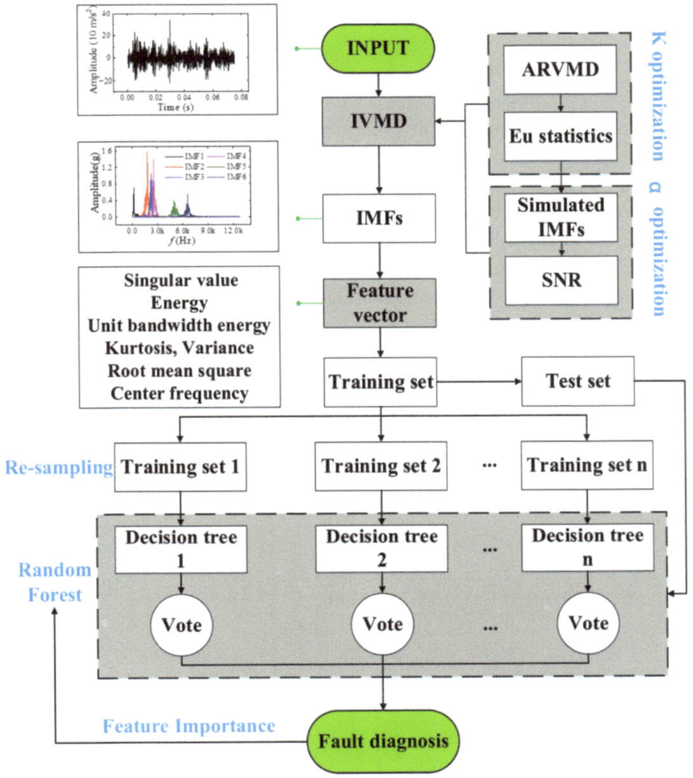

Figure 8. IVMD-RF fault detection process.

Table 6. Accuracy comparison of different algorithms (%).

Methods	1H-2000 rpm	1H-2300 rpm	1B-2300 rpm	3H-2300 rpm	1H-Multi-Speed
SMO-SVM	99.06	97.50	93.75	92.91	92.59
MLP	97.81	98.13	92.36	92.71	96.79
1DCNN	94.06	99.06	92.89	94.06	96.56
LSTM-RNN	59.06	75.16	78.12	68.75	56.23
ResNet	91.09	96.88	88.28	86.72	92.19
IVMD-RF	98.75	99.38	92.91	93.54	97.32

Note: "3H" represents the third cylinder head, "1B" represents the first cylinder block.

In addition, the proposed method requires less training time to achieve high accuracy. For comparison, all algorithms are run in the same environment (Python 3.8, Windows 11, Intel Core i7-10700 CPU @ 2.9 GHz), and the running time is recorded in Table 7 The results show that SMO-SVM has the highest training efficiency, followed closely by IVMD-RF. ResNet has the longest training time due to the deep network layers. Therefore, the proposed method has high efficiency and high accuracy. It is worth noting that deep learning may provide better diagnostic results for the original raw signal. However, the significant increase in data dimensionality leads to an increase in computation time and higher hardware requirements, which deviates from the purpose of this paper. No diagnostics were performed on the original raw data to keep the variables consistent.

Next, the 1300 rpm, 1600 rpm, 2000 rpm, and 2300 rpm data were mixed into one dataset. The data were labeled into 12 categories according to Table 4, regardless of the speed change. The cross-speed datasets include the 1H dataset, 1B dataset, and 3H dataset. The above methods are still used for classification, and the results are shown in Table 8. The results show that the accuracy of each algorithm has a certain drop compared to

Table 6, especially the SMO-SVM. The 1DCNN and IVMD-RF still maintain a relatively high accuracy rate. The overall accuracy of the 3H dataset is low because the sensor in the third cylinder head is far from the cylinder where some failures occurred. The proposed method still has some advantages over other algorithms. Table 8 also shows the classification precision, recall, and f1-score. These indicators are weighted averages, where the weights are determined by the proportion of each class sample distribution. The results show that the proposed method also performs well on these metrics. For the 3H dataset with relatively poor diagnostic results, Figure 9 shows the comparison of recall and precision of each algorithm for the 12 classes. The results showed low recall and accuracy for reduced valve clearance (condition II) and abnormal injection advance angles (condition IX to XII). The reason for the low precision and recall of Fault II is the slight increase in valve clearance and the long distance of the sensor from the cylinder where the fault occurred. Faults IX to XII, on the other hand, are due to small changes in injection advance angle, causing only minor differences in combustion conditions. The training efficiency of the proposed method is much higher than that of the deep learning method and slightly lower than that of SMO-SVM. Figure 10 shows the confusion matrix for the 1H dataset, indicating that most of the misclassified samples are data of the same type but with different failure levels, which proves the effectiveness of the proposed method.

Table 7. Comparison of training time of various algorithms (s).

Methods	1H-2000 rpm	1H-2300 rpm	1B-2300 rpm	3H-2300 rpm	1H-Multi-Speed
SMO-SVM	0.05	0.06	0.05	0.07	2.84
MLP	9.60	9.91	9.55	9.53	274.21
1DCNN	13.02	14.10	14.65	14.87	167.38
LSTM-RNN	17.85	18.21	20.26	20.48	224.95
ResNet	49.32	47.66	54.53	51.66	446.92
IVMD-RF	0.26	0.28	0.31	0.29	4.30

Note: "3H" represents the third cylinder head, "1B" represents the first cylinder block.

Table 8. Comparison of fault diagnosis results of each algorithm for cross-speed dataset.

/	Methods	SMO-SVM	MLP	1DCNN	LSTM-RNN	ResNet	IVMD-RF
First cylinder head Y-direction (1H)	Accuracy	0.81	0.92	0.93	0.68	0.87	0.97
	Precision	0.82	0.92	0.94	0.65	0.89	0.97
	Recall	0.81	0.92	0.92	0.63	0.88	0.96
	F1-score	0.82	0.92	0.92	0.62	0.88	0.97
	Time (s)	0.99	70.19	130.02	189.29	219.77	2.20
First cylinder block Y-direction (1B)	Accuracy	0.81	0.89	0.93	0.67	0.86	0.92
	Precision	0.81	0.89	0.94	0.75	0.87	0.92
	Recall	0.80	0.89	0.91	0.72	0.85	0.92
	F1-score	0.80	0.89	0.91	0.72	0.86	0.92
	Time (s)	1.13	70.69	145.99	192.59	224.21	2.74
Third cylinder head Y-direction (3H)	Accuracy	0.65	0.75	0.83	0.47	0.78	0.94
	Precision	0.66	0.76	0.83	0.42	0.72	0.93
	Recall	0.65	0.76	0.80	0.42	0.71	0.94
	F1-score	0.65	0.76	0.81	0.40	0.71	0.93
	Time (s)	1.03	70.28	143.31	203.47	221.34	2.92

The datasets with different training/testing ratios are set up for classification to verify the diagnostic effectiveness of various methods for the small sample case. Figure 11 shows each algorithm's accuracy and time consumption curves for the 1H dataset at different training test ratios (0.1 to 4). When the training test ratio <0.25, the accuracy of 1DCNN, RNN, and MLP significantly decrease, while SMO-SVM and IVMD-RF decrease more smoothly. When the training test ratio is 0.1, IVMD-RF has the highest accuracy of 88.77%. Figure 11b shows that the training efficiency of each algorithm increases as the training/test ratio decreases. The efficiency of IVMD-RF and SMO-SVM remains higher than the other methods. The results illustrate the good diagnostic effect of the proposed method for small samples of single-sensor data. Figure 12 shows the accuracy and training loss curves when the three deep learning methods are applied to the 1H dataset. Figure 12 indicates that the 1DCNN has converged while the RNN clearly shows over-fitting, which is the reason for its low accuracy. Continuing to train ResNet may improve the accuracy, but the training efficiency is too low compared to other methods. Therefore, IVMD-RF has a high fault diagnosis accuracy and high efficiency for cross-speed data. It is worth noting that deep learning methods still have more advantages and potential when the amount of labeled data and computational resources are sufficient.

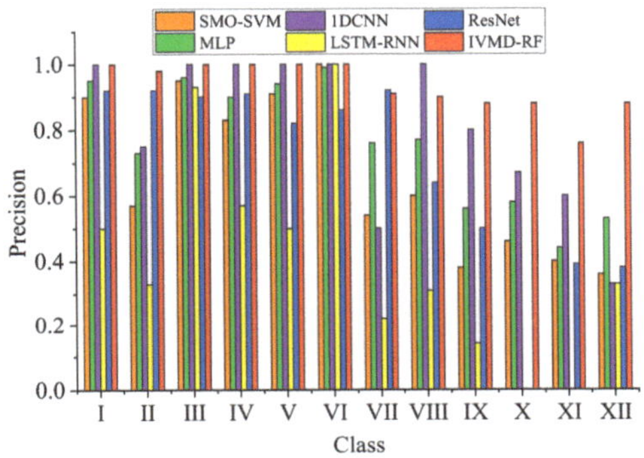

(a)

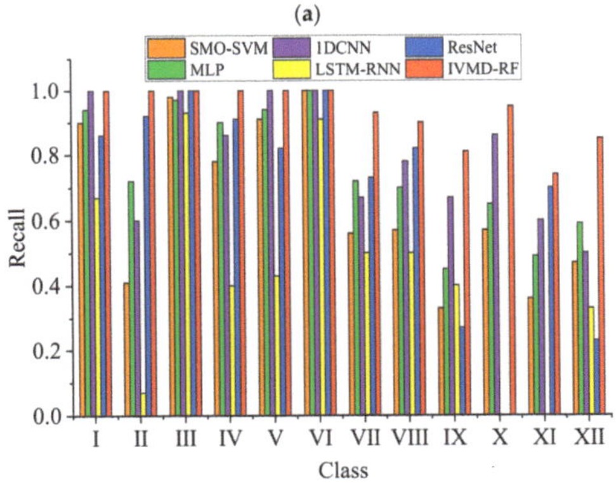

(b)

Figure 9. Precision and recall results of each algorithm for the 3H dataset. (**a**) Precision result. (**b**) Recall result.

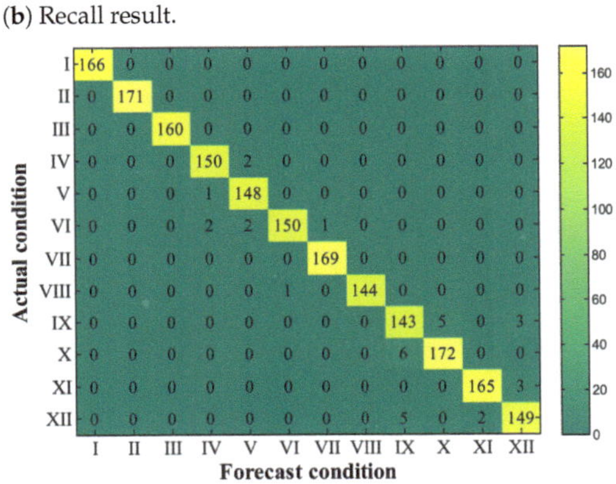

Figure 10. Confusion matrix for 1H dataset of IVMD-RF.

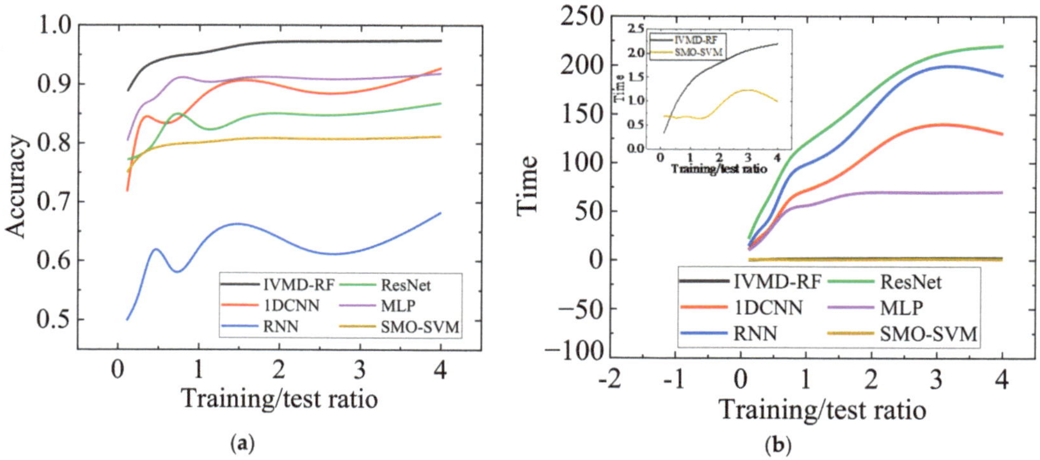

Figure 11. Comparison of each algorithm with different training/testing ratios (0.1 to 4). (**a**) Comparison of accuracy. (**b**) Comparison of the training time.

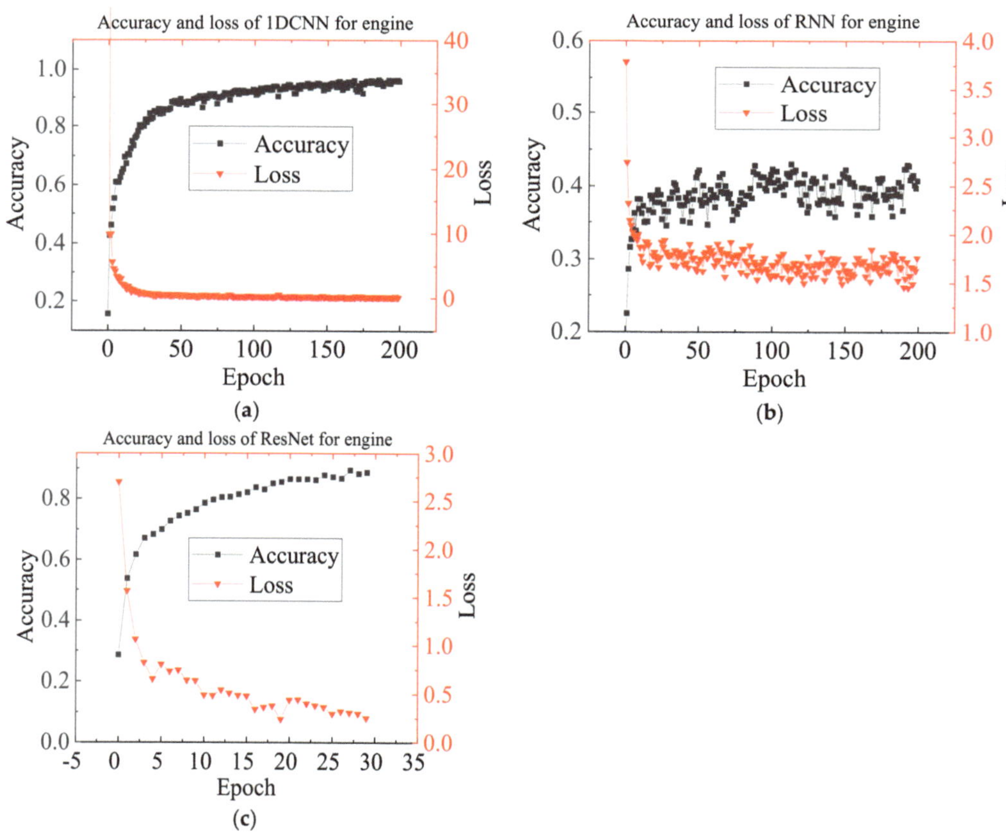

Figure 12. Accuracy and training loss of deep learning methods for 1H dataset. (**a**) Accuracy and training loss of 1DCNN. (**b**) Accuracy and training loss of LSTM-RNN. (**c**) Accuracy and training loss of ResNet.

The study of feature importance can further improve the performance of the method. Using the RF to rank the importance of features, the results for the 1H dataset are shown in Figure 13. Singular values, energy, and center frequency contributed more to the classification, followed by variance and unit bandwidth energy. However, kurtosis has little contribution to the classification results. Figure 13 shows that features with suffixes 1 and 4 contribute significantly to the classification, i.e., IMF1 and IMF4 contribute the most to the classification, followed by IMF5 and IMF6. For different working conditions, the difference in the body surface vibration is mainly reflected in the low-frequency (IMF1) and high-frequency components (IMF4~6). IMF2 and IMF3 have high energy but weak contribution. This conclusion is valuable for the study of unsupervised engine fault diagnosis. Figure 14 shows the impact of using different numbers of features in order of importance on training time and accuracy. Finally, we found an optimal point. When using fifteen features, it only takes 1.23 s to train and can achieve 97% diagnostic accuracy as marked in Figure 14. The selected fifteen categories of features are marked in Figure 13. Feature selection significantly improves training time with little change in accuracy.

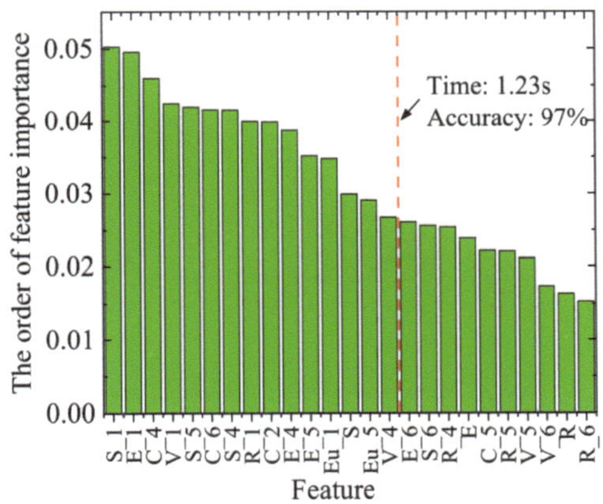

Figure 13. The order of feature importance.

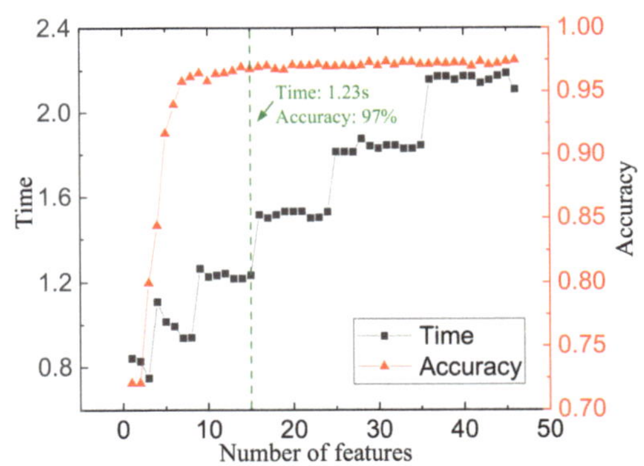

Figure 14. The impact of changing the number of features used.

5.2. Case 2: Gasoline Engine Fault Diagnosis

To verify the effectiveness of the proposed method on different engines, the gasoline engine fault data will be diagnosed in the following. The fault data came from a two-cylinder, two-stroke gasoline engine with the specific engine parameters shown in Table 9. The sensor locations and coordinate system for the engine are shown in Figure 15. The data used are from the Y-direction of cylinder 1 and the X-direction of cylinder 2 (the two sensors connected by the white wire in Figure 15). Three common faults were simulated:

abnormal injection advance angle, abnormal air-fuel ratio, and misfire. Among them, the abnormal injection advance angle occurs in both cylinders, while the other two failures occur only in cylinder 1. The specific fault level settings are shown in Table 10. Use Roman numerals I through VII to indicate individual faults. Signals from 5000 rpm and 7000 rpm were collected. Each working condition contains 200 samples, each containing data from one working cycle.

Table 9. Parameters of gasoline engine.

Items	Parameters
Displacement	0.294 L
Rated power/Rated speed	35.5 kW/8500 rpm
Maximum torque/Speed range	44.5 Nm/7000 rpm

Figure 15. Sensor positions and coordinate direction of the engine. (**a**) Position of sensor 1Y. (**b**) Position of sensor 2X.

Table 10. Fault type and degree parameter setting.

Mark	Injection Advance Angle	Air/Fuel Ratio	Misfire Rate
I	10 °CA	1	0
II	5 °CA	1	0
III	15 °CA	1	0
IV	10 °CA	1.1	0
V	10 °CA	1.2	0
VI	10 °CA	1	0.05
VII	10 °CA	1	0.1

Note: The shaded green marks the location of the faulty parameter.

The 5000 and 7000 rpm data were mixed to form the cross-speed dataset. Various algorithms diagnose the fault data of 1Y and 2X sensors separately. The settings of each algorithm are shown in Section 5.1. A comparison of the diagnostic results for each algorithm is shown in Table 11. The results show an overall decrease in the diagnostic accuracy of each algorithm due to the increase in signal noise as the two-cylinder, two-stroke engine vibrates more than the diesel engine. The higher speed is also one of the reasons. SMO-SVM is still the fastest, but its accuracy is low. The proposed method works best for fault diagnosis of 1Y data, and 1DCNN works best for 2X data. The difference in accuracy between the two is not significant. The proposed method is more efficient and suitable for low hardware conditions. The results show that IVMD-RF can be used for gasoline engine fault diagnosis.

Table 11. Comparison of fault diagnosis results of each algorithm for gasoline engine data.

	Methods	SMO-SVM	MLP	1DCNN	LSTM-RNN	ResNet	IVMD-RF
First cylinder head Y-direction (1Y)	Accuracy	0.67	0.75	0.82	0.52	0.78	0.85
	Precision	0.68	0.77	0.81	0.69	0.78	0.85
	Recall	0.67	0.75	0.82	0.51	0.79	0.84
	F1-score	0.67	0.72	0.82	0.52	0.78	0.84
	Time (s)	0.33	15.83	9.45	6.44	11.02	1.16
Second cylinder head X-direction (2X)	Accuracy	0.72	0.74	0.82	0.59	0.79	0.79
	Precision	0.72	0.74	0.82	0.58	0.77	0.78
	Recall	0.72	0.74	0.81	0.58	0.76	0.79
	F1-score	0.72	0.74	0.82	0.58	0.74	0.79
	Time (s)	0.25	15.80	9.61	6.51	11.05	1.12

6. Conclusions and Discussion

This article proposes an IVMD-RF for single-sensor multi-fault detection of the engine. In IVMD, the engine data spectral energy distribution is obtained through multiple decompositions and statistics. The alpha value was chosen based on the spectral distribution and the SNR. By presetting the center frequency and the optimal K and α values, the efficiency is improved, the mode aliasing is reduced, and the feature size is unified. The effectiveness of IVMD is proved by decomposing the engine signals. Seven types of attributes are calculated to form a feature group for IMFs, which is input into RF for classification. Compared with various machine learning and deep learning algorithms, it is proved that the proposed method has advantages in training efficiency and accuracy. Through the feature importance study, it is found that the high-frequency and low-frequency IMFs contribute more to the classification. Fifteen optimal features have been selected to improve the efficiency of RF. The IVMD-RF method has application prospects in engine single-sensor multi-class fault detection.

Author Contributions: Conceptualization, D.T. and F.B.; software, J.C.; validation, D.T., X.Y. and P.S.; writing—original draft preparation, D.T.; writing—review and editing, J.C. and X.B.; project administration, F.B.; funding acquisition, X.B. All authors have read and agreed to the published version of the manuscript.

Funding: This research was funded in part by the Science and Technology Research Project of Higher Education in Hebei province of China under Grant QN2022159, and in part by National Key Research and Development Program of China under Grant 2021YFD2000303.

Institutional Review Board Statement: Not applicable.

Informed Consent Statement: Not applicable.

Data Availability Statement: The data presented in this study are not publicly available at this time but may be obtained upon reasonable request from the authors.

Conflicts of Interest: The authors declare no conflict of interest.

References

1. Zhong, J.H.; Wong, P.K.; Yang, Z.X. Fault diagnosis of rotating machinery based on multiple probabilistic classifiers. *Mech. Syst. Signal Process.* **2018**, *108*, 99–114. [CrossRef]
2. Delvecchio, S.; Bonfiglio, P.; Pompoli, F. Vibro-acoustic condition monitoring of Internal Combustion Engines: A critical review of existing techniques. *Mech. Syst. Signal Process.* **2018**, *99*, 661–683. [CrossRef]
3. Goyal, D.; Pabla, B.S. The Vibration Monitoring Methods and Signal Processing Techniques for Structural Health Monitoring: A Review. *Arch. Comput. Methods Eng.* **2016**, *23*, 585–594. [CrossRef]
4. Ma, Y.L.; Cheng, J.S.; Wang, P.; Wang, J.; Yang, Y. A novel Lanczos quaternion singular spectrum analysis method and its application to bevel gear fault diagnosis with multi-channel signals. *Mech. Syst. Signal Process.* **2022**, *168*, 108679. [CrossRef]
5. Ribeiro, R.F.; Areias, I.A.D.; Campos, M.M.; Teixeira, C.E.; da Silva, L.E.B.; Gomes, G.F. Fault detection and diagnosis in electric motors using 1d convolutional neural networks with multi-channel vibration signals. *Measurement* **2022**, *190*, 110759.

6. Bhowmik, B.; Panda, S.; Hazra, B.; Pakrashi, V. Feedback-driven error-corrected single-sensor analytics for real-time condition monitoring. *Int. J. Mech. Sci.* **2022**, *214*, 106898. [CrossRef]
7. Ayati, M.; Shirazi, F.A.; Ansari-Rad, S.; Zabihihesari, A. Classification-Based Fuel Injection Fault Detection of a Trainset Diesel Engine Using Vibration Signature Analysis. *J. Dyn. Syst. Meas. Control* **2020**, *142*, 051003. [CrossRef]
8. Wang, H.F.; Jiang, W.; Deng, X.Y.; Geng, J. A new method for fault detection of aero-engine based on isolation forest. *Measurement* **2021**, *185*, 110064. [CrossRef]
9. Ellefsen, A.L.; Han, P.H.; Cheng, X.; Holmeset, F.T.; Æsøy, V.; Zhang, H.X. Online Fault Detection in Autonomous Ferries: Using Fault-Type Independent Spectral Anomaly Detection. *IEEE Trans. Instrum. Meas.* **2020**, *69*, 8216–8225. [CrossRef]
10. Liu, J.X.; Yang, L.D.; Xu, M.J.; Zhang, Q.; Yan, R.Q.; Chen, X.F. Model-based detection of soft faults using the smoothed residual for a control system. *Meas. Sci. Technol.* **2021**, *32*, 015107. [CrossRef]
11. Wang, Z.J.; Zhang, J.J.; Jiang, Z.N.; Mao, Z.W.; Chang, K.; Wang, C.G. Quantitative misalignment detection method for diesel engine based on the average of shaft vibration and shaft shape characteristics. *Measurement* **2021**, *181*, 109527. [CrossRef]
12. Naderi, E.; Khorasani, K. Data-driven fault detection, isolation and estimation of aircraft gas turbine engine actuator and sensors. *Mech. Syst. Signal Process.* **2018**, *100*, 415–438. [CrossRef]
13. Chao, M.A.; Adey, B.T.; Fink, O. Implicit supervision for fault detection and segmentation of emerging fault types with Deep Variational Autoencoders. *Neurocomputing* **2021**, *454*, 324–338. [CrossRef]
14. Elmasry, W.; Wadi, M. EDLA-EFDS: A Novel Ensemble Deep Learning Approach for Electrical Fault Detection Systems. *Electr. Power Syst. Res.* **2022**, *207*, 107834. [CrossRef]
15. Feng, Y.; Liu, Z.J.; Chen, J.L.; Lv, H.X.; Wang, J.; Zhang, X.W. Unsupervised Multimodal Anomaly Detection with Missing Sources for Liquid Rocket Engine. *IEEE Trans. Neural Netw. Learn. Syst.* **2022**. [CrossRef]
16. Bi, F.R.; Ma, T.; Wang, X. Development of a novel knock characteristic detection method for gasoline engines based on wavelet-denoising and EMD decomposition. *Mech. Syst. Signal Process.* **2019**, *117*, 517–536.
17. Wang, X.; Cai, Y.P.; Li, A.H.; Zhang, W.; Yue, Y.J.; Ming, A.B. Intelligent fault diagnosis of diesel engine via adaptive VMD-Rihaczek distribution and graph regularized bi-directional NMF. *Measurement* **2021**, *172*, 108823. [CrossRef]
18. Dragomiretskiy, K.; Zosso, D. Variational Mode Decomposition. *IEEE Trans. Signal Process.* **2014**, *62*, 531–544. [CrossRef]
19. Huang, S.L.; Sun, H.Y.; Wang, S.; Qu, K.F.; Zhao, W.; Peng, L.S. SSWT and VMD Linked Mode Identification and Time-of-Flight Extraction of Denoised SH Guided Waves. *IEEE Sens. J.* **2021**, *21*, 14709–14717. [CrossRef]
20. Miao, Y.H.; Zhao, M.; Lin, J. Identification of mechanical compound-fault based on the improved parameter-adaptive variational mode decomposition. *ISA Trans.* **2019**, *84*, 82–95. [CrossRef]
21. Nazari, M.; Sakhaei, S.M. Successive variational mode decomposition. *Signal Process.* **2020**, *174*, 107610. [CrossRef]
22. Lian, J.J.; Liu, Z.; Wang, H.J.; Dong, X.F. Adaptive variational mode decomposition method for signal processing based on mode characteristic. *Mech. Syst. Signal Process.* **2018**, *107*, 53–77. [CrossRef]
23. Mei, L.; Li, S.Y.; Zhang, C.; Han, M.X. Adaptive Signal Enhancement Based on Improved VMD-SVD for Leak Location in Water-Supply Pipeline. *IEEE Sens. J.* **2021**, *21*, 24601–24612. [CrossRef]
24. Li, J.M.; Wang, H.; Zhang, J.F.; Yao, X.F.; Zhang, Y.G. Impact fault detection of gearbox based on variational mode decomposition and coupled underdamped stochastic resonance. *ISA Trans.* **2019**, *95*, 320–329. [CrossRef]
25. Diao, X.; Jiang, J.C.; Shen, G.D.; Chi, Z.Z.; Wang, Z.R.; Ni, L.; Mebarki, A.; Bian, H.T.; Hao, Y.M. An improved variational mode decomposition method based on particle swarm optimization for leak detection of liquid pipelines. *Mech. Syst. Signal Process.* **2020**, *143*, 106787. [CrossRef]
26. Sahani, M.; Dash, P.K. Deep Convolutional Stack Autoencoder of Process Adaptive VMD Data with Robust Multikernel RVFLN for Power Quality Events Recognition. *IEEE Trans. Instrum. Meas.* **2021**, *70*, 9001912. [CrossRef]
27. Luo, X.J.; Fong, K.F.; Sun, Y.J.; Leung, M.K.H. Development of clustering-based sensor fault detection and diagnosis strategy for chilled water system. *Energy Build.* **2019**, *186*, 17–36. [CrossRef]
28. Ni, Y.R.; Zeng, X.J.; Liu, Z.L.; Yu, K.; Xu, P.B.; Wang, Z.; Zhuo, C.; Huang, Y. Faulty feeder detection of single phase-to-ground fault for distribution networks based on improved K-means power angle clustering analysis. *Int. J. Electr. Power Energy Syst.* **2022**, *142*, 108252. [CrossRef]
29. Shahid, S.M.; Ko, S.; Kwon, S. Real-time abnormality detection and classification in diesel engine operations with convolutional neural network. *Expert Syst. Appl.* **2022**, *192*, 116233. [CrossRef]
30. Zhang, B.; Zhang, S.H.; Li, W.H. Bearing performance degradation assessment using long short-term memory recurrent network. *Comput. Ind.* **2019**, *106*, 14–29. [CrossRef]
31. Lee, J.; Lee, Y.C.; Kim, J.T. Fault detection based on one-class deep learning for manufacturing applications limited to an imbalanced database. *J. Manuf. Syst.* **2020**, *57*, 357–366. [CrossRef]
32. Li, H.F.; Hu, G.Z.; Li, J.Q.; Zhou, M.C. Intelligent Fault Diagnosis for Large-Scale Rotating Machines Using Binarized Deep Neural Networks and Random Forests. *IEEE Trans. Autom. Sci. Eng.* **2022**, *19*, 1109–1119. [CrossRef]
33. Breiman, L. Random forests. *Mach. Learn.* **2001**, *45*, 5–32. [CrossRef]
34. Nahim, H.M.; Younes, R.; Shraim, H.; Ouladsine, M. Oriented review to potential simulator for faults modeling in diesel engine. *J. Mar. Sci. Technol.* **2016**, *21*, 533–551. [CrossRef]
35. Yu, S.W.; Ma, J.W. Complex Variational Mode Decomposition for Slop-Preserving Denoising. *IEEE Trans. Geosci. Remote Sens.* **2018**, *56*, 586–597. [CrossRef]

36. Tang, D.J.; Bi, F.R.; Lin, J.W.; Li, X.; Yang, X.; Bi, X.Y. Adaptive Recursive Variational Mode Decomposition for Multiple Engine Faults Detection. *IEEE Trans. Instrum. Meas.* **2022**, *71*, 3513111. [CrossRef]
37. The Bearing Data Source. Available online: https://csegroups.case.edu/bearingdatacenter (accessed on 10 April 2018).
38. He, K.M.; Zhang, X.Y.; Ren, S.Q.; Sun, J. Deep Residual Learning for Image Recognition. In Proceedings of the IEEE Conference on Computer Vision and Pattern Recognition, Seattle, WA, USA, 27–30 June 2016; pp. 770–778.

sensors

MDPI

Article

Short-Training Damage Detection Method for Axially Loaded Beams Subject to Seasonal Thermal Variations

Marta Berardengo [1], Francescantonio Lucà [2,*], Marcello Vanali [3] and Gianvito Annesi [3]

1 Department of Mechanical, Energy, Management and Transportation Engineering, Università degli Studi di Genova, Via all'Opera Pia, 15A, 16145 Genoa, Italy
2 Department of Mechanical Engineering, Politecnico di Milano, Via La Masa, 34, 20156 Milan, Italy
3 Department of Engineering and Architecture, Università degli Studi di Parma, Parco Area delle Scienze, 181/A, 43124 Parma, Italy
* Correspondence: francescantonio.luca@polimi.it

Abstract: Vibration-based damage features are widely adopted in the field of structural health monitoring (SHM), and particularly in the monitoring of axially loaded beams, due to their high sensitivity to damage-related changes in structural properties. However, changes in environmental and operating conditions often cause damage feature variations which can mask any possible change due to damage, thus strongly affecting the effectiveness of the monitoring strategy. Most of the approaches proposed to tackle this problem rely on the availability of a wide training dataset, accounting for the most part of the damage feature variability due to environmental and operating conditions. These approaches are reliable when a complete training set is available, and this represents a significant limitation in applications where only a short training set can be used. This often occurs when SHM systems aim at monitoring the health state of an already existing and possibly already damaged structure (e.g., tie-rods in historical buildings), or for systems which can undergo rapid deterioration. To overcome this limit, this work proposes a new damage index not affected by environmental conditions and able to properly detect system damages, even in case of short training set. The proposed index is based on the principal component analysis (PCA) of vibration-based damage features. PCA is shown to allow for a simple filtering procedure of the operating and environmental effects on the damage feature, thus avoiding any dependence on the extent of the training set. The proposed index effectiveness is shown through both simulated and experimental case studies related to an axially loaded beam-like structure, and it is compared with a Mahalanobis square distance-based index, as a reference. The obtained results highlight the capability of the proposed index in filtering out the temperature effects on a multivariate damage feature composed of eigenfrequencies, in case of both short and long training set. Moreover, the proposed PCA-based strategy is shown to outperform the benchmark one, both in terms of temperature dependency and damage sensitivity.

Keywords: structural health monitoring; unsupervised learning; environmental variations; principal component analysis; short baseline; tie-rods; beam-like structures; mahalanobis squared distance

Citation: Berardengo, M.; Lucà, F.; Vanali, M.; Annesi, G. Short-Training Damage Detection Method for Axially Loaded Beams Subject to Seasonal Thermal Variations. *Sensors* **2023**, *23*, 1154. https://doi.org/10.3390/s23031154

Academic Editor: Theodore E. Matikas

Received: 15 December 2022
Revised: 6 January 2023
Accepted: 13 January 2023
Published: 19 January 2023

1. Introduction

Structures are naturally subject to deterioration and material degradation, which can lead to critical damage conditions. When the structural integrity is compromised, system current or future performances are affected. Thus, being able to detect damage at an early stage plays a key role in order to carry out prompt maintenance actions, preventing structural failure. This aspect has a relevant impact, first and foremost in terms of safety for the users, but also from an economic point of view. Indeed, carrying out effective maintenance actions, acting only when required, allows a better use of the maintenance resources.

The research area aiming at defining automatic damage detection strategies goes by the name of structural health monitoring (SHM) [1]. Due to the availability of advanced sensing techniques, data acquisition, computing, and information management, these strategies are mainly data-driven, i.e., they exploit data acquired by sensors on the monitored structure. Since no device directly measures damage, a crucial point is the extraction of damage sensitive quantities, or damage features, from the signals acquired by the sensors [2].

Vibration-based approaches are among the most commonly adopted approaches, as reported by many exhaustive review papers in the literature (e.g., [3–7]). According to these approaches, damage sensitive features are extracted from the dynamic response of the monitored structure by adopting, e.g., time series models [8–10] or modal analysis [11], relying on a simple assumption: damage manifests itself as a change in structural properties (e.g., a change of mass, stiffness, constraint characteristics or structural connectivity) that reflects in changes of modal parameters (i.e., eigenfrequencies, mode shapes and damping coefficients) [12]. Vibration-based approaches are also called global approaches [12], since the information that can be extracted from the response of a structure is related to the overall structural condition. This aspect comes with two significant advantages. Firstly, as opposite to local techniques, vibration-based techniques can be successfully adopted to detect damage without knowing the expected damage location in advance. Secondly, vibration-based techniques often use a limited number of sensors and the instrumentation required can be easily integrated in the monitored structure [5,13]. Vibration-based techniques, together with their practical advantages, are crucial for all those structures whose dynamic behaviour is significantly affected by damage, such as tie-rods, which are the main focus of this study.

Tie-rods are axially-loaded metallic beams used to balance lateral forces in arches and vaults of civil structures. Due to their characteristics, these slender elements undergo significant vibration levels under operational conditions, which make the adoption of vibration-based SHM techniques particularly suitable. Considering real operating tie-rods, they show a high uncertainty, generally associated to geometrical and material properties, loading conditions and constraint characteristics. Moreover, many different damage scenarios are possible and, in most cases, damage-related data are not available at the beginning of the monitoring phase. These factors make the use of supervised methods difficult and unreliable. Thus, an unsupervised learning approach becomes interesting, since damage is assessed when a statistically significant variation of the adopted vibration-based damage features is observed, with respect to a reference condition [2].

However, the main obstacle to the adoption of unsupervised learning approaches to real structures is related to the effects of environmental and operational variations [14]. Indeed, changes of environmental variables, e.g., temperature, cause changes to structural properties that can significantly increase the variability associated to vibration-based damage features [15,16]. For the specific case of tie-rods, it has been observed that this high variability can mask the effects of damage at an early stage, hampering a prompt damage detection [17,18].

In the literature of SHM, different approaches have been proposed to face the problems related to environmental and operational variations. A family of approaches is that of input-output models, which require measurements of both the environmental variables (the input) and the structural response (the output) to filter out the environmental effects through the adoption of, e.g., linear correlation models [19–22], neural networks [23–25] or support vector machines [26,27]. However, often not all the relevant environmental and operational variables are measured or known. For this reason, output-only approaches can be adopted to compensate environmental and operational changes, without relying on any additional measurement related to these changes.

When output-only techniques are considered, a possible approach to filter out the temperature effects is to actually include the normal variability of environmental factors in the training data and to use multivariate data with enough redundancy to remove the unwanted effects, using the data correlation structure [28]. Some recent examples of such approach can be found in the literature, based on Kalman filtering [29], Bayesian virtual sensing [30,31] and principal component analysis (PCA) [32–34]. One of the most popular tool is the multivariate metrics known as Mahalanobis squared distance (MSD) [35]. The MSD is used to assess when a new observation of a multivariate damage feature is an outlier with respect to a reference data set, called the baseline set. The MSD can naturally filter out the environmental variability, provided that a proper baseline set, containing the full range of environmental conditions, is adopted and a high-enough number of variables is used, to ensure some separability between the damage effects and the environmental effects [35]. Therefore, a critical aspect for MSD-based damage detection, and in general for any method relying on an exhaustive baseline, is the amount of time needed to build such a complete baseline set, representative of most of the natural variability. There are several cases, indeed, where this aspect prevents a reliable use of monitoring systems, and where methods not sensitive to changes of operational and environmental conditions are necessary to properly detect structural damage. This paper aims at solving this problem by proposing an SHM method able to filter out any change of the considered damage feature due to environmental effects, and able to work even when short training set, which is inevitably lacking in information, must be used.

Many different cases fall in this category and would benefit of an SHM method with these peculiarities; some examples are listed below:

- when a new structure is considered, the reference data acquired at beginning of the monitoring campaign refers to the healthy condition of the structure. In this case, damage detection cannot be effectively carried out until all the temperature conditions are observed, due to long-term seasonal effects. This can imply excessively long time before being able to start the actual monitoring of the structure, also resulting in the impossibility of detecting early damages;
- another critical scenario could be that of a case where an already operating structure shows a suspicious structural behaviour that suggests the installation of an SHM system, such as in the case of tie-rods of historical buildings. In this case, since damage can potentially be already ongoing, the goal would be detecting the possible evolution of the deterioration process. In such a situation, the need for a long training set represents a clear limit;
- even when a long and exhaustive training set is possible, there could be cases where the structure finds itself working in rare operating and environmental conditions, not accounted for in the training set (e.g., extreme meteorological events, different climate conditions). In these situations, an SHM method unable to filter out the effects of these changes on the damage feature would detect a structural damage/alteration, leading to a false positive.

In these scenarios, the SHM approach here proposed has a great impact with implications in many fields such as safety, maintenance and system reliability.

It is worth mentioning that another possible approach, which can be used as an alternative to the one proposed here, is that of adopting damage features which are not sensitive to environmental and operational variations [36,37]. This approach is attractive, since it directly tackles the cause of the problem. However, it is also challenging and difficult to apply, since it is hard to find vibration-based damage features showing a high sensitivity to damage and, at the same time, a low sensitivity to environmental effects. This is especially true for the structures considered here, i.e., tie-rods. Indeed, during their normal operational conditions, temperature variations cause changes in the mechanical and geometrical properties of both the tie-rod and the structure, which reflects into changes of the axial load and, thus, of the dynamic response properties. However, at the same time, other tension variations are due to deformation and displacement of the connecting walls, that may be caused by terrain crawl, subsidence of foundations or seismic events [17,38].

Tie-rods are, thus, challenging structures for SHM procedure. Most of the works in the literature related to SHM of tie-rods regard the axial-load identification (e.g., [39–48]); however none of these works considers the presence of damage in the beam. Moreover, as already mentioned, a change of the axial load cannot be directly related to the presence of a crack in the tie-rod, due to the axial load sensitivity to physical variables, not correlated to the state of health of the tie-rod, and due to environmental effects, e.g., temperature. Only recently, the problem of detecting damage in tie-rods has been faced, with a focus on cracks [17] or corrosion [49,50], and this is an important aspect when SHM of larger structures where tie-rods are in use must be carried out (e.g., [51]). Lucà et al. showed that tie-rod eigenfrequencies can be used as synthetic damage features that are representative of all physical variables which affect the system behaviour, included the axial load. At the same time, they can be used for MSD-based damage detection, when a long-term baseline set is available [18]. However, as mentioned, there are cases when short time baseline is needed.

The novel approach presented in this paper represents a solution to this kind of problems since it adopts a technique allowing for filtering out the temperature effects from the damage index which thus results effective, even in presence of an incomplete set of environmental conditions. This is done by relying on the PCA, which is a well known multivariate analysis technique, often adopted in data representation or data compression [52]. This tool allows projecting the original data set into a new space, defined by the principal components (PCs). The PCs are new variables that are sorted such that the majority of the variability in the original data set is explained by the first few PCs. Since under normal operational conditions the majority of the variability of a multivariate damage feature set is due to environmental effects, it is reasonable to expect that the first few PCs will be representative of these effects [19,34,53]. The idea behind the damage detection algorithm developed in this work is to exclude these PCs and, then, to use the remaining ones to define a damage index which is, thus, insensitive to environmental effects. To show the effectiveness and the reliability of this novel PCA-based procedure, it will be compared with one of the most used approaches in this field, which is the MSD-based method presented in [18]. The comparison will be carried out both on simulated and experimental data of axially-loaded beams.

The article is organized as it follows: in Section 2, both the MSD-based and the PCA-based damage detection algorithms are explained. Moreover, the simulated data and the experimental set-up are described. In Section 3, the results of the simulations are showed and discussed. The experimental results are presented and commented in Section 4. Finally, the conclusions are drawn in Section 5.

2. The New PCA-Based SHM Approach and the Validation Plan

In this section, the two methods that are compared in this paper are introduced. Furthermore, a description of the simulated and experimental data is provided.

Before entering into details of the two compared approaches, it is worth mentioning that the initial damage feature is a collection of eigenfrequencies of the monitored tie-rod. This starting point comes from previous research works where it has been proved that the eigenfrequencies of an axially-loaded beam-like structure, used as a multivariate damage feature, can be effectively adopted to spot damage in operating tie-rods [18,49,50,54].

Indeed, eigenfrequencies can be used to synthetically represent the state of the monitored tie-rod, since they are representative of the physical variables that mostly influence its dynamic behaviour (e.g., the axial load). Moreover, the effect of environmental changes is different from that of damage, if multiple eigenfrequencies are considered as a multivariate damage feature. As an example, the eigenfrequencies of the first four bending vertical modes of a healthy tie-rod are considered: a decrease of temperature would cause an increase in the values of all four eigenfrequencies and the lower the vibration mode considered, the higher the effect [18]. If, instead, the temperature does not change but damage (e.g., a reduction of flexural stiffness) is present at midspan, only the eigenfrequencies of the first and third vibration modes would change, since midspan is a vibration node for the even vibration modes. Furthermore, the higher the vibration mode considered, the higher the effect [18].

If a number m of vibration modes are considered, the associated eigenfrequency values are referred to as $f_1, f_2, \dots, f_i, \dots, f_m$, with $i = 1, 2, \dots, m$ (according to this notation, the eigenfrequencies are sorted in ascending order and $i = 1$ simply indicates the lowest eigenfrequency value among those considered, not necessarily that associated to the first vibration mode). The eigenfrequency values can be arranged in a column vector $\mathbf{f}$, defined as it follows:

$$\mathbf{f} = [f_1, f_2, \ldots, f_i, \ldots, f_m]^{\mathrm{T}} \tag{1}$$

where the superscript "T" means the transpose.

The vector $\mathbf{f}$ constitutes the damage feature and it is used to represent the state of the structure with few variables, with respect to the raw acceleration data. When the structure is monitored over time, the feature vector can be estimated several times. In this case, a generic number r of feature vectors $\mathbf{f}_1, \mathbf{f}_2, \ldots, \mathbf{f}_j, \ldots, \mathbf{f}_r$, with $j = 1, 2, \ldots, r$ can be arranged in a matrix $\mathbf{F}$ as it follows:

$$\mathbf{F} = \begin{bmatrix} \mathbf{f}_1^{\mathrm{T}} \\ \mathbf{f}_2^{\mathrm{T}} \\ \vdots \\ \mathbf{f}_j^{\mathrm{T}} \\ \vdots \\ \mathbf{f}_r^{\mathrm{T}} \end{bmatrix}. \tag{2}$$

In the following, the symbol $\mathbf{F}^0$ will be adopted to indicate the baseline set, i.e., a matrix containing a number b of observations (i.e., $r = b$) of the damage feature when the structure is in the reference initial condition and which will be used for training the methods. The symbol $\mathbf{f}^*$ will be adopted to indicate a generic new observation of the damage feature which does not belong to the baseline set, thus, it is associated to an unknown health condition. Finally, the symbol $\mathbf{F}^*$ will be adopted to indicate a set containing a number n of observations of the damage feature (i.e., $r = n$) that do not belong to the baseline set, thus, $\mathbf{F}^*$ can potentially include damage-related data.

2.1. The Benchmark MSD-Based Approach

In this paper, the benchmark is represented by an MSD-based damage index. The MSD is a well-known multivariate metrics, often adopted in the field of SHM to define damage indexes. In the considered case, the MSD between the new vector $\mathbf{f}^*$ and the baseline set $\mathbf{F}^0$ can be evaluated according to the following expression:

$$d^{\mathrm{MSD}} = \left(\mathbf{f}^* - \boldsymbol{\mu}^0\right)^{\mathrm{T}} \left(\boldsymbol{\Sigma}^0\right)^{-1} \left(\mathbf{f}^* - \boldsymbol{\mu}^0\right) = MSD(\mathbf{f}^*, \mathbf{F}^0) \tag{3}$$

where $\boldsymbol{\mu}^0$ is a $m \times 1$ vector where every i-th element is the mean of the i-th column of $\mathbf{F}^0$, $\boldsymbol{\Sigma}^0$ is the covariance matrix of $\mathbf{F}^0$ and "-1" means the inverse. The notation $MSD(\mathbf{f}^*, \mathbf{F}^0)$ is used here to indicate the result of the application of the MSD operator to the vector $\mathbf{f}^*$ with respect to $\mathbf{F}^0$. It is also noticed that the equivalent vector operator $MSD(\mathbf{F}^*, \mathbf{F}^0)$ used further in the paper indicates the MSD operator applied to each observation contained in $\mathbf{F}^*$ with respect to $\mathbf{F}^0$, resulting in a $n \times 1$ vector. The result of Equation (3), the MSD, is a scalar number and constitutes the damage feature of the benchmark method.

To detect possible structural changes, the scalar value d^{MSD} has to be checked against a threshold to state whether the vector $\mathbf{f}^*$ can be considered as an outlier with respect to the set $\mathbf{F}^0$. The threshold can be set according to a procedure based on the Monte Carlo method explained in [2,55], briefly described in the following:

- construct a matrix of size $b \times m$, where every element is a random number generated from a zero mean and unit standard deviation normal distribution;
- calculate the MSD between the transpose of each row of the matrix and the the matrix itself;
- store the maximum of the b obtained distances;

- repeat the operation for a large number of trials, e.g., 1000, and then sort all the maxima in terms of magnitude;
- the inclusive threshold t is then defined as the 95th percentile of the distribution of the MSD maxima (the term inclusive refers to a case when the baseline may also contain damaged or altered data which will be, thus, considered as outliers);
- if the baseline set does not include outliers, the exclusive threshold t^{MSD} must be adopted. The threshold t^{MSD} can be calculated according to the following equation:

$$t^{\mathrm{MSD}} = \frac{(b-1)(b+1)^2 t}{b(b^2-(b+1)t)}. \quad (4)$$

Summarizing, the main steps required by the MSD-approach used as a benchmark in this work are shown in the flowchart reported in Figure 1.

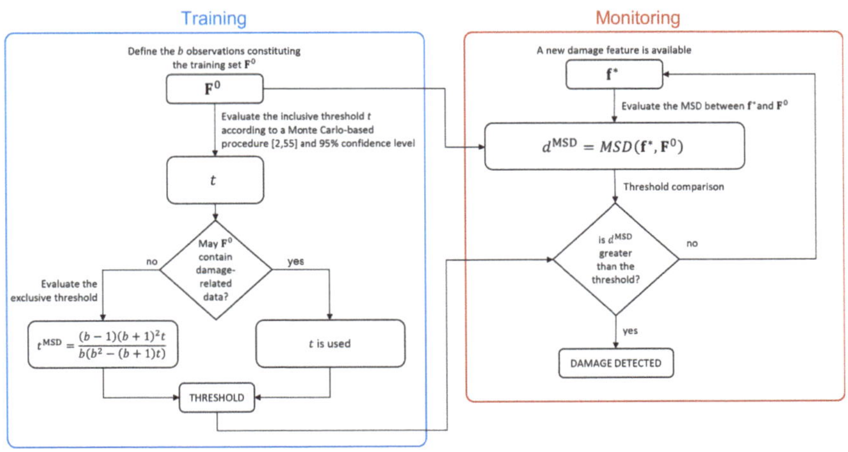

Figure 1. Flowchart of the MSD-based approach.

The MSD is very popular in the field of SHM since this metric naturally filters out the variability associated with the environmental effects while keeping a high sensitivity to structural changes [35]. However, it is known that to properly filter out environmental effects, a full range of environmental conditions must be included in the baseline set to describe the whole variability of the considered feature in operational conditions. In real applications, which are usually characterized not only by short-term trends but also seasonal ones, this aspect translates in the need for long baseline sets. In the following section, a new approach is proposed to try to overcome this limit.

2.2. The PCA-Based Approach

The new proposed approach is obtained through the adoption of the PCA. The PCA is a multivariate analysis technique that allows an orthogonal projection of a given data set onto a different coordinate system, where each of the new coordinates (the PCs) accounts for a decreasing amount of the variance of the original data set. The PCs are uncorrelated each other and they are sorted so that the first few components retain most of the variability present in the original data set.

This new description of the data set is usually adopted when a dimensionality reduction is needed. Considering just the very first PCs allows, indeed, to retain most part of the data set variability with a few number of variables. Here, instead, PCA is used for a different purpose. Its aim will be the removal of the data variability due to operating and environmental changes, as will be clarified in the following.

In this case, the data set $\mathbf{F}^0$ (of size $b \times m$) must be centred by subtracting the mean of each column from each value in that column, obtaining the matrix $\mathbf{C}^0$. Then, the

PCA transforms the data in $\mathbf{C}^0$ into a new set $\mathbf{Z}^0$ (the score matrix) through a rotational transformation according to the following equation:

$$\mathbf{Z}^0 = \mathbf{C}^0\mathbf{R} \tag{5}$$

where $\mathbf{R}$ is an $m \times m$ matrix (the loading matrix). The matrix $\mathbf{Z}^0$ contains the scores in the principal directions of $\mathbf{C}^0$, arranged such as the first column contains the scores related to the PC accounting for the largest variance, the second column contains the scores related to the PC accounting for the second largest variance, and so on. The matrix $\mathbf{R}$ can be evaluated, for example, by adopting the singular value decomposition. The reader can refer to [52] for a complete theory on the topic.

In the proposed framework, the PCA is used to project the centred baseline matrix $\mathbf{C}^0$ and obtain the scores associated to the PCs $\mathbf{Z}^0$. As it will be shown in the following sections, in the baseline data set, where no damage occurs (i.e., the baseline data set is considered as the reference structural condition), the majority of the variance in the eigenfrequencies is associated with temperature effects. The idea is, then, to remove the first p columns of the matrix $\mathbf{Z}^0$, associated to the first p PCs, to filter out the temperature effect from the baseline dataset. Once the first p columns of the matrix $\mathbf{Z}^0$ are removed, the matrix $\hat{\mathbf{Z}}^0$ is obtained (in the following, the hat symbol is used to indicate score matrices after the removal of the first p columns).

The key idea of the new SHM procedure proposed here is that, when new observations of the feature vector are available, if they are still referring to the same structural condition as the baseline, the PCA should project the data in the same principal directions (i.e., the transformation matrix $\mathbf{R}$ is still the same). Let's consider the matrix $\mathbf{F}^*$, containing n new feature vectors $\mathbf{f}^*$ which are not included in the baseline. A matrix $\mathbf{F}^{0*}$ can be assembled as it follows:

$$\mathbf{F}^{0*} = \begin{bmatrix} \mathbf{F}^0 \\ \mathbf{F}^* \end{bmatrix}. \tag{6}$$

Following the same steps previously described for $\mathbf{F}^0$, the matrix $\mathbf{F}^{0*}$ is centred and the PCA is applied obtaining $\mathbf{Z}^{0*}$. Then, the first p columns are, again, removed from the score matrix, obtaining the matrix $\hat{\mathbf{Z}}^{0*}$.

Now, the MSD is calculated between each element of $\hat{\mathbf{Z}}^{0*}$ and $\hat{\mathbf{Z}}^0$, i.e.,:

$$\mathbf{d} = MSD(\hat{\mathbf{Z}}^{0*}, \hat{\mathbf{Z}}^0) \tag{7}$$

and the result is a vector $\mathbf{d}$, containing the MSD of the transpose of each row of $\hat{\mathbf{Z}}^{0*}$ with respect to $\hat{\mathbf{Z}}^0$.

The vector $\mathbf{d}$ is a $(b+n) \times 1$ column vector. The first b distances contained in $\mathbf{d}$ are considered, and the number o of these b distances which exceed a reference value (further indicated as $P_{0.95}$, see below) is counted. The new damage index is defined as it follows:

$$d^{\mathrm{PCA}} = \frac{o}{b}. \tag{8}$$

In order to calculate the damage detection threshold t^{PCA}, the procedure described in the following is adopted:

- Only the baseline is considered and the MSDs are calculated between the transpose of each row of $\hat{\mathbf{Z}}^0$ and the matrix $\hat{\mathbf{Z}}^0$, i.e.,:

$$\mathbf{d}^0 = MSD(\hat{\mathbf{Z}}^0, \hat{\mathbf{Z}}^0). \tag{9}$$

- A probability density function is estimated, by fitting a Gamma distribution [56] to the elements in $\mathbf{d}^0$.

- The 95th percentile $P_{0.95}$ and its lower and upper 95% confidence bounds, $P_{0.95,\mathrm{UB}}$ and $P_{0.95,\mathrm{LB}}$, are extracted.
- The number of the first b elements of $\mathbf{d}^0$ exceeding $P_{0.95}$, $P_{0.95,\mathrm{UB}}$ and $P_{0.95,\mathrm{LB}}$ are counted, obtaining respectively c, u and l.
- c, u and l are then normalized with respect to the number b of elements in the baseline, obtaining the damage threshold t^{PCA} and its 95% confidence bounds, i.e.,:

$$t^{\mathrm{PCA}} = c/b \tag{10}$$

$$t^{\mathrm{PCA,up}} = u/b \tag{11}$$

$$t^{\mathrm{PCA,lo}} = l/b. \tag{12}$$

Finally, the possible presence of a damage is assessed if d^{PCA} exceeds $t^{\mathrm{PCA,up}}$. This indeed means that more than 5% of the first b elements of $\mathbf{d}$ exceed the 95th percentile $P_{0.95}$ (with a confidence level of 95%), implying that the new $\mathbf{d}$ does not belong to the Gamma distribution fitted on $\mathbf{d}^0$, thus suggesting the presence of damage. Finally, the main steps required by the proposed PCA-based approach are shown in the flowchart reported in Figure 2.

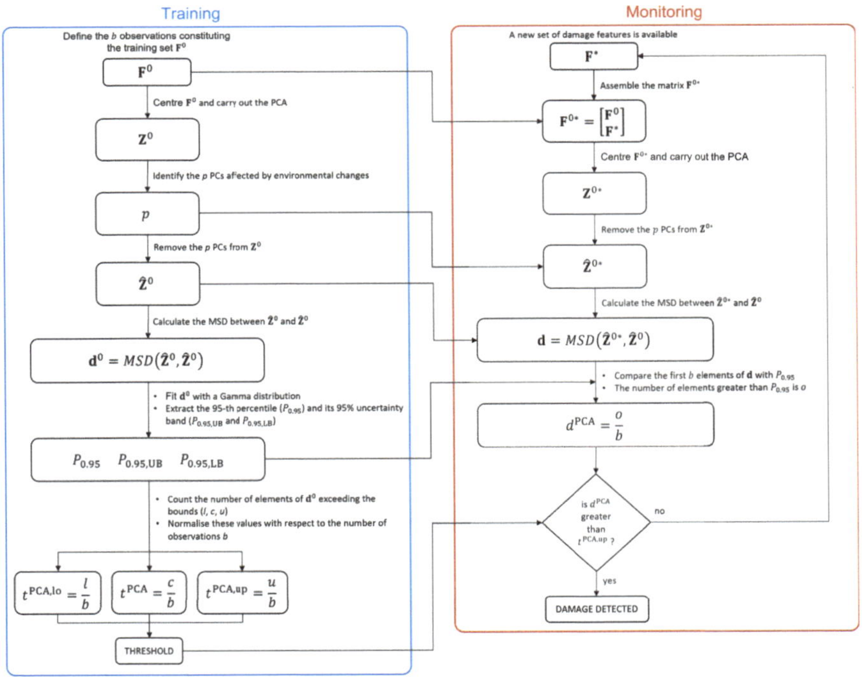

Figure 2. Flowchart of the PCA-based approach.

A difference between the new PCA-based approach and the MSD-based one is that d^{MSD} compares a single observation $\mathbf{f}^*$ with the baseline set, while d^{PCA} requires a set of new samples $\mathbf{F}^*$ to be assembled with the baseline matrix $\mathbf{F}^0$. Thus, after the baseline data set, each time a new observation of the damage feature is available, the matrix $\mathbf{F}^{0*}$ will be increased of one row, until the number of new observations is equal to n. From that moment onward, the matrix $\mathbf{F}^{0*}$ will always have size $b+n$, meaning that, every time a new observation is available, it is included in $\mathbf{F}^*$ and the least recent one is discarded, proceeding as a travelling window.

The length of the data set $\mathbf{F}^*$ (i.e., n) defines the sensitivity and the readiness of the method in detecting the damage, as will be mentioned later in the paper. Indeed, if n is much lower than b, and $\mathbf{F}^*$ contains data referring to an altered condition, their weight in the coordinate transformation of $\mathbf{F}^{0*}$ will be low. If n is much higher than b, when an alteration occurs the method will show the damage effect only when a certain number of damaged samples will replace the undamaged ones in $\mathbf{F}^*$. This translates in a transient effect and the higher n with respect to b is, the slower the transient is. In this application we, thus, choose to use $n = b$ as a compromise.

Finally, it should be noted that the steps required by the proposed method during the monitoring phase (see Figure 2) can be carried out through computationally inexpensive numerical algorithms (e.g., the above mentioned singular value decomposition to carry out the PCA). This means that the health condition of the considered structure can be evaluated in near-real time, every time a new observation of the damage feature is available.

2.3. PCA-Based Method Validation: Simulations and Experiments

The two methods presented in Sections 2.1 and 2.2 will be compared using both simulated and experimental data. Two aspects will be investigated: the effectiveness in filtering out the effects of environmental variables and in detecting damage of different severity. To this purpose, different situations were simulated:

1. cases with no damage and with a cyclic (sinusoidal) temperature trend, simulating its daily or seasonal variations;
2. cases with no damage and two cyclic temperature trends, simulating both daily and seasonal variations;
3. cases with damage and two cyclic temperature trends;
4. cases with no damage and temperature trends coming from experimental measurements (i.e., real temperature variations);
5. cases with damage and temperature trends coming from experimental data.

Cases 1 and 2 allow comparing the effectiveness of the two methods in filtering out the temperature effects when considering a whole temperature cycle (i.e., one period of the main sine) or a fraction of it in the training set. Case 3 allows the assessment of both the ability of the methods in filtering out the environmental effects and their sensitivity to damage of different severities. Finally, cases 4 and 5 remove the constraint of pure cyclic trends using real temperature data, thus allowing for an evaluation of the robustness of the methods to generic temperature variations.

Furthermore, again with the same aim of validating the proposed method in different situations and comparing its results with a benchmark SHM method, experimental tests were then performed. The tests were conducted on a sample structure placed in a room with monitored but uncontrolled temperature conditions. Data were acquired both without damage and with a purposely introduced damage with different severity levels, thus allowing for a validation of the simulation results, in terms of method behaviour.

This section will present in detail the simulations carried out and the experimental set-up, while the comparison results will be presented in Sections 3 and 4.

2.3.1. The Simulations

The simulations are meant to investigate the sensitivity of the two methods to environmental changes and to estimate their effectiveness in separating temperature and damage effects. The case of a simply supported axially-loaded beam is considered, for which the eigenfrequency values for the bending vertical modes can be analytically estimated, according to the following equation [57–59]:

$$f_i = \frac{i}{2L}\sqrt{\frac{S + EJ(\frac{i^2\pi^2}{L^2})}{q}}. \quad (13)$$

In Equation (13), L is the tie-rod length, S is the axial load, E is the Young's modulus, J is the momentum of inertia of the cross section and q is the mass per unit length. The simulations were carried out on a beam with rectangular cross-section with height h and width w. Thus, $J = (w\,h^3)/12$ and $q = w\,h\,\rho$, where ρ is the material density.

Eigenfrequency time-trends are generated by changing the axial load value, with respect to an initial reference value S_0, which corresponds to a generic initial temperature T_0. A linear relationship between the axial load and the temperature T is assumed, i.e., $S = S_0 + k\,(T - T_0)$, where k is a constant (i.e., the slope of the line that describes the axial load as a function of the temperature). For this reason, in Section 3, temperature trends for simulated data will always be represented as the difference with respect to the initial temperature T_0, i.e., $T - T_0$. The reference values adopted for the simulations are reported in Table 1.

Table 1. Parameters of the simulated tie-rod.

L [m]	S_0 [N]	E [GPa]	ρ [kg/m^3]	w [m]	h [m]	k [N/°C]
4	8×10^3	69	2.7×10^3	1.5×10^{-2}	2.5×10^{-2}	−60

Temperature trends, made by either a single sinusoidal trend or two sinusoidal trends, are simulated. If the latter case is considered, both long-term and short-term cyclic trends are present, to mimic seasonal and daily temperature fluctuations, respectively. A simple sine function with amplitude equal to 8 °C and mean equal to 0 °C is used for the long-term temperature trend, which represents the seasonal trend of the mean daily temperature. A series of sinusoidal functions characterized by a shorter period are used to represent the cyclic daily fluctuations. Each of the short-term sinusoidal functions has mean equal to 0 °C and amplitude which is randomly extracted from uniformly distributed numbers in the interval between 1.5 and 4 °C, to simulate that the thermal excursion may change from day to day. The two trends, i.e., the long-term sine function and the series of short-term trends, are summed up, to obtain the simulated temperature with two cyclic components. Conversely, simulated temperature trends with a single cyclic component are pure sines with amplitude equal to 8 °C, as the seasonal trend described above. The temperature values adopted to define the amplitudes of short-term and long-term trends are similar to those registered by meteorological outdoor stations, located in the north of Italy. Finally, the possibility to simulate eigenfrequency trends as function of the temperature allows also to use real temperature data as an input (see Sections 3.4 and 3.5). Also in this case, temperature data are represented as variations with respect to a reference mean value.

The effect of damage is, then, introduced as a reduction of Young's Modulus of a portion of the tie-rod of extent equal to 1% of L, at midspan. The way to introduce the effect of damage is by reducing each i-th eigenfrequency value, provided by Equation (13), by a certain percentage Δf_i. The values for Δf_i are obtained through finite element simulations carried out considering a three-dimensional axially-loaded beam model. The reader can refer to [18], where complete details on the finite element simulations are provided. In this work, two different damage levels are considered, i.e., 10% and 30% of Young's modulus reduction (for both damage conditions, the values for Δf_i for the first five tie-rod eigenfrequencies are reported in Table 2). A summary of all the simulated test cases is shown in Table 3.

Table 2. List of Δf_i values used to simulate damage.

Young's Modulus Reduction [%]	Δf_1 [%]	Δf_2 [%]	Δf_3 [%]	Δf_4 [%]	Δf_5 [%]
10	1.0×10^{-2}	1.4×10^{-4}	5.4×10^{-2}	4.4×10^{-4}	8.6×10^{-2}
30	3.9×10^{-2}	5.6×10^{-4}	2.0×10^{-1}	1.7×10^{-3}	3.2×10^{-1}

Table 3. Simulated test cases.

Test Case	Damage	*T* Cycle	*b*	Total Number of Samples
sim 1	No	Long	4320 (b_1)	8640
sim 2	No	Long	1008 (b_2)	8640
sim 3	No	Long + short	1008 (b_2)	8640
sim 4	Yes	Long + short	4320 (b_1)	21600
sim 5	No	Real	1008 (b_2)	12960
sim 6	No	Real	4320 (b_1)	12960
sim 7	Yes	Real	4320 (b_1)	12960
sim 8	Yes	Real	1008 (b_2)	12960

The outcome of the simulations is discussed in Section 3. In the next subsection, the experimental set-up is described.

2.3.2. The Experiments

The experimental data come from a test bench (see Figure 3) installed in the Mechanical Engineering laboratory of Politecnico di Milano, in Italy. A full-scale aluminium tie-rod is considered, characterized by a free length of 4 m and a cross-section equal to 0.015×0.025 m^2.

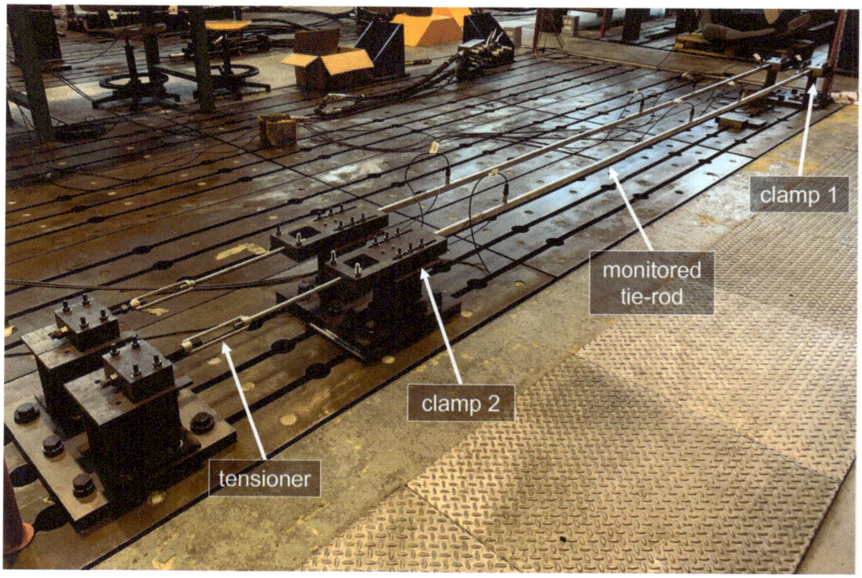

Figure 3. The experimental set-up.

Clamps made from steel plates are located at the two ends of the beam, to provide the constraints. The plates are in contact with the upper and lower faces of the tie-rod and they are held together by bolted joints. During the installation, the bolted joints of one of the two clamps (clamp 1 in Figure 3) were tightened, while the ones of the other clamp (clamp 2 in Figure 3) were left loose. In this way, since the beam was not fully constrained along the axial direction, a tension was applied through a tensioner. When a tension of 8000 N was applied to the tie-rod, also the bolted joints of clamp 2 were tightened up, to finally obtain a tensioned beam with a clamped-clamped constraint configuration.

Preliminary tests revealed that the broadband excitation provided by the environment, under normal conditions, significantly decreases for frequencies higher than 200 Hz and that the vibration modes which are mostly excited by the operational environment are the first six bending vertical modes (the eigenfrequency values for the first six bending vertical modes, identified through an impact hammer test carried out immediately after the tensioning procedure, are reported in Table 4).

Table 4. Tie-rod eigenfrequencies of the first six bending vertical modes, identified after the tensioning procedure.

f_1 [Hz]	f_2 [Hz]	f_3 [Hz]	f_4 [Hz]	f_5 [Hz]	f_6 [Hz]
13.89	30.98	53.36	81.82	116.55	157.95

The choices related to the sensor layout were aimed to obtain a sufficient spatial resolution to distinguish the mode shapes associated with the first six bending vertical modes, using as few sensors as possible, in order to reduce the load effect and to mimic real applications. Indeed, the use of as few sensors as possible is often desirable in field applications, for both practical and economic reasons. Many different layouts were evaluated based on the autoMAC matrix [60], to finally select a layout composed of four uniaxial accelerometers, fixed on the top face of the tie-rod, at distances of $\frac{1}{20}L, \frac{1}{3}L, \frac{1}{2}L$ and $\frac{9}{10}L$ from clamp 1. However, it should be noted that the choice of considering only bending modes in the vertical plane is specific of this experimental campaign. Indeed, by using, e.g., triaxial accelerometers, also other vibration modes, as the bending lateral ones, can be included in the analysis.

More in detail, the adopted accelerometers are general-purpose industrial piezoelectric accelerometers, model PCB603C01 (sensitivity of 10.2 $mV/(m/s^2)$, full scale of ± 490 m/s^2). The choice for general-purpose industrial accelerometers comes from the decision to not adopt high-end sensors, which are typical of laboratory environments and not representative of real applications. Moreover, axially-loaded beam-like structures are usually subject to significant vibration levels in operational conditions, due to their slenderness, making possible the use of, e.g., industrial piezoelectric accelerometers or accelerometers based on microelectromechanical systems (MEMS). Regarding the acquisition system, it is composed by NI 9234 modules with anti-aliasing filter on board and the sampling frequency is set to 512 Hz, obtaining a bandwidth of approximately 200 Hz that includes the range of frequency significantly excited by the operational environment.

It must be pointed out that neither the temperature nor the excitation are controlled, thus, even though it is a laboratory experiment, acquired data are similar to those of real monitoring applications. The temperature reaches minimum values approximately equal to 5 °C, during winter, and maximum values approximately equal to 30 °C during summer. Daily thermal excursion ranges from ± 1.5 °C to ± 4 °C.

The characteristics of the operational environment allow for a stable modal identification of the first four bending vertical modes, through the adoption of the polyreference least-square complex frequency-domain method [61]. Thus, the eigenfrequencies used to calculate the damage indexes in Section 4 are those of the first four bending vertical modes. However, the proposed strategy is of general validity and can also be used when other output-only modal identification algorithms are adopted to extract the required number of modal parameters. Furthermore, since only the eigenfrequency values are used to calculate either the MSD-based or the PCA-based damage indexes, also the adoption of a single accelerometer and simple single-degree-of-freedom output-only techniques is possible [50].

The damage effect has been introduced through the addition of a concentrated mass, to alter the dynamic properties of the tie-rod with a simple and reversible strategy, often used in literature (e.g., in [62–65]). Damage is simulated close to the constraints, at a distance equal to $\frac{1}{10}L$, which represents a challenging scenario for eigenfrequency-based damage detection [18]. Two different masses are used, equal to 1% and 3% of the total mass of the beam, to test different damage severity.

3. Results: Simulations

In this section, the results of the simulations are presented. The different subsections discuss the results of the simulations 1 and 2, 3, 4, 5 and 6, 7 and 8, respectively, described in Table 3 and associated to different temperature and damage conditions.

3.1. Long-Term Temperature Trends and No Damage

At first, the temperature profile reported in Figure 4 is considered, which is composed by 8640 samples. In this set of simulated data, the eigenfrequency changes are only associated to the temperature change and the tie-rod is always in the same healthy condition.

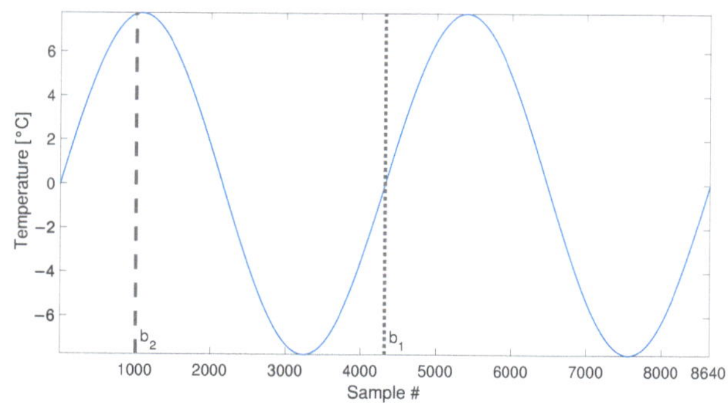

Figure 4. Simulated temperature trend: long-term trend only. Vertical dotted and dashed lines identify the number of samples used as baseline in sim 1 and sim 2, respectively.

The eigenfrequency trends for the first five vertical bending modes of the simulated tie-rod are reported in Figure 5.

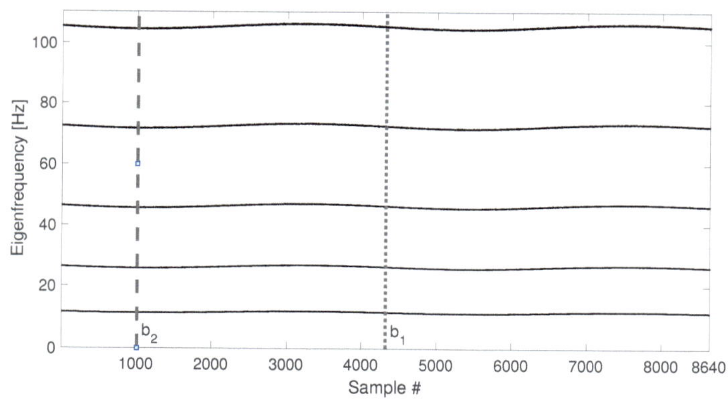

Figure 5. Simulated eigenfrequency trends due to long-term temperature trend. Vertical dotted and dashed lines identify the number of samples used as baseline in sim 1 and sim 2, respectively.

The temperature follows a simple sine function and it completes two identical cycles, covering the range −8 to +8 °C with respect to the initial temperature value. As it is reasonable to expect, also the eigenfrequency trends follow the cyclical trend of temperature.

In order to compare the MSD-based and the PCA-based strategies in their capability to filter out the environmental effects, first, a number $b_1 = 4320$ of observations of the damage feature are considered (see Table 3, sim 1), i.e., half of the total number of samples shown in Figure 5 (the limit of the baseline set is represented as a vertical dotted line in Figures 4 and 5). In this way, the baseline set $\mathbf{F}^0$ includes data referring to an entire temperature cycle, i.e., all the environmental conditions to which the tie-rod is subject.

For MSD-based strategy, the damage index d^{MSD} is evaluated by calculating the MSD of each observation subsequent to the baseline (i.e., samples after b_1) and compared with the threshold t^{MSD}. As for the PCA-based strategy, Figure 6 shows the PC scores for the baseline set of eigenfrequencies $\mathbf{F}^0$, i.e., the columns of matrix $\mathbf{Z}$ (see Section 2.2). As it is possible to see, the scores in the first principal direction show a deterministic trend that is strictly related to the temperature trend (compare the first plot of Figure 6 with that of Figure 4). Conversely, the scores in the other principal directions do not show deterministic trends. Since the first PC seems to be highly correlated with temperature, it is removed from the damage feature, before the evaluation of d^{PCA} ($p = 1$, according to Section 2.2).

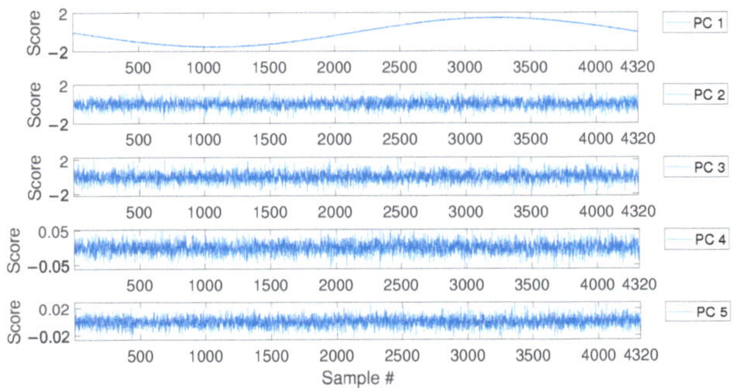

Figure 6. PC scores for the baseline set of eigenfrequencies containing a number $b = 4320$ of observations, considering a long-term temperature trend.

Figure 7 shows the comparison of the two approaches on the data which are not included in the baseline (i.e., from sample 4321 to sample number 8640). To allow for a direct comparison of the two approaches, from now on, the two indexes d^{MSD} (blue dotted line) and d^{PCA} (red solid line) will always be plotted as normalized on the respective damage detection threshold (t^{MSD} and t^{PCA}, respectively). For this reason, the damage detection threshold is represented by a black horizontal dot-dashed line of value 1 (from now on, referred to as the unitary threshold) for both the methods. In the same way, the upper and lower bounds for the PCA-based threshold (see Equations (11) and 12), $t^{PCA,up}$ and $t^{PCA,lo}$, respectively, will be presented as normalized on the damage detection threshold t^{PCA} and indicated by black horizontal dashed lines.

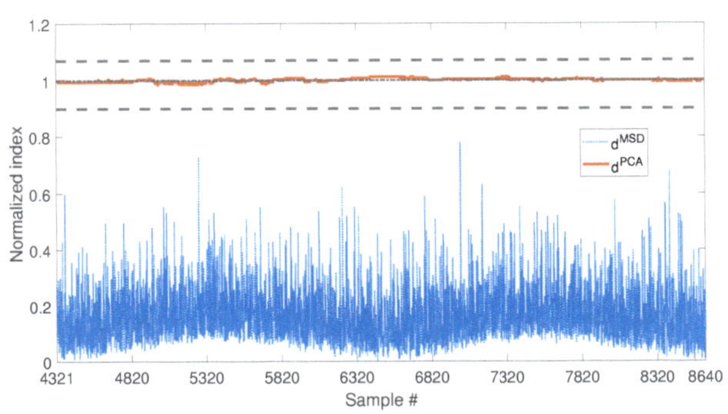

Figure 7. Comparison between d^{MSD} (blue dotted line) and d^{PCA} (red solid line), with $b = 4320$, considering a long-term temperature trend.

As it is possible to see by observing the results presented in Figure 7, both the strategies are effective in filtering out the temperature effect. Indeed, d^{MSD} is below the damage detection threshold and d^{PCA} is inside the range defined by $t^{PCA,up}$ and $t^{PCA,lo}$. Thus, no false positives are detected due to the environmental effects, which are correctly filtered out since all the temperature conditions from sample 4321 to 8640 were included in the baseline set.

The second case discussed considers a shorter baseline set. In this case, the baseline includes a number $b_2 = 1008$ of samples (see Table 3, sim 2), which is approximately one quarter of the entire temperature cycle (see the black vertical dashed lines in Figures 4 and 5, which indicate the end of the baseline set). In more detail, in this case $\mathbf{F}^0$ contains only the eigenfrequencies associated to temperatures in the range 0 to +8 °C.

Figure 8 shows the comparison of the two approaches. In this case, also the temperature is plotted on the right axis of the figure with a black thin line, to facilitate the interpretation of the results. As it is possible to see, the PCA-based strategy is still filtering out the temperature effect correctly. Indeed, d^{PCA} is always inside the range defined by $t^{PCA,up}$ and $t^{PCA,lo}$. This confirms that most of the variability of the data, which is associated to temperature, is explained by the first PC. Therefore, removing the first PC from the damage feature allows to filter out any change due to temperature effects. On the contrary, d^{MSD} clearly shows a deterministic trend, with values that increase when data outside of the training set are considered. This can be stated by observing that d^{MSD} increases when the temperature is in the range 0 to -8 °C, which is not included in the baseline (e.g., compare d^{MSD} and the temperature trend from sample 7008 to sample 8008 in Figure 8). The influence of temperature causes the index d^{MSD} to exceed the damage detection threshold even if no damage is present, thus producing false positives.

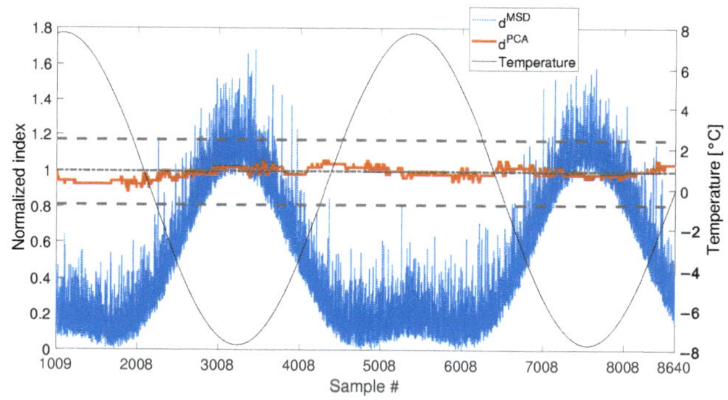

Figure 8. Comparison between d^{MSD} (blue dotted line) and d^{PCA} (red solid line), with $b = 1008$, considering a long-term temperature trend (black thin line).

3.2. Short-Term and Long-Term Temperature Trends with No Damage

The second set of simulations considers a different temperature profile, characterized by two cyclical trends: a long-term trend (which is the same as the first set of simulations) and a short-term trend. This data set is meant to mimic the presence of both seasonal and daily temperature trends. Indeed, the long-term trend again covers a range of temperature from -8 to +8 °C in 4320 samples, and it simulates the seasonal trend of the daily mean temperature. The short-term trend, instead, completes an entire cycle in 144 samples. For every daily cycle, the range of temperatures around the daily mean temperature is generated as described in Section 2.3.1.

The temperature trend shown in Figure 9 is used to simulate the eigenfrequency trends which are reported in Figure 10. As it is expected, also the eigenfrequency trends show both daily and long term trends.

Also in this case, the first $b_2 = 1008$ samples are used as a baseline (see Table 3, sim 3), as indicated by a black vertical dashed line, both in Figures 9 and 10.

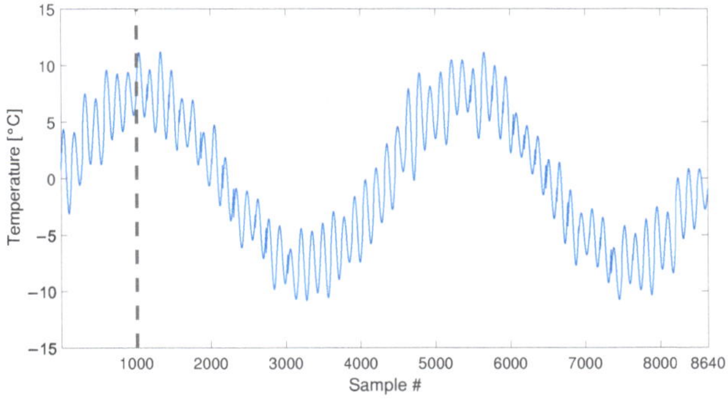

Figure 9. Simulated temperature trend: long-term and short-term trends. The vertical dashed line identifies the number of samples used as baseline in sim 3.

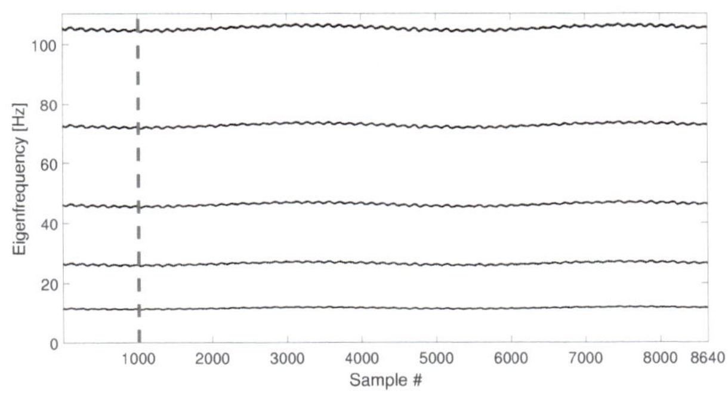

Figure 10. Simulated eigenfrequency trends related to long-term and short-term temperature trends. The vertical dashed line identifies the number of samples used as baseline in sim 3.

As in the case of the first set of simulations, the PCA confirms that the first PC has a clear deterministic trend which is highly correlated with the temperature (compare the plot labelled as PC 1 in Figure 11 with the first 1008 samples in Figure 9). Thus, also in this case, the first PC is removed before calculating d^{PCA}.

The comparison of the two approaches is reported in Figure 12 and it shows results which are similar to those commented in Figure 8. The PCA-based strategy is able to filter out the effects of temperature, also in presence of both short-term and long-term temperature trends. The index d^{PCA} is always in the range defined by $t^{\mathrm{PCA,up}}$ and $t^{\mathrm{PCA,lo}}$. The MSD-based index, instead, shows the same deterministic trend observed in Figure 8, i.e., it increases when the temperature ranges from 0 °C to −8 °C, thus exceeding the damage detection threshold. Moreover, it is possible to notice that also a short-term trend is present in the damage index, which has the same periodicity of the short-term trends in temperature (e.g., compare d^{MSD} and the temperature trend between samples 3008 and 4008, in Figure 12).

The outcome of these first simulations (sim 1, 2 and 3 of Table 3), where the effect of damage is not accounted for, is that both the strategies are potentially able to be insensitive to temperature effects in the data. However, a strong difference emerged: while the MSD-based strategy requires a complete set of environmental effects to filter them out, the PCA-based strategy can provide a temperature-insensitive damage index without needing for a complete set of environmental conditions. This aspect is relevant in situations where a short baseline set is available, e.g., at the beginning of a monitoring campaign.

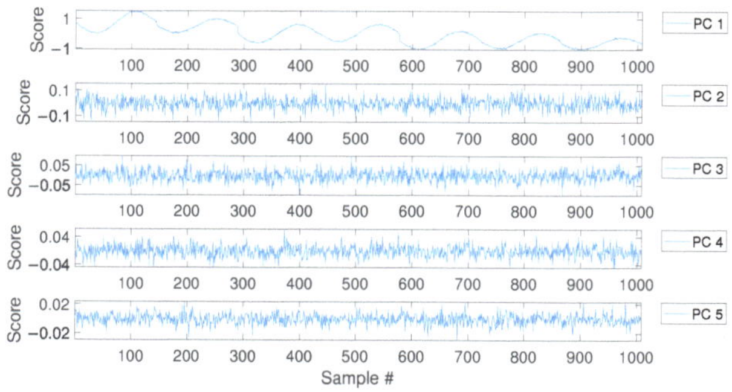

Figure 11. PC scores for the baseline set of eigenfrequencies containing a number $b = 1008$ of observations, considering both long-term and short-term temperature trends.

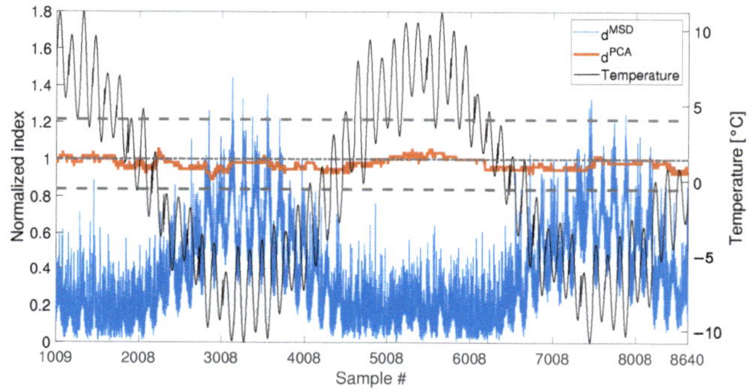

Figure 12. Comparison between d^{MSD} (blue dotted line) and d^{PCA} (red solid line), with $b = 1008$, considering both long-term and short-term temperature trends (black thin line).

3.3. Short-Term and Long-Term Temperature Trends with Simulated Damage

This set of simulations aims at answering a central question: are the damage indexes insensitive enough to temperature to allow for damage detection? The simulations discussed in the following, thus, consider the presence of damage.

As mentioned in Section 2.3.1, damage is simulated as a reduction of Young's modulus of a portion of the tie-rod of extent equal to 1% of the free-length. The portion of the tie-rod which is affected by damage is located at mid-span and two levels of damage are considered: 10% and 30% of Young's modulus reduction. In order to simulate damage, a change of eigenfrequency value is introduced, using the corresponding eigenfrequency decrease Δf_i (see Section 2.3.1).

In this case, the total number of samples is equal to 21600, which includes five entire long-term temperature trends (see Figure 13). A number equal to $b_1 = 4320$ samples (see Table 3, sim 4) is used to define the baseline, in order to include a complete long-term temperature trend (see the black vertical dotted line in Figure 13).

Damage is introduced after two and a half temperature cycles and the beginning of the sample set containing damage-related data is indicated by a red vertical dot-dashed line in Figure 13.

The performances of d^{MSD} and d^{PCA} in presence of damage can be compared, for the two damage levels 10% and 30%, in Figures 14 and 15, respectively.

In both cases, the two indexes are below the respective threshold, when the samples before the beginning of damage are considered, thus they are not producing false positives due to temperature fluctuations (same conclusions of Sections 3.1 and 3.2). However, the two indexes perform differently when damage occurs: d^{PCA} is always able to detect damage, exceeding the upper limit of the range defined by $t^{\text{PCA,up}}$ and $t^{\text{PCA,lo}}$, both with low and high damage severity. Moreover it is sensitive to different levels of damage, as proved by the higher level reached by d^{PCA} in Figure 15 (around 12) than in Figure 14 (around 1.75). In both cases, a transient can be observed that finishes approximately b_1 samples after the beginning of damage. This is because, due to the travelling window used to calculate d^{PCA} (see Section 2.2), for the first b_1 samples after the introduction of damage, $\mathbf{F}^*$ still contains data referring to the healthy structure.

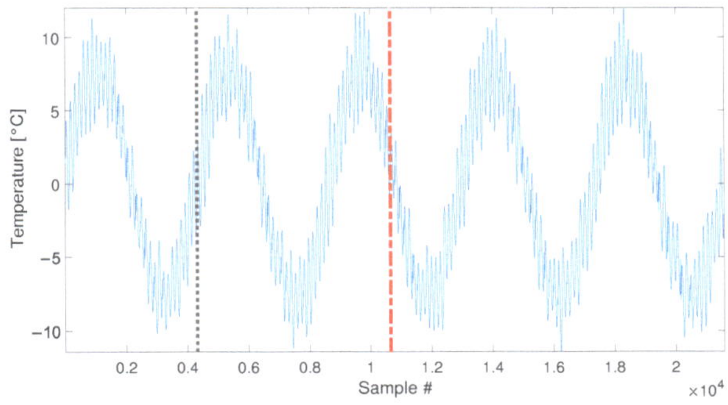

Figure 13. Simulated temperature trend: long-term and short-term temperature trends. The vertical dotted line identifies the number of samples used as baseline in sim 4. The beginning of the damage-related data is indicated by a red vertical dot-dashed line.

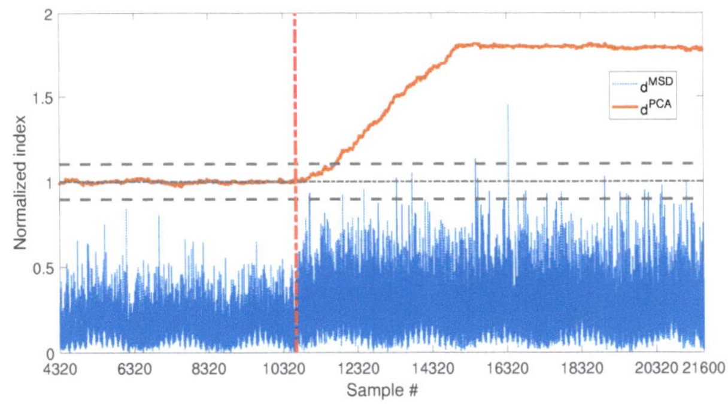

Figure 14. Comparison between d^{MSD} (blue dotted line) and d^{PCA} (red solid line), with $b = 4320$, considering long and short-term temperature trends and damage (10% reduction of Young's modulus).

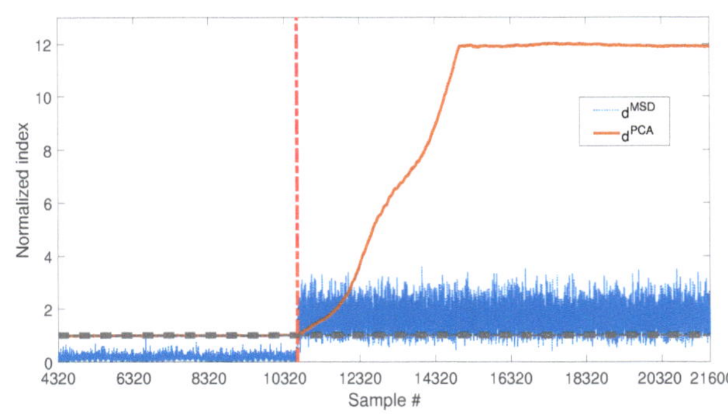

Figure 15. Comparison between d^{MSD} (blue dotted line) and d^{PCA} (red solid line), with $b = 4320$, considering long and short-term temperature trends and damage (30% reduction of Young's modulus).

As for d^{MSD}, the MSD-based damage index is not able to detect the lowest simulated damage, as proved by the fact that d^{MSD} stays below the unitary threshold in Figure 14. Only the most severe simulated damage is detected (d^{MSD} almost always above the unitary threshold in Figure 15). However, the conclusion is less clear, if compared with the index d^{PCA}, on the same conditions (compare the blue dotted line with the red-solid line in Figure 15).

Results proved that, when the PCA-based strategy is used, removing the first principal component filters out the temperature effect, while preserving sensitivity to damage. Moreover, d^{PCA} has a higher sensitivity than d^{MSD}.

3.4. *Real Temperature Trends without Damage*

Before moving to the experimental results, a last set of simulations is discussed. In this case, temperatures are not numerically defined but real temperature values are used. In more detail, the temperature trend comes from the experimental data, collected by a thermocouple in the laboratory where the experimental set-up, described in Section 2.3.2, is located. This set of simulations is meant to check the conclusions of previous Sections 3.1–3.3, where simple temperature trends were adopted, to easily separate the effects.

The temperature trend used is presented in Figure 16 and refers to the acquisition of the temperature every ten minutes, for a total number of samples equal to 12960 (90 days of data). Data are presented as the difference with respect to the average temperature value. The temperature shows both short-term and long-term trends. The short-term trends show a cyclical behaviour and they are related to the daily temperature trends. Moreover, it is possible to see that the mean daily temperature drifts during the observation window, from values approximately around +4 °C to values approximately around −4 °C.

Two different baselines will be adopted in the following: a short baseline, containing $b_2 = 1008$ samples, see Table 3, sim 5, (the end of the short baseline is indicated by a black vertical dashed line in Figure 16), and a long baseline, containing $b_1 = 4320$ samples, see Table 3, sim 6 (the end of the long baseline is indicated by a black vertical dotted line in Figure 16). As opposite to the previous simulations, it must be noted that even when the longest baseline is considered, it is not enough to include all the temperature values that characterize the remaining part of data. Indeed, the long baseline will include only temperature higher than, approximately, −2 °C, while, in the remaining part of the data, the temperature reaches lower levels.

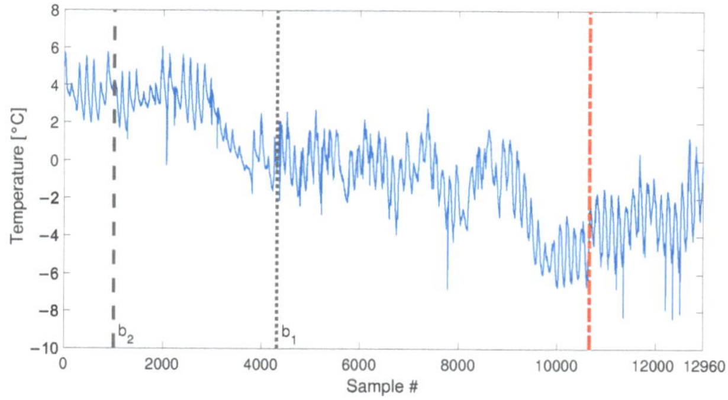

Figure 16. Real temperature trend, including daily trends and long-term drift. Black vertical dotted and dashed lines identify the number of samples used as baseline in sim 6 and sim 5, respectively. The beginning of the damage related data of sim 7 and 8 of Table 3 is indicated by a red vertical dot-dashed line.

The results of the PCA of the baseline matrix $\mathbf{F}^0$ again confirmed that the first PC is that presenting a deterministic trend which is highly correlated with that of temperature (see Figure 17). For this reason, again the index d^{PCA} is calculated after removing the first principal component.

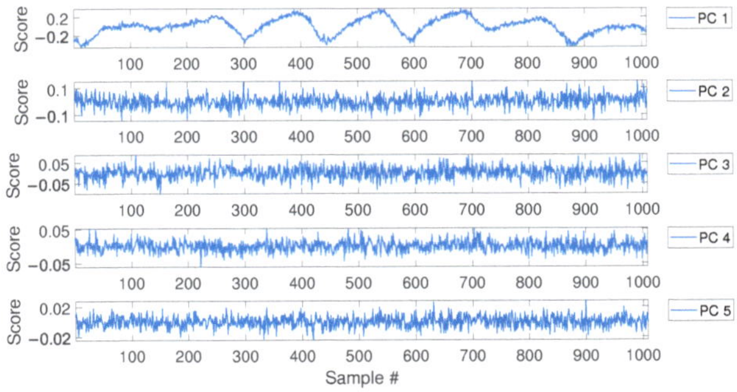

Figure 17. PC scores for the baseline set of eigenfrequencies containing a number $b = 1008$ of observations, considering real temperature trends.

Cases where no damage is present are here discussed. The results are presented in Figure 18 for sim 5 (short baseline), and in Figure 19 for sim 6 (long baseline).

With respect to the results presented in Sections 3.1 and 3.2, the insensitivity of d^{PCA} to temperature is confirmed: when either 1008 or 4320 samples are considered, d^{PCA} never exceeds the range defined by $t^{\mathrm{PCA,up}}$ and $t^{\mathrm{PCA,lo}}$, i.e., no false positives are produced. Furthermore, also the performances of the MSD-based strategy are confirmed. Indeed, d^{MSD} significantly exceeds the damage threshold with the baseline containing 1008 samples, causing false positives. As an example, when the mean trend of temperature decreases around sample 9008 (see Figure 18), the mean trend of d^{MSD} increases and stays constantly above the threshold. In this case, damage would be detected even if the structure is in healthy condition. Toward the end of the observation window, while temperature increases, d^{MSD} decreases, coming back to threshold level. Moreover, despite the effect is reduced by adopting a larger baseline (see Figure 19), it is still possible to notice that d^{MSD} sometimes exceeds the threshold and shows cyclic trends due to daily temperature variations.

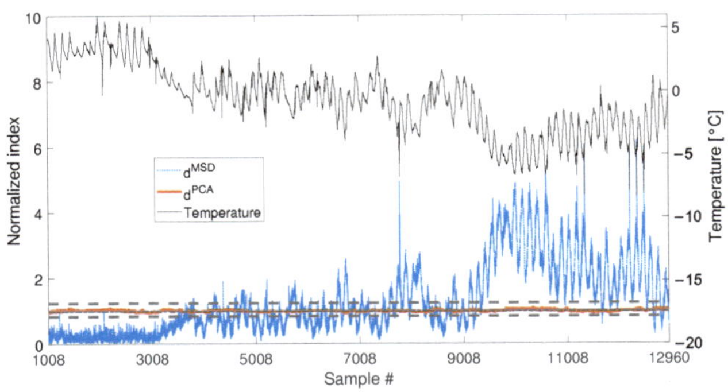

Figure 18. Comparison between d^{MSD} (blue dotted line) and d^{PCA} (red solid line), with $b = 1008$, in case of a real temperature trend (black thin line).

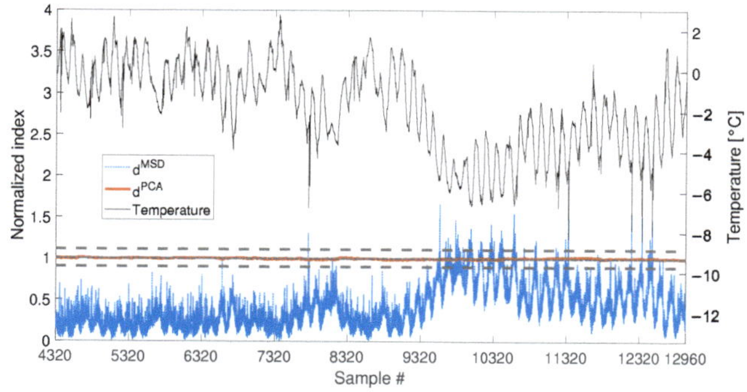

Figure 19. Comparison between d^{MSD} (blue dotted line) and d^{PCA} (red solid line), with $b = 4320$, in case of a real temperature trend (black thin line).

3.5. Real Temperature Trends With Damage

Finally, the performances in presence of damage are discussed. The results are presented in Figure 20, for the long baseline (see Table 3, sim 7), and in Figure 21, for the short baseline (see Table 3, sim 8). The damage simulated in this case is a 30% reduction of Young's modulus at midspan and it is indicated by the red vertical dot-dashed line, in Figures 16, 20 and 21.

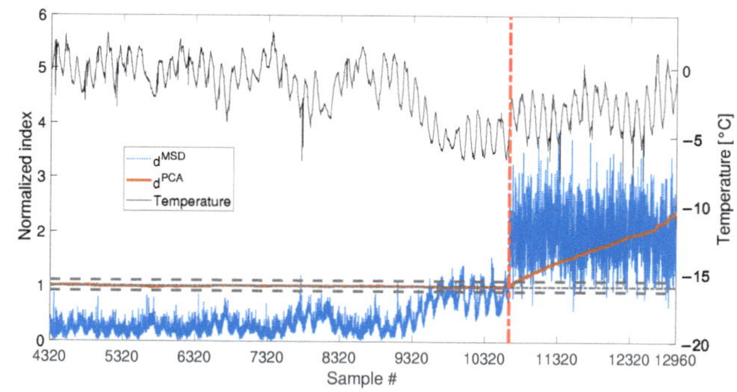

Figure 20. Comparison between d^{MSD} (blue dotted line) and d^{PCA} (red solid line), with $b = 4320$, considering a real temperature trend (black thin line) and damage (30% reduction of Young's modulus).

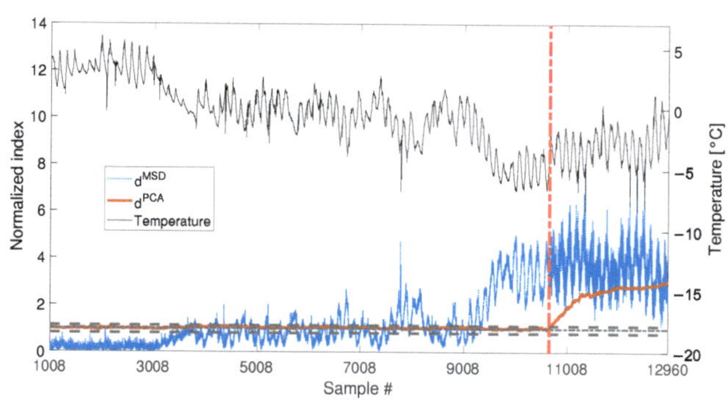

Figure 21. Comparison between d^{MSD} (blue dotted line) and d^{PCA} (red solid line), with $b = 1008$, considering a real temperature trend (black thin line) and damage (30% reduction of Young's modulus).

When a baseline of $b_1 = 4320$ samples are adopted (see Figure 20 and Table 3, sim 7), both strategies are able to detect damage. However, d^{PCA} shows a clear increasing trend unlike d^{MSD}. The decreasing trend of d^{MSD} due to temperature, previously observed in Figure 19 (i.e., from about sample 10700 to sample 12960), seems to be mitigated by the effect of damage; however, this effect is still present.

When 1008 samples are adopted (see Figure 21 and Table 3, sim 8) d^{PCA} is able to clearly detect damage, exceeding the range defined by $t^{\mathrm{PCA,up}}$ and $t^{\mathrm{PCA,lo}}$, confirming that not only the damage index is insensitive to temperature, but it is sensitive to damage. Conversely, the trend of d^{MSD} is similar to that of Figure 18, where no damage is present. Indeed, damage is detected even when the tie-rod is healthy since d^{MSD} is above the threshold before damage is introduced (i.e., d^{MSD} exceeds the threshold approximately at sample 9008). Moreover, the trend of d^{MSD} immediately before the introduction of damage is similar to that after the introduction of damage. This observation confirms that the increase of d^{MSD} is mainly due to temperature and not to damage.

4. Results: Experiments

In this section, the results obtained by considering real data coming from the experimental set-up (see Section 2.3.2) are presented. Two damage scenarios are considered, where the effect of damage is obtained through the addition of concentrated masses of 1% and 3% of the total mass of the tie-rod, close to one of the two fixed ends.

A set of 1008 samples is used to define the baseline matrix $\mathbf{F}^0$ (i.e., $b = 1008$), composed by the experimentally identified eigenfrequencies for the first four vibration modes (see Section 2.3.2). Considering that an estimate of the four eigenfrequencies is available every 10 min, the baseline set includes 7 days. The temperature trend related to the baseline set is reported in Figure 22, and the temperature trends of the validation and damage sets are reported in the following Figures 23 and 24, together with the damage indexes. It is noticed that, in all the figures related to the experiments, the temperature T is plotted in place of $T - T_0$. The gap of temperature data noticeable in Figures 23 and 24 is due to missing data caused by a malfunctioning of the temperature sensor. As it is possible to observe, the daily temperature cycles can be clearly noted. Furthermore, a drift in the daily mean temperature is also present. The available baseline set approximately covers the range of temperatures 10.5 to 17.5 °C.

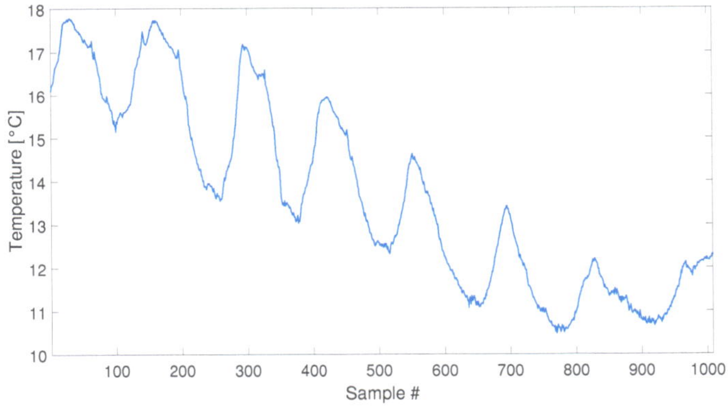

Figure 22. Temperature trend for the baseline set of real data.

The comparison between the two approaches is presented in Figures 23 and 24, for an added mass of 1% and 3% of the total mass, respectively. The indexes (blue-dotted trend for d^{MSD} and red-solid trend for d^{PCA}) are normalized on the respective damage detection threshold, as done in the simulations. The horizontal dot-dashed line represents the threshold for d^{MSD}, while the two horizontal dashed lines, below and above the unitary threshold, indicate the range defined by $t^{\text{PCA,up}}$ and $t^{\text{PCA,lo}}$, for d^{PCA} (see Section 2.2). The right y-axis is used to represent the temperature (black-thin line).

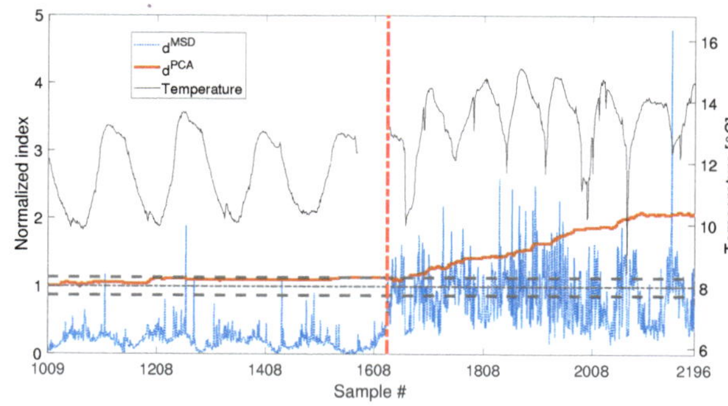

Figure 23. Comparison between d^{MSD} (blue dotted line) and d^{PCA} (red solid line), with $b = 1008$, for added mass equal to 1% of the total mass. A black thin line identifies the temperature.

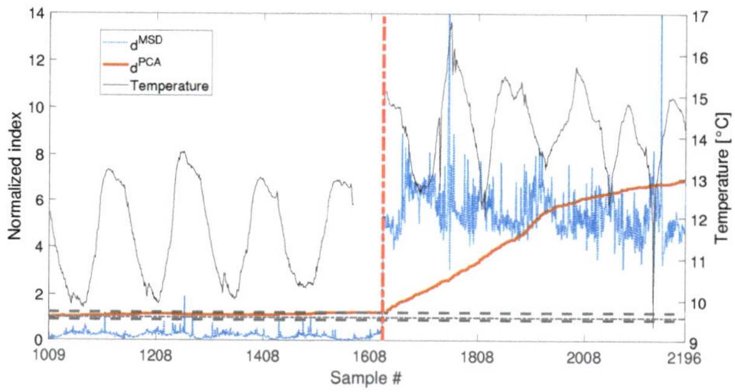

Figure 24. Comparison between d^{MSD} (blue dotted line) and d^{PCA} (red solid line), with $b = 1008$, for added mass equal to 3% of the total mass. A black thin line identifies the temperature.

As for the PCA-based strategy, the effects of temperature on the variance of $\mathbf{F}^0$ is retained by the first two PCs. They indeed show deterministic trends and, thus, were removed from the damage feature. The proposed procedure proved to be able to effectively filter out the temperature effect, as it can be seen from Figures 23 and 24. Indeed, d^{PCA} does not show any temperature-correlated trend (compare the red and black curves) and, when no damage is present, it does not exceed the range limited by $t^{\text{PCA,up}}$ and $t^{\text{PCA,lo}}$.

In both cases, d^{PCA} shows a clear growing trend when damage is introduced, thus the PCA-based damage index is able to promptly detect damage. By comparing the trends of d^{PCA} in Figures 23 and 24, it is possible to observe that the PCA-based damage index is sensitive to different magnitudes of damage: indeed, when damage is 3% the index grows faster than when damage is 1% (compare the values of the red solid trends in Figure 23 with those of Figure 24).

It is worth noticing that only the most severe damage condition (i.e., 3% of added mass) is clearly detected by d^{MSD}. For the case of 1% of added mass, instead, it remains below the threshold for most of the samples and just a slight damage index increase can be deduced, not allowing for a clear damage detection.

The experimental results confirmed what observed on the simulated data: when just few temperature conditions are available to define the baseline data set, the PCA-based strategy can provide a damage index which is robust with respect to the environmental effects, while the MSD-based index is still sensitive to temperature. Moreover, not only d^{PCA} is less sensitive to temperature than d^{MSD}, but d^{PCA} has a higher sensitivity to damage than d^{MSD}. Thus, the novel approach, based on the PCA, is expected to outperform the traditional approach, based on the MSD, in applications where few baseline data are available.

5. Conclusions

This paper presented an unsupervised learning vibration-based damage detection strategy for SHM applications where only few data are available to build the training set. In these cases, indeed, the whole variability of the damage feature due to operational and environmental conditions is not described in the training set. This leads to changes of the damage feature which can possibly either mask a damage or lead to false positives. The proposed SHM approach is based on the use of a damage index obtained through the PCA of the selected damage features. Indeed, relying on the assumption that under healthy reference conditions the variability of the collected damage features is only due to environmental and operational variations, these variations will affect the first few PCs, which explain most of the variability in the data. Thus, by discarding these few PCs, the remaining ones are not correlated to the environmental effects and can be used to define a temperature-insensitive damage index.

The effectiveness of the proposed approach was proved on both simulated and experimental data related to an axially loaded beam-like structure and considering the first bending natural frequencies as a multivariate damage feature. In both the cases, the proposed approach was compared with the MSD-based outlier detection method, widely adopted in unsupervised learning SHM literature. The simulations allowed highlighting the behaviour of the method when seasonal temperature trends are present. Both strategies showed similar performances when a complete temperature cycle is contained in the baseline set. Conversely, by reducing the baseline, and thus limiting the temperature conditions included in the training set, the PCA-based damage index outperformed the MSD-based one. It, indeed, did not produce any false positive and showed a higher sensitivity to damage, even when only a quarter of the simulated seasonal trend was included in the training set. Moreover, unlike the MSD-based approach, the PCA-based one successfully identified the smallest damage which was intentionally introduced in the experimental set-up. The experimental campaign proved the PCA-based method robustness, sensitivity and effectiveness in presence of real and uncontrolled temperature conditions.

It is worth mentioning that, when a damage is introduced in the structure, a transient of the PCA-based damage index is noticed. Although the effect of the damage can be clearly detected even during the transient, it may represent a limit of the approach. Thus, future studies will be devoted to the investigation of the effect of some parameters (e.g., the length of the new data added to the training set and used to calculate the damage index) on the transient duration and on the method sensitivity. Moreover, also the number of PCs to discard in the damage index evaluation is worthy of a deeper analysis. Future studies could, indeed, allow for an automated strategy able to define the PCs which have to be neglected in the damage index evaluation. The proposed approach, together with the future studies on its optimisation, will constitute a step forward in the monitoring of all those structures where long training is not possible and whose most relevant damage features are also the most sensitive to environmental and operating conditions.

Author Contributions: Conceptualization, M.B. and F.L.; methodology, M.B., F.L. and G.A.; software, F.L. and G.A.; validation, M.B., F.L. and G.A.; formal analysis, M.B. and F.L.; investigation, M.B., F.L., M.V. and G.A.; resources, M.V.; data curation, M.B., F.L., M.V. and G.A.; writing—original draft preparation, M.B. and F.L.; writing—review and editing, M.B., F.L. and M.V.; visualization, M.V. and G.A.; supervision, M.B. All authors have read and agreed to the published version of the manuscript.

Funding: This research received no external funding.

Institutional Review Board Statement: Not applicable.

Data Availability Statement: The data presented in this study are available on request from the corresponding author.

Conflicts of Interest: The authors declare no conflict of interest.

References

1. Farrar, C.R.; Worden, K. An introduction to structural health monitoring. *Philos. Trans. R. Soc. A Math. Phys. Eng. Sci.* **2007**, *365*, 303–315. [CrossRef] [PubMed]
2. Farrar, C.R.; Worden, K. *Structural Health Monitoring: A Machine Learning Perspective*; John Wiley and Sons: Hoboken, NJ, USA, 2012. [CrossRef]
3. Fan, W.; Qiao, P. Vibration-based damage identification methods: A review and comparative study. *Struct. Health Monit.* **2011**, *10*, 83–111. [CrossRef]
4. Hou, R.; Xia, Y. Review on the new development of vibration-based damage identification for civil engineering structures: 2010–2019. *J. Sound Vib.* **2021**, *491*, 115741. [CrossRef]
5. Avci, O.; Abdeljaber, O.; Kiranyaz, S.; Hussein, M.; Gabbouj, M.; Inman, D.J. A review of vibration-based damage detection in civil structures: From traditional methods to Machine Learning and Deep Learning applications. *Mech. Syst. Signal Process.* **2021**, *147*, 107077. [CrossRef]
6. Brownjohn, J.M.; de Stefano, A.; Xu, Y.L.; Wenzel, H.; Aktan, A.E. Vibration-based monitoring of civil infrastructure: Challenges and successes. *J. Civ. Struct. Health Monit.* **2011**, *1*, 79–95. [CrossRef]
7. Limongelli, M.P.; Manoach, E.; Quqa, S.; Giordano, P.F.; Bhowmik, B.; Pakrashi, V.; Cigada, A. Vibration Response-Based Damage Detection. In *Springer Aerospace Technology*; Springer: Cham, Switzerland, 2021; pp. 133–173. [CrossRef]
8. Entezami, A.; Shariatmadar, H.; Karamodin, A. Data-driven damage diagnosis under environmental and operational variability by novel statistical pattern recognition methods. *Struct. Health Monit.* **2019**, *18*, 1416–1443. [CrossRef]
9. Entezami, A.; Sarmadi, H.; Behkamal, B.; Mariani, S. Big data analytics and structural health monitoring: A statistical pattern recognition-based approach. *Sensors* **2020**, *20*, 2328. [CrossRef]
10. Razavi, B.S.; Mahmoudkelayeh, M.R.; Razavi, S.S. Damage identification under ambient vibration and unpredictable signal nature. *J. Civ. Struct. Health Monit.* **2021**, *11*, 1253–1273. [CrossRef]
11. Tran, T.T.; Ozer, E. Automated and model-free bridge damage indicators with simultaneous multiparameter modal anomaly detection. *Sensors* **2020**, *20*, 4725. [CrossRef]
12. Farrar, C.R.; Doebling, S.W.; Nix, D.A. Vibration-based structural damage identification. *Philos. Trans. R. Soc. A Math. Phys. Eng. Sci.* **2001**, *359*, 131–149. [CrossRef]
13. Chen, H.P.; Ni, Y.Q. Vibration-Based Damage Identification Methods. *Struct. Health Monit. Large Civ. Eng. Struct.* **2018**, 155–193. [CrossRef]
14. Sohn, H. Effects of environmental and operational variability on structural health monitoring. *Philos. Trans. R. Soc. A Math. Phys. Eng. Sci.* **2007**, *365*, 539–560. [CrossRef] [PubMed]
15. Peeters, B.; Maeck, J.; De Roeck, G. Vibration-based damage detection in civil engineering: Excitation sources and temperature effects. *Noise Vib. Worldw.* **2004**, *35*, 33. [CrossRef]
16. Alampalli, S. Effects of testing, analysis, damage, and environment on modal parameters. *Mech. Syst. Signal Process.* **2000**, *14*, 63–74. [CrossRef]
17. Collini, L.; Garziera, R.; Riabova, K. Detection of cracks in axially loaded tie-rods by vibration analysis. *Nondestruct. Test. Eval.* **2020**, *35*, 121–138. [CrossRef]
18. Lucà, F.; Manzoni, S.; Cigada, A.; Frate, L. A vibration-based approach for health monitoring of tie-rods under uncertain environmental conditions. *Mech. Syst. Signal Process.* **2022**, *167*, 108547. [CrossRef]
19. Pereira, S.; Magalhães, F.; Gomes, J.P.; Cunha, Á.; Lemos, J.V. Vibration-based damage detection of a concrete arch dam. *Eng. Struct.* **2021**, *235*, 112032. [CrossRef]
20. Hu, W.H.; Tang, D.H.; Teng, J.; Said, S.; Rohrmann, R.G. Structural health monitoring of a prestressed concrete bridge based on statistical pattern recognition of continuous dynamic measurements over 14 years. *Sensors* **2018**, *18*, 4117. [CrossRef]
21. Peeters, B.; Roeck, G.D. One-year monitoring of the Z24Bridge: Environmental effects versus damage events. *Earthq. Eng. Struct. Dyn.* **2015**, *30*, 149–171. [CrossRef]
22. Cross, E.J.; Koo, K.Y.; Brownjohn, J.M.; Worden, K. Long-term monitoring and data analysis of the Tamar Bridge. *Mech. Syst. Signal Process.* **2013**, *35*, 16–34. [CrossRef]
23. Torzoni, M.; Rosafalco, L.; Manzoni, A.; Mariani, S.; Corigliano, A. SHM under varying environmental conditions: An approach based on model order reduction and deep learning. *Comput. Struct.* **2022**, *266*, 106790. . [CrossRef]
24. Zhou, H.F.; Ni, Y.Q.; Ko, J.M. Constructing input to neural networks for modeling temperature-caused modal variability: Mean temperatures, effective temperatures, and principal components of temperatures. *Eng. Struct.* **2010**, *32*, 1747–1759. [CrossRef]
25. Shan, W.; Wang, X.; Jiao, Y. Modeling of Temperature Effect on Modal Frequency of Concrete Beam Based on Field Monitoring Data. *Shock Vib.* **2018**, *2018*, 8072843. [CrossRef]
26. Ni, Y.Q.; Hua, X.G.; Fan, K.Q.; Ko, J.M. Correlating modal properties with temperature using long-term monitoring data and support vector machine technique. *Eng. Struct.* **2005**, *27*, 1762–1773. [CrossRef]
27. Kromanis, R.; Kripakaran, P. Support vector regression for anomaly detection from measurement histories. *Adv. Eng. Inform.* **2013**, *27*, 486–495. [CrossRef]

28. Kullaa, J. Distinguishing between sensor fault, structural damage, and environmental or operational effects in structural health monitoring. *Mech. Syst. Signal Process.* **2011**, *25*, 2976–2989. [CrossRef]
29. Erazo, K.; Sen, D.; Nagarajaiah, S.; Sun, L. Vibration-based structural health monitoring under changing environmental conditions using Kalman filtering. *Mech. Syst. Signal Process.* **2019**, *117*, 1–15. [CrossRef]
30. Kullaa, J. Damage detection and localization under variable environmental conditions using compressed and reconstructed bayesian virtual sensor data. *Sensors* **2022**, *22*, 306. [CrossRef]
31. Kullaa, J. Robust damage detection in the time domain using Bayesian virtual sensing with noise reduction and environmental effect elimination capabilities. *J. Sound Vib.* **2020**, *473*, 115232. [CrossRef]
32. Maes, K.; Van Meerbeeck, L.; Reynders, E.P.; Lombaert, G. Validation of vibration-based structural health monitoring on retrofitted railway bridge KW51. *Mech. Syst. Signal Process.* **2022**, *165*, 108380. [CrossRef]
33. Sen, D.; Erazo, K.; Zhang, W.; Nagarajaiah, S.; Sun, L. On the effectiveness of principal component analysis for decoupling structural damage and environmental effects in bridge structures. *J. Sound Vib.* **2019**, *457*, 280–298. [CrossRef]
34. Soo Lon Wah, W.; Chen, Y.T.; Roberts, G.W.; Elamin, A. Separating damage from environmental effects affecting civil structures for near real-time damage detection. *Struct. Health Monit.* **2018**, *17*, 850–868. [CrossRef]
35. Deraemaeker, A.; Worden, K. A comparison of linear approaches to filter out environmental effects in structural health monitoring. *Mech. Syst. Signal Process.* **2018**, *105*, 1–15. [CrossRef]
36. Cross, E.J.; Manson, G.; Worden, K.; Pierce, S.G. Features for damage detection with insensitivity to environmental and operational variations. *Proc. R. Soc. A Math. Phys. Eng. Sci.* **2012**, *468*, 4098–4122. [CrossRef]
37. Surace, C.; Bovsunovsky, A. The use of frequency ratios to diagnose structural damage in varying environmental conditions. *Mech. Syst. Signal Process.* **2020**, *136*, 106523. [CrossRef]
38. Martakis, P.; Reuland, Y.; Imesch, M.; Chatzi, E. Reducing uncertainty in seismic assessment of multiple masonry buildings based on monitored demolitions. *Bull. Earthq. Eng.* **2022**, *20*, 4441–4482. [CrossRef]
39. Campagnari, S.; Di Matteo, F.; Manzoni, S.; Scaccabarozzi, M.; Vanali, M. Estimation of axial load in tie-rods using experimental and operational modal analysis. *J. Vib. Acoust. Trans. ASME* **2017**, *139*, 041005. [CrossRef]
40. Kernicky, T.; Whelan, M.; Al-Shaer, E. Dynamic identification of axial force and boundary restraints in tie rods and cables with uncertainty quantification using Set Inversion Via Interval Analysis. *J. Sound Vib.* **2018**, *423*, 401–420. [CrossRef]
41. Rainieri, C.; Fabbrocino, G. Development and validation of an automated operational modal analysis algorithm for vibration-based monitoring and tensile load estimation. *Mech. Syst. Signal Process.* **2015**, *60*, 512–534. [CrossRef]
42. Resta, C.; Chellini, G.; Falco, A.D. Dynamic assessment of axial load in tie-rods by means of acoustic measurements. *Buildings* **2020**, *10*, 23. [CrossRef]
43. De Falco, A.; Resta, C.; Sevieri, G. Sensitivity analysis of frequency-based tie-rod axial load evaluation methods. *Eng. Struct.* **2021**, *229*, 111568. [CrossRef]
44. Coïsson, E.; Collini, L.; Ferrari, L.; Garziera, R.; Riabova, K. Dynamical Assessment of the Work Conditions of Reinforcement Tie-Rods in Historical Masonry Structures. *Int. J. Archit. Herit.* **2019**, *13*, 358–370. [CrossRef]
45. Cescatti, E.; Da Porto, F.; Modena, C. Axial Force Estimation in Historical Metal Tie-Rods: Methods, Influencing Parameters, and Laboratory Tests. *Int. J. Archit. Herit.* **2019**, *13*, 317–328. [CrossRef]
46. Garziera, R.; Amabili, M.; Collini, L. A hybrid method for the nondestructive evaluation of the axial load in structural tie-rods. *Nondestruct. Test. Eval.* **2011**, *26*, 197–208. [CrossRef]
47. Tullini, N.; Rebecchi, G.; Laudiero, F. Reliability of the tensile force identification in ancient tie-rods using one flexural mode shape. *Int. J. Archit. Herit.* **2019**, *13*, 402–410. [CrossRef]
48. Makoond, N.; Pelà, L.; Molins, C. Robust estimation of axial loads sustained by tie-rods in historical structures using Artificial Neural Networks. *Struct. Health Monit.* **2022**, 14759217221123326. [CrossRef]
49. Lucà, F.; Manzoni, S.; Cerutti, F.; Cigada, A. A Damage Detection Approach for Axially Loaded Beam-like Structures Based on Gaussian Mixture Model. *Sensors* **2022**, *22*, 8336. [CrossRef]
50. Lucà, F.; Manzoni, S.; Cigada, A.; Barella, S.; Gruttadauria, A.; Cerutti, F. Automatic Detection of Real Damage in Operating Tie-Rods. *Sensors* **2022**, *22*, 1370. [CrossRef]
51. Gentile, C.; Poggi, C.; Ruccolo, A.; Vasic, M. Vibration-Based Assessment of the Tensile Force in the Tie-Rods of the Milan Cathedral. *Int. J. Archit. Herit.* **2019**, *13*, 402–415. [CrossRef]
52. Jolliffe, I.T. *Principal Component Analysis*; Springer: Berlin/Heidelberg, Germany, 2002.
53. Datteo, A.; Lucà, F.; Busca, G.; Cigada, A. Long-time monitoring of the G. Meazza stadium in a pattern recognition prospective. *Procedia Eng.* **2017**, *199*, 2040–2046. [CrossRef]
54. Lucà, F.; Manzoni, S.; Cigada, A. Data Driven Damage Detection Strategy Under Uncontrolled Environment. In *European Workshop on Structural Health Monitoring*; Springer: Cham, Switzerland, 2023; Volume 2, pp. 764–773. [CrossRef]
55. Worden, K.; Manson, G.; Fieller, N.R. Damage detection using outlier analysis. *J. Sound Vib.* **2000**, *229*, 647–667. [CrossRef]
56. Gallego, G.; Cuevas, C.; Mohedano, R.; García, N. On the mahalanobis distance classification criterion for multidimensional normal distributions. *IEEE Trans. Signal Process.* **2013**, *61*, 4387–4396. [CrossRef]
57. Cheli, F.; Diana, G. *Advanced Dynamics of Mechanical Systems*; Springer: Cham, Switzerland, 2015; pp. 1–818. . [CrossRef]
58. Valle, J.; Fernández, D.; Madrenas, J. Closed-form equation for natural frequencies of beams under full range of axial loads modeled with a spring-mass system. *Int. J. Mech. Sci.* **2019**, *153–154*, 380–390. [CrossRef]

59. Galef, A.E. Bending Frequencies of Compressed Beams. *J. Acoust. Soc. Am.* **1968**, *44*, 643. [CrossRef]
60. Ewins, D.J. *Modal Testing: Theory, Practice and Application*; John Wiley & Sons: Hoboken, NJ, USA, 2001; p. 562.
61. Peeters, B.; Van Der Auweraer, H.; Guillaume, P.; Leuridan, J. The PolyMAX frequency-domain method: A new standard for modal parameter estimation? *Shock Vib.* **2004**, *11*, 395–409. [CrossRef]
62. Sakaris, C.S.; Sakellariou, J.S.; Fassois, S.D. Random-vibration-based damage detection and precise localization on a lab–scale aircraft stabilizer structure via the Generalized Functional Model Based Method. *Struct. Health Monit.* **2017**, *16*, 594–610 [CrossRef]
63. Sarrafi, A.; Mao, Z.; Niezrecki, C.; Poozesh, P. Vibration-based damage detection in wind turbine blades using Phase-based Motion Estimation and motion magnification. *J. Sound Vib.* **2018**, *421*, 300–318. [CrossRef]
64. Banerjee, S.; Ricci, F.; Monaco, E.; Mal, A. A wave propagation and vibration-based approach for damage identification in structural components. *J. Sound Vib.* **2009**, *322*, 167–183. [CrossRef]
65. Yi, W.J.; Zhou, Y.; Kunnath, S.; Xu, B. Identification of localized frame parameters using higher natural modes. *Eng. Struct.* **2008**, *30*, 3082–3094. [CrossRef]

sensors

MDPI

Article

The IBA-ISMO Method for Rolling Bearing Fault Diagnosis Based on VMD-Sample Entropy

Deyu Zhuang [1], Hongrui Liu [1], Hao Zheng [2], Liang Xu [2], Zhengyang Gu [2], Gang Cheng [2] and Jinbo Qiu [1,*]

[1] China Coal Technology and Engineering Group Shanghai Company Ltd., Shanghai 200030, China
[2] School of Mechatronic Engineering, China University of Mining and Technology, Xuzhou 221116, China
* Correspondence: jinboqiu@sh163.net

Abstract: Rolling bearings are important supporting components of large-scale electromechanical equipment. Once a fault occurs, it will cause economic losses, and serious accidents will affect personal safety. Therefore, research on rolling bearing fault diagnosis technology has important engineering practical significance. Feature extraction with high price density and fault identification are two keys to overcome in the field of fault diagnosis of rolling bearings. This study proposes a feature extraction method based on variational modal decomposition (VMD) and sample entropy and also designs an improved sequence minimization algorithm with optimal parameters to identify the fault. Firstly, a variational modal decomposition system based on vibration signals is designed, and the sample entropy of the components is extracted as the eigenvalue of the signal. Secondly, in order to improve the accuracy of fault diagnosis, the sequence minimum optimization algorithm optimized by the bat algorithm is used as the classifier. Certainly, the traditional bat algorithm (BA) and the sequence minimum optimization algorithm (SMO) are improved, respectively. Therefore, a fault diagnosis algorithm based on IBA-ISMO is obtained. Finally, the experimental verification is designed to prove that the algorithm model has a good state recognition rate for bearings.

Keywords: variational mode decomposition; sample entropy; sequence minimum optimization algorithm; fault diagnosis

Citation: Zhuang, D.; Liu, H.; Zheng, H.; Xu, L.; Gu, Z.; Cheng, G.; Qiu, J. The IBA-ISMO Method for Rolling Bearing Fault Diagnosis Based on VMD-Sample Entropy. *Sensors* **2023**, *23*, 991. https://doi.org/10.3390/s23020991

Academic Editors: Yongbo Li and Bing Li

Received: 20 December 2022
Revised: 5 January 2023
Accepted: 13 January 2023
Published: 15 January 2023

1. Introduction

The common faults of electromechanical equipment can be divided into electrical faults and mechanical faults. Electrical faults are often caused by circuit aging, component damage, etc., while mechanical faults refer to the phenomenon that electromechanical equipment loses or reduces its specified functions and cannot continue to operate due to some inevitable damage. Mechanical faults have a serious impact on the safety state of electromechanical equipment [1–3]. On the one hand, the root cause of faults is complex and the evolution time is long; on the other hand, once the mechanical faults lead to an accident, the impact and consequences are unpredictable.

Many mechanical faults are reflected in the form of vibration, and the vibration signals contain rich information, which can quickly and directly reflect the operating status of critical parts in major equipment such as bearings [4–6]. It is very necessary to carry out vibration monitoring and fault diagnosis. However, extracting feature information with high valence density and designing a classification space that is suitable for strong nonlinear and non-stationary information are urgent problems to be solved in the field of fault diagnosis, and even in the field of machine learning. Reference [7] proposed an adaptive boundary determination method based on empirical wavelet transform and applied it to fault detection of high-speed train wheelset bearings. Park et al. [8] proposed a minimum variance cepstrum based on cepstral analysis, which avoided the influence of the system frequency and the selection of the resonance band, and realized the detection of early faults of the rotating parts. Borghesani proposed a method of whitening the signal using cepstrum [9]. [10] proposed an improved empirical mode decomposition method to effectively extract the fault features of rolling bearings. In view of the nonlinear and non-stationary characteristics of the vibration signal of planetary gears, [11] proposed a feature extraction method of initial fault based on ensemble empirical mode decomposition (EEMD) and adaptive stochastic resonance (ASR), which provided the strong initial fault diagnosis of planetary gears against noise background. However, in practical application, the wavelet decomposition method has problems such as difficulty in selecting the wavelet basis function [12], and the empirical mode method often has the problem of end-point effects and mode mixing, which presents some challenges regarding the extraction of fault features. As a time-frequency domain analysis method, variational mode decomposition (VMD) has better adaptability than other analysis methods. This method combines the classic Wiener filtering, Hilbert transform and frequency mixing in mathematical theory. Based on these advantages, the number of self-determined modal components and the lower time complexity are realized, and the non-stationary original signal can be decomposed into relatively stationary subsequences containing multiple frequency domains by VMD. Wang et al. [13] proposed the characteristic parameter of spectral kurtosis entropy (SKE) and combined it with VMD to realize the feature extraction of the bogie vibration signal under variable working conditions. [14] proposed a sparse VMD (sparsity-oriented VMD) method, which effectively extracted encoder information and realized gear fault diagnosis.

Fault identification is also an important step for establishing the correlation between fault features and class labels. Fan et al. [15] proposed a high-performance SVM multi-feature fusion and self-tuning particle swarm optimization algorithm. The method extracted multi-dimensional fault features by EMD. Then, the multi-dimensional parameters of the high-performance SVM were configured by adjusting the particle swarm optimization algorithm, which has improved the effectiveness in bearing fault detection and classification. David E. Runelhart et al. [16] proposed the back propagation (BP) neural network algorithm, which constituted a multi-layer feedforward perceptron to solve the problem of connection weight learning in the hidden layer of the multi-layer neural network. Lu et al. [17] proposed an improved feature selection and neural network classification algorithm for the problem of rotating machinery fault diagnosis. The study extracted the time domain and frequency domain features of the whole machine under multiple working conditions and used an optimized backpropagation neural network algorithm for fault diagnosis. He et al. [18] proposed a bearing fault diagnosis method based on a Gaussian constrained Boltzmann machine, which takes the envelope spectrum of the resampled data directly as a feature vector to represent the bearing fault. Wang et al. [19] proposed an intelligent bearing fault diagnosis method that combined the symmetric point pattern representation and the compressed excitation convolutional neural network model for the problems of fault visualization and automatic feature extraction.

Based on the above analysis, this paper intends to use variational modal decomposition (VMD) and sample entropy for signal decomposition and feature extraction of vibration signals. The improved sequence minimum optimization algorithm has been chosen as the pattern recognition method in this study.

The rest of the article is organized as follows: Section 2 focuses on the feature extraction method based on VMD-Sample Entropy and the sequence minimum optimization algorithm after optimizing parameters. Section 3 verifies the validity of the fault diagnosis model proposed in this study through experiments and conducts necessary analysis on the experiments. The fourth part summarizes the research of the full study.

2. The Enhanced Fault Diagnosis Method

The study plans to decompose the original signal by VMD to obtain the intrinsic mode component (IMF) and to extract the sample entropy of each component as the eigenvalue of the bearing fault. The training set is input into the improved sequence minimum optimization algorithm (ISMO) model for training. At the same time, the penalty factor and Gaussian kernel parameters of the ISMO are optimized by an improved bat algorithm (IBA). The fault diagnosis model established in this study is shown in Figure 1.

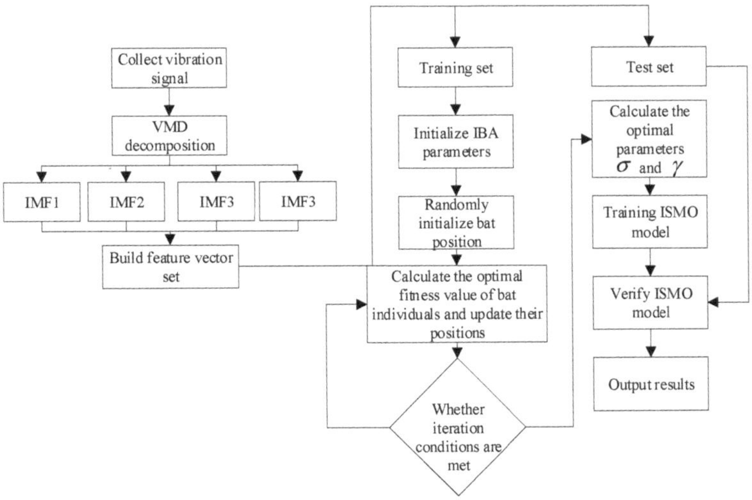

Figure 1. The model of fault diagnosis.

2.1. Feature Extraction Method Based on VMD-Sample Entropy

Signal denoise and feature extraction often play a vital role, respectively. In this study, the variable scale processing method, known as variational mode decomposition (VMD), is proposed. VMD satisfies the self-adaptability of the decomposition model in a non-recursive way and transforms the complex signal decomposition problem in the time domain and frequency domain into a mathematical model for the solution so as to avoid the end effect and restrain the mode mixing caused by noise, and an ideal optimal result of signal decomposition is naturally obtained.

After the optimal result of signal decomposition is obtained through the variational model, in order to better carry out fault diagnosis, it is necessary to extract the most prominent feature information from the signal decomposition results. In this study, the sample entropy is selected as the feature of the vibration signal. As a new algorithm based on approximate entropy algorithm, the physical meaning of sample entropy can be expressed as calculating the probability of the change of time series caused by the change of data bits.

2.1.1. Variational Mode Decomposition

1. Constructing a constrained variational model.

Firstly, the analytic signal u_k of the original signal is obtained by Hilbert transform of the real mode function u_k^+.

$$u_k^+(t) = \left(\delta(t) + \frac{j}{\pi t}\right) * u_k(t) \tag{1}$$

In Formula (1), t and $\delta(t)$ denote time and influence function, respectively.

The analytical signal u_k^+ is mixed with each estimated center frequency, and the spectrum of each mode is modulated to the corresponding fundamental frequency band as follows:

$$u_k^m(t) = \left(\delta(t) + \frac{j}{\pi t}\right) * u_k(t)e^{-jw_k t} \tag{2}$$

Finally, by calculating the L_2 norm of the time gradient, the effective value of the modal component bandwidth can be calculated as follows:

$$\Delta w = \|\partial_t\left[\left(\delta(t) + \frac{j}{\pi t}\right) * u_k(t)e^{-jw_k t}\right]\|_2^2 \tag{3}$$

Therefore, the bandwidth of the modal components of each frequency can be expressed as Formula (4):

$$\begin{gathered}\min_{\{u_k\},\{w_k\}}\left\{\sum_{K=1}^{K}\|\partial_t\left[\left(\delta(t)+\frac{j}{\pi t}\right) * u_k(t)e^{-jw_k t}\right]\|_2^2\right\}\\ \sum_{K=1}^{K} u_k(t) = f(t)\end{gathered} \tag{4}$$

In Formula (4), $\{u_k\} = \{u_1, \ldots, u_k\}$ represents the IMF components obtained by VMD; $\{w_k\} = \{w_1, \ldots, w_k\}$ represents the central frequency of IMF, $f(t)$ is the original input signal.

(2) Solving constrained variational model

Formula (4) is constructed as a Lagrangian expression by adding the quadratic penalty factor and the Lagrange operator $\lambda(t)$.

$$\begin{gathered}L_{(\{u_k\},\{w_k\},\lambda)} := \alpha\left\{\sum_k \|\partial_t[(\delta(t)+\frac{j}{\pi t}) \times u_k(t)]e^{-jw_k t}\|_2^2\right\} + \|f(t)-\sum_k u_k(t)\|_2^2\\ +\left\langle \lambda(t), f(t) - \sum_k u_k(t)\right\rangle\end{gathered} \tag{5}$$

In Formula (5), the appropriate penalty factor is selected to ensure that the reconstruction accuracy is high enough under variable working conditions, and the Lagrangian operator $\lambda(t)$ is introduced to make the solution of Formula (5) theoretical and rigorous.

The alternating direction multiplier algorithm is introduced to solve the above variational problems. The main idea is to obtain the saddle point of the extended Lagrangian expression by alternately updating the parameters $u_k(t)$, $w_k(t)$ and $\lambda_k(t)$. The updated Formula is as follows (6):

$$\begin{gathered}u_k^{n+1}(t) = \underset{u_k \in X}{\operatorname{argmin}}\left\{\alpha\sum_{K=1}^{K}\|\partial_t\left[\left(\delta(t)+\frac{j}{\pi t}\right) * u_k(t)e^{-jw_k t}\right]\|_2^2\right.\\ \left. + \|f(t) - \sum_i u_i(t) + \frac{\lambda(t)}{2}\|_2^2\right\}\end{gathered} \tag{6}$$

Under the condition of L_2 norm, Equation (6) is transformed into the frequency domain by Fourier isometric transform, and the variable w in the equation is replaced by the updated $w - w_k$. According to the Hermitain symmetry theorem, the expression of the k eigenmode function (IMF) is obtained as follows:

$$\hat{u}_k^{n+1}(w) = \frac{\hat{f}(w) - \sum_{i<k}\hat{u}_i^{n+1}(w) - \sum_{i>k}\hat{u}_i^{n}(w) + \frac{\hat{\lambda}(w)}{2}}{1 + 2\alpha(w - w_n)^2} \tag{7}$$

The central frequency expression of the updated modal IMF is:

$$\hat{w}_k^{n+1} = \frac{\int_0^\infty w\left|\hat{u}_k^{n+1}(w)\right|^2 dw}{\int_0^\infty \left|\hat{u}_k^{n+1}(w)\right|^2 dw} \tag{8}$$

The updated expression of all non-negative center frequencies is $w \geq 0$. and the updated expression of operator λ^{n+1} is:

$$\hat{\lambda}^{n+1}(w) = \hat{\lambda}^{n}(w) + \tau\left[\hat{f}(w) - \sum_{k} \hat{u}_k^{n+1}(w)\right] \tag{9}$$

Summing up the above description, the decomposition process of VMD algorithm can be summarized as follows:

(1) Initialize the value of $u_k^1(t)$, $w_k^1(t)$ and $\lambda_k^1(t)$, n is 0.
(2) Set the out-of-loop condition: $n = n + 1$.
(3) Update $u_k(t)$ and $w_k(t)$ until the number of intrinsic mode decomposition of the original sample meets the preset number of the decomposition, ending the current internal cycle.
(4) Get a new $\lambda_k(t)$ license.
(5) Give the jump condition ε as the operator precision, and the $\frac{\sum_k \|u_k^{n+1} - u_k^n\|_2^2}{\|u_k^n\|_2^2} < \varepsilon$ as the stop condition, when the condition is satisfied the loop ends. If not, the outer loop operation is performed again (step 2).

From the solving process of the above VMD algorithm, it can be concluded that the VMD algorithm adaptively decomposes the characteristic frequency of the original signal to get its frequency bandwidth. Through the termination condition to control the IMF and the center frequency to calculate repeatedly in the time-frequency domain of the signal. The adaptive decomposition process ends when the stop condition is satisfied.

2.1.2. Sample Entropy

Sample entropy, which can better measure the complexity of time series, is widely used in signal analysis and processing.

Suppose that there are N pieces of data, and the time series of data sampling is defined as $X = [x(n), n = 1, 2, \ldots, N]$. The theoretical derivation of the definition of sample entropy is as follows:

(1) According to the sampling time of the signal, a vector sequence based on time series is constructed, and the dimension of the vector sequence is m, $X_m(1), \ldots, X_m(N - m + 1)$. Each element in the vector sequence can be represented by the following array: $X_m(i) = \{x(i), x(i+1), \ldots, x(i+m-1)\}, 1 \leq i \leq N - m + 1$. The array represents the continuous x values of the time series from i to $m + i$;
(2) Define the distance between $X_m(i)$ and $X_m(j)$: $d[X_m(i), X_m(j)]$ is the absolute value of the difference between $X_m(i)$ and $X_m(j)$.

$$d[X_m(i), X_m(j)] = \max_{k=0,\ldots,m-1}(|x(i+k) - x(j+k)|) \tag{10}$$

(3) For the constructed $d[X_m(i), X_m(j)]$, the number of j $(1 \leq j \leq N - m, j \neq i)$ is calculated and marked as B_i, $1 \leq i \leq N - m$, B_i is defined as follows:

$$B_i^m(r) = \frac{1}{N - m - 1} B_i \tag{11}$$

(4) Average $B^{(m)}(r)$ as Formula (12):

$$B^{(m)}(r) = \frac{1}{N - m} \sum_{i=1}^{N-m} B_i^m(r) \tag{12}$$

(5) Update the vector dimension to $m+1$, and recalculate the number of distances and $d[X_m(i), X_m(j)] \leq r$ bands, where $(1 \leq j \leq N-m, j \neq i)$ benchmark is marked as A_i. Define $A_i^m(r)$ and $A^m(r)$ as the following expressions:

$$A_i^m(r) = \frac{1}{N-m-1} A_i \tag{13}$$

$$A^m(r) = \frac{1}{N-m} \sum_{i=1}^{N-m} A_i^m(r) \tag{14}$$

From the above steps, $B^{(m)}(r)$ is the probability of two sequences matching m points under the similar tolerance r, while $A^m(r)$ is the probability of two sequences matching $m+1$ points. Therefore, the definition of sample entropy is:

$$SampEn(m,r) = \lim_{N\to\infty} \left\{ -\ln\left[\frac{A^m(r)}{B^m(r)}\right] \right\} \tag{15}$$

When N is a finite value, the following Formula can be used:

$$SampEn(m,r,N) = -\ln\left[\frac{A^m(r)}{B^m(r)}\right] \tag{16}$$

As can be seen from the above description, the sample entropy has the following characteristics:

(1) This feature quantity can avoid the disadvantage of approximate entropy, prevent the data length from being compared by itself and can make the operation results more accurate and consistent.
(2) Comparing the two sequences, no matter what the scale of the two sequences is, if the m and r values are changed, the calculation results will not change.
(3) In the process of signal acquisition, it is inevitable to lose some frames. For the sample entropy algorithm, the loss of a small part of data has no great impact on the overall structure. Sample entropy can restore the operation results of real data to the maximum.

In any algorithm involving parameter selection, the influence of parameters cannot be ignored. When calculating the sample entropy of the signal, the value of the parameters has the same important influence on the result of the sample entropy operation. According to the theoretical derivation in the previous section, the main parameters of sample entropy include embedding dimension m, similarity tolerance r and data points N., the indexes of these parameters are as follows:

(1) The embedded dimension m represents the dimension of the window function in the sample entropy algorithm, which is similar to the size of the window function in the Fourier transform. In most cases, $m = 1.2$. When $m > 2$, the deviation of the parameter value will result in the following: first, a large number of original data sets will be needed to increase the computational complexity of the algorithm; second, a too large m will affect the value of r, and there is a positive correlation between the two. When m is larger, r is larger, r will remove too much useful information.
(2) Similarity capacity r is usually obtained based on $(0.15 \sim 0.25)\delta(x)$, where $\delta(x)$ represents the standard deviation of sampling. The r value is too large, resulting in invalid data redundancy; the r value is too small, resulting in a reduction in the amount of data in similar patterns.
(3) N indicates the number of sampled data points, which is usually obtained from 100 to 6000.

2.2. Fault Identification Method Based on IBA-ISMO

The penalty factor ζ in ISMO and the parameter σ of Gaussian kernel function have a considerable influence on the classification result and running time. In this paper, the neural network algorithm based on the bat algorithm is selected to optimize the parameters. The algorithm has a good local search ability. By optimizing the bat algorithm, the shortcomings of the algorithm in the process of global optimization are improved and the global optimal solution of the parameters is obtained.

2.2.1. The Improved Bat Algorithm

The bat algorithm is used to solve the optimal solution by simulating the feeding habits of bats through echolocation. The main idea of the algorithm is that each bat represents a solution in the feasible region, imitating the method of identifying the direction of bat sound waves. Bat individuals constantly emit pulses of a fixed range of frequencies and capture the sound waves reflected after the pulse collides with the target. The distance and position of the target are obtained according to the difference in pulse frequency and the time difference of senses to feel the pulse.

Let the dimension of search space be d-dimensional, and the relevant parameters emitted by bat i in the process of finding the optimal solution are pulse frequency f_i, velocity v_i, position x_i, transmitted pulse frequency $[f_{\min}, f_{\max}]$ and the maximum number of iterations max$T$. Therefore, the update Formula for the position of the bat at t moment is as follows:

$$f_i = f_{\min} + \beta(f_{\max} - f_{\min}) \tag{17}$$

$$v_i^t = v_i^{t-1} + (x_i^t - x_*)f_i \tag{18}$$

$$x_i^t = x_i^{t-1} + v_i^t \tag{19}$$

where β is a random number in $[0, 1]$, and x_* is the optimal position of the current population.

In the process of searching for prey, each bat will adjust the loudness and pulse frequency of its sound wave according to the location of the target to improve the capture probability. In the process of getting closer to the target, the search area of the bat will gradually decrease. Therefore, when the loudness decreases below a certain fixed value, the frequency is rapidly increased to facilitate the faster acquisition of prey, and the changes of loudness and pulse in the process of catching prey can be obtained, as shown in the following Formula:

$$A_i^{t+1} = \beta A_i^t \tag{20}$$

$$r_i^{t+1} = r_i^0[1 - \exp(-\gamma t)] \tag{21}$$

where A represents the pulse loudness, $\gamma > 0$ pulse represents the pulse frequency enhancement coefficient, and r_i^0 represents the initial pulse frequency.

As can be seen from the above Formula, when there is $t \to \infty$, there is $A_i^t \to 0$. When $A_i^t \to 0$, it means that the bat has found its prey at this time, and the iteration ends and no longer sends out pulses.

Bat algorithm has obvious advantages over other parameter optimization algorithms in global search ability and convergence speed, but it also has the disadvantage that individuals of the population are easy to fall into the local optimal solution. In order to solve this problem, this paper proposes an improved bat optimization algorithm by introducing a new variable w; namely, the adaptive weight factor, to measure the difference between the current position and the global optimal solution. In order to avoid the final solution vector falling into the local optimal situation to the greatest extent.

The calculation Formula of adaptive weight factor w is as follows:

$$w_i = \frac{(x_i - x_*)}{t + 1} \tag{22}$$

By updating Formula (22) to:

$$v_i^t = v_i^{t-1} w_i + (x_i^t - x_*) f_i \tag{23}$$

To sum up, the flow of the IBA algorithm is shown in Figure 2.

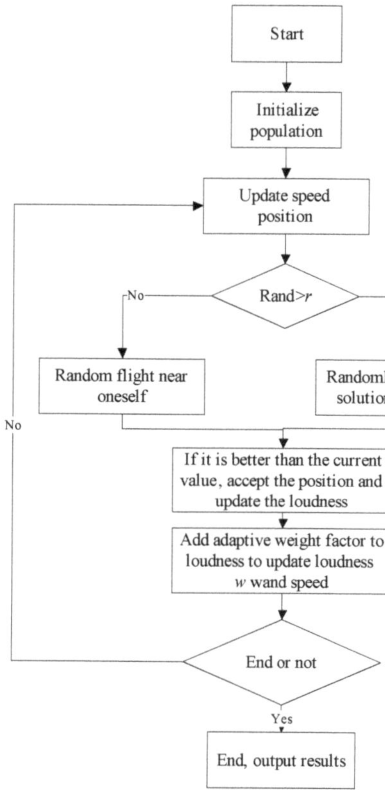

Figure 2. The flowchart of IBA algorithm.

According to the characteristics of IBA algorithm, it is found that the parameters will affect the convergence speed of the algorithm itself and the accuracy of the optimal solution. For example, the parameters of this kind of group optimization algorithm need to be selected through strong experiment and experience, and either too large or too small parameters will affect the results, so the selected parameters are as follows:

Pulse enhancement coefficient $\gamma = 0.9$, pulse frequency $f_{\max} = 2$, $f_{\min} = 0$; loudness coefficient $A_0 = 1.5$, initial pulse intensity $r_0 = 0.5$, algorithm population size $n = 50$, dimension $d = 5$, maximum iterations $M = 1000$, adaptive weight factor $w_{\max} = 0.9$,$w_{\min} = 0.2$.

Because the IBA algorithm avoids the disadvantage of falling into the local optimal solution compared with the traditional BA algorithm, in order to verify whether the parameter selection of the IBA algorithm is reasonable, this section selects the Rastrigin function to test the global optimization performance of the IBA algorithm and selects the Ackley function to test the global convergence ability of the IBA algorithm.

(1) Rastrigin function

$$f(x) = \sum_{i=1}^{n} \left[x_i^2 - 10\cos(2\pi x_i) + 10 \right] \tag{24}$$

Among them, $x \in [-5.12, 5.12]$, $i = 1, 2$, and the overall shape of the function is similar to that of the hills, which proves that the algorithm has a good ability for global optimization. The function image is shown in Figure 3:

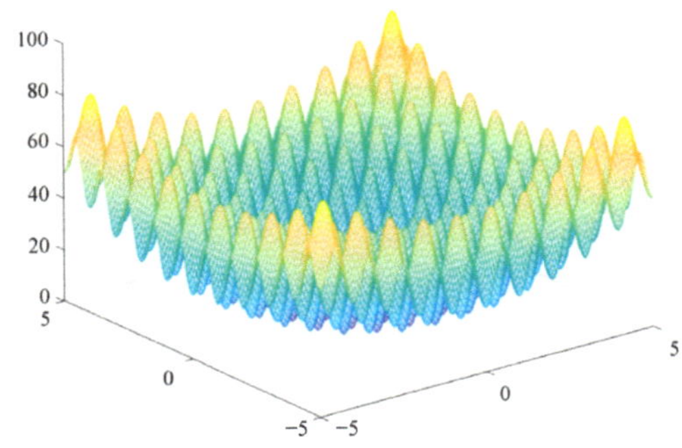

Figure 3. Rastrigin function.

(2) Ackley function

$$f(x) = 20 + e - 20\exp\left(-0.2\sqrt{\frac{1}{n}\sum_{i=1}^{n} x_i^2}\right) - \exp\left(\frac{1}{n}\sum_{i=1}^{n}\cos(2\pi x_i)\right) \tag{25}$$

Among them, $x \in [-32.768, 32.768]$, $i = 1, 2$. The closer $f(x)$ is to 0, the stronger the global convergence ability of the algorithm is. The image of the IBA algorithm after applying this function is shown in Figure 4.

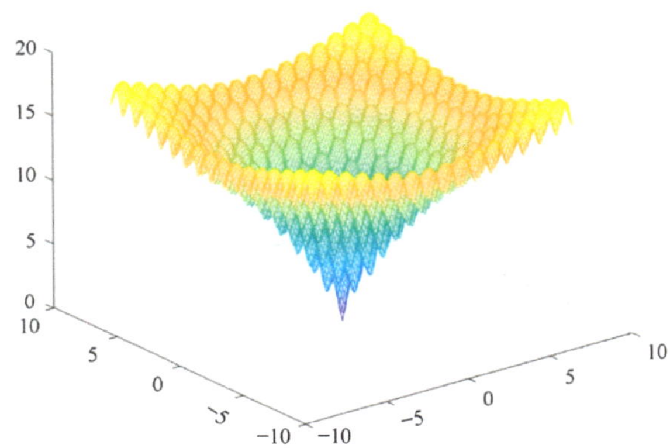

Figure 4. Ackley function image.

In order to digitize the image and show the global optimization ability and global convergence ability of the improved bat algorithm more intuitively, the IBA algorithm after parameter selection is run 15 times independently, and the test results shown in Table 1 are obtained.

Table 1. Test function results.

Function	Algorithm	Optimal Value	Average Value	Standard Deviation
Rastrigin	IBA	0	7.11×10^{-16}	1.50×10^{-15}
Ackley	IBA	4.26×10^{-14}	5.97×10^{-13}	8.33×10^{-13}

As can be seen from the table, the standard deviation and average of the Rastrigin function and the Ackley function are both close to 0. Therefore, it has been proven that the IBA algorithm overcomes the disadvantages of the traditional BA algorithm and has a significant improvement in global convergence and global optimization.

2.2.2. Improved Sequence Minimization Algorithm (ISMO)

As an algorithm in the SVM model, SMO algorithm essentially uses a very important functional relationship-kernel function. In this study, the Gaussian kernel function is improved to improve the efficiency of the SMO algorithm.

The accuracy of sample classification predicted by SMO should meet the following expression:

$$E_N[P(error)] \leq \frac{E_{N[SV]}}{N} \tag{26}$$

where N represents the total number of training set samples and E_N represents the expected value calculated through the training set samples. It can be seen from the Formula that when the number of samples in the training set is N, we can choose to reduce the number of support vectors to reduce the probability of operational errors and improve the application range of support vector machines. The control of the number of support vectors depends on the mapping relationship of the algorithm and the selection of algorithm parameters.

Based on the type of kernel function determined in the previous section, the Gaussian kernel function and the coefficient $(1 + m)(m > 0)$ are as follows:

$$K(x, x_i) = (1 + m) * \exp\left(-\gamma * \|x - x_i\|^2\right) \tag{27}$$

According to Formula (27), the Gaussian kernel coefficient is magnified by $(1 + m)$ times, and the number of support vectors and the number of samples on the boundary are reduced by increasing the absolute value of the quadratic coefficient in Q^v, which can effectively reduce the classification error rate. By reducing the solution vectors in data samples that meet the KKT boundary conditions, the time complexity of the algorithm is reduced, and the SMO classification accuracy and application range are improved. The improved SMO algorithm is named ISMO.

Based on the application background of the system in engineering practice, the acquisition and analysis system take the vibration signal as the original signal, and the original signal has the characteristics of small sample and non-linearity. The ISMO model is selected to classify the vibration signal eigenvector obtained in the previous chapter, and the mapping of eigenvector from linear inseparable to linear separable is completed, which enhances the classification accuracy and application range of the algorithm.

3. Experiment and Analysis

3.1. Extracted Features of Rolling Bearing Signals

A variational mode decomposition algorithm is an adaptive signal decomposition algorithm. By using this method, not only can part of the noise signal be removed, but also the information of the signal will not be lost, and the characteristic components of the original signal can be preserved as much as possible, so the VMD algorithm is chosen to preprocess the original signal. Sample entropy is a kind of eigenvalue used to measure the complexity of time series, which is improved on the basis of other entropy values, so it also has the characteristics of anti-noise, so sample entropy is chosen as the eigenvalue of the

IMF signal. Therefore, this paper proposes a method of feature extraction by combining the VMD algorithm with sample entropy. The detailed flow chart is shown in Figure 5.

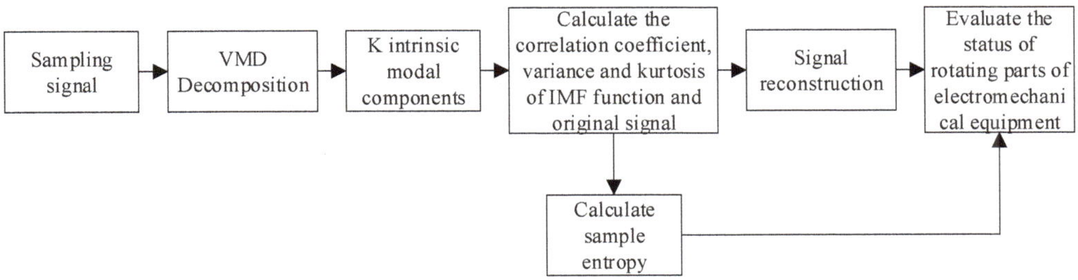

Figure 5. Flow chart of feature extraction.

In order to verify the effectiveness of the feature extraction algorithm based on VMD and sample entropy proposed in this study, this section uses the bearing data in the CRWU database for related experiments [20]. The data set is mainly composed of the following data: drive acceleration data, fan segment acceleration data, basic acceleration data and speed data. The experimental system consists of test bearings, torque sensors, control motors with different functions and programmable controllers. The test bench is shown in Figure 6.

Figure 6. Bearing simulation failure test bench of Western Reserve University.

In the experiments, the sampling frequency is 12 kHz, the motor speed is 1797 r/min, and the fault state bearing damage diameter is 0.1778 mm. The bearing states selected in this experiment include normal state, inner ring fault, outer ring fault and roller fault, and the number of sampling points of each sample is 6000. The original sampling signals of the four states of the bearing are shown in Figure 7:

From the original vibration signal shown in Figure 7, it can be seen that there are great differences in the vibration period and amplitude of the bearing in different states. The vibration signals of the three fault states all confirm the above analysis of the vibration signal that there is a periodic abnormal signal, and there is little difference in amplitude in different periods in the same state.

The vibration signals of four states in Figure 7 are decomposed by variational mode decomposition. Because the variational mode algorithm has the advantage of removing some redundant component information, as shown in Figure 8, four IMF component informations are obtained according to different time–frequency domain characteristics.

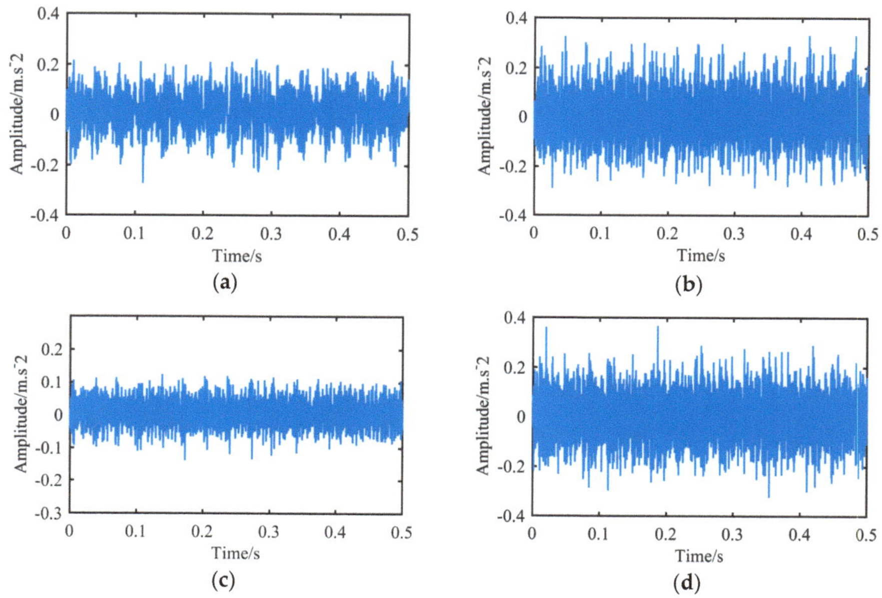

Figure 7. Vibration signals of rolling bearings in four states. (**a**) Normal state. (**b**) Inner ring fault. (**c**) Rolling body fault. (**d**) Outer ring fault.

The sample entropy of the decomposed components is calculated, and four groups of data are randomly selected from each state, as shown in Table 2. As can be seen from Table 2, there are obvious differences in the sample entropy of each modal component after the VMD decomposition of vibration signals in different states. Therefore, the sample entropy index based on VMD decomposition can be used as the eigenvalue of the bearing. The total number of samples obtained according to the above process is $350 \times 4 = 1400$.

Table 2. Sample entropy of some samples.

Status	Sample Entropy Features			
	IMF1	**IMF2**	**IMF3**	**IMF4**
Normal	0.270606	0.538289	0.266344	0.756758
	0.278523	0.547741	0.261999	0.741383
	0.27207	0.556223	0.234584	0.815750
	0.271301	0.543783	0.229493	0.520792
Inner ring fault	0.583038	0.507441	0.245161	0.276670
	0.592259	0.510691	0.239027	0.287463
	0.585477	0.483586	0.304042	0.234883
	0.586028	0.487997	0.317364	0.219320
Rolling element fault	0.427306	0.609396	0.267466	0.155322
	0.398368	0.512955	0.240769	0.177347
	0.589774	0.482116	0.304627	0.163799
	0.582455	0.473968	0.281762	0.206347
Outer ring fault	0.427306	0.609396	0.267466	0.155322
	0.398368	0.595303	0.228638	0.157244
	0.578411	0.495655	0.192259	0.119652
	0.579525	0.506227	0.204141	0.142848

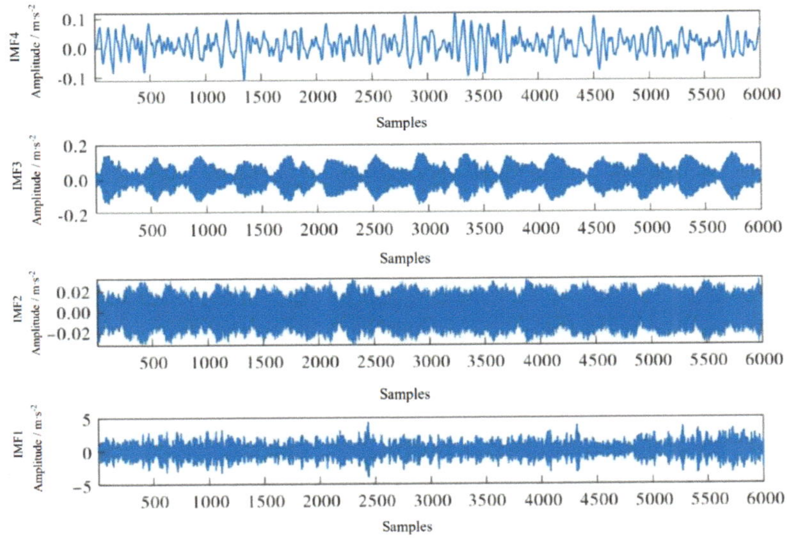

(a)

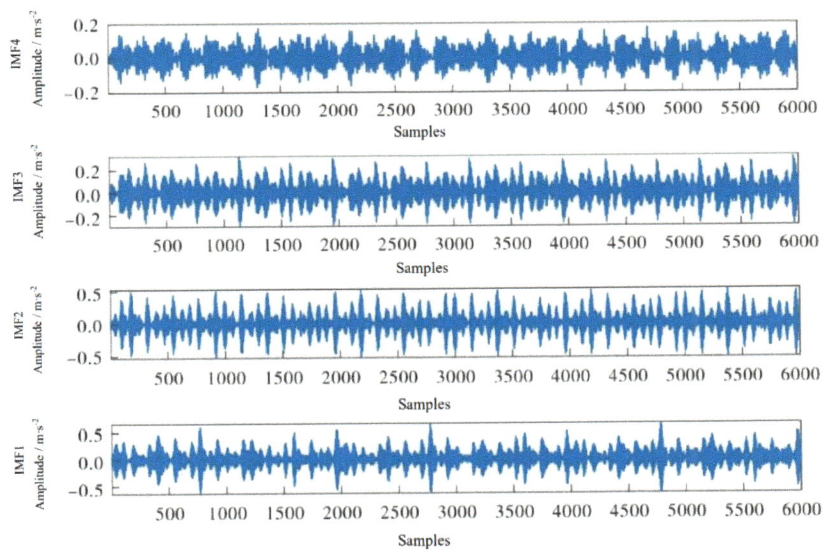

(b)

Figure 8. *Cont.*

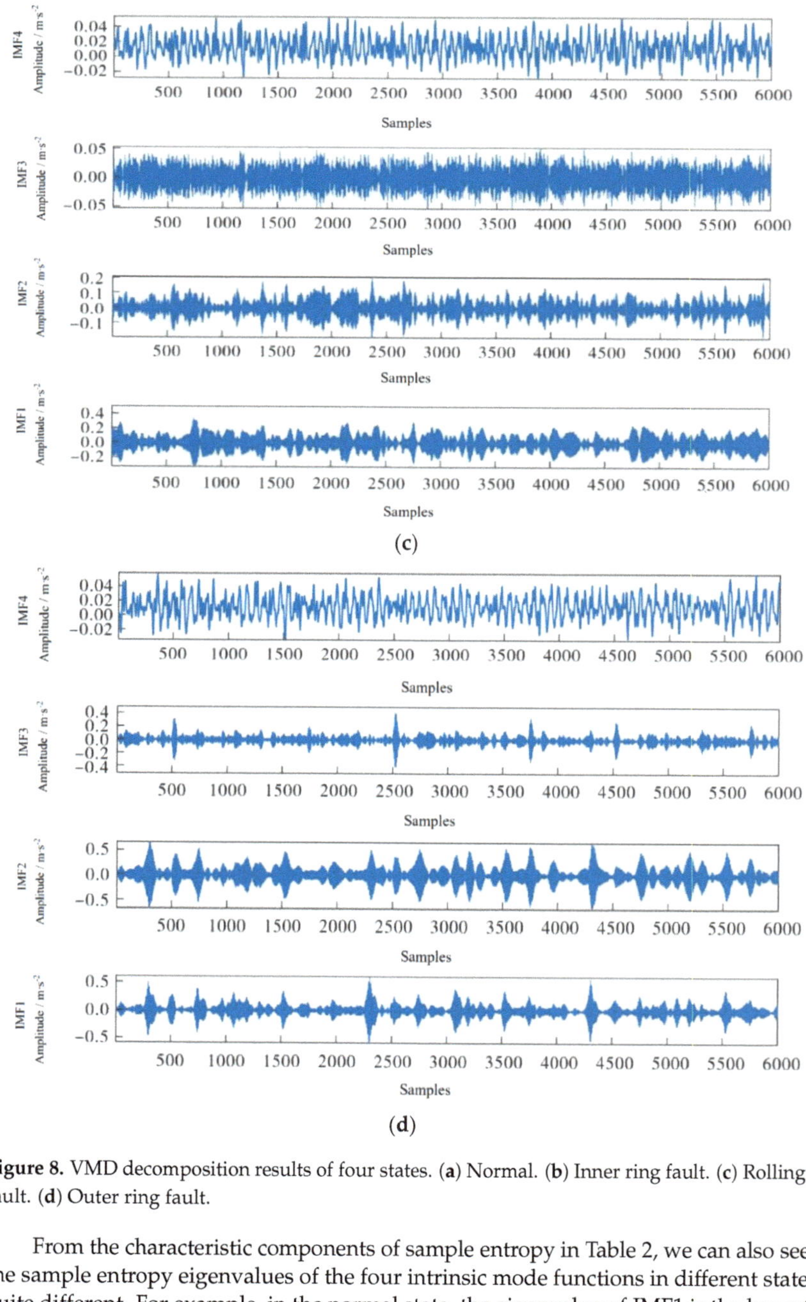

Figure 8. VMD decomposition results of four states. (**a**) Normal. (**b**) Inner ring fault. (**c**) Rolling body fault. (**d**) Outer ring fault.

From the characteristic components of sample entropy in Table 2, we can also see that the sample entropy eigenvalues of the four intrinsic mode functions in different states are quite different. For example, in the normal state, the eigenvalue of IMF1 is the lowest and IMF4 is the highest among the four states; the IMF1 component has the highest eigenvalue in the inner ring fault state; the IMF1 and IMF2 have relatively high eigenvalues in the rolling body fault state; in the outer ring fault state, the eigenvalue is the lowest among the four states and the highest in the four states of IMF2. From the above analysis, it can be concluded that the vibration signal is decomposed into the IMF component by the VMD algorithm, and the sample entropy characteristic value of the IMF component has a high degree of identification and discrimination. Therefore, the sample entropy characteristic index based on VMD decomposition can be used as the eigenvalue of the bearing.

3.2. Result of Fault Identification Based on IBA-SMO Algorithm

In the experiment, the feature extracted sample set is divided into a training set and verification set, and the training set is input to the ISMO model for training. According to the improved bat algorithm (IBA), the optimal penalty factor and "Gaussian kernel function parameter" of the ISMO model are obtained while training the ISMO model parameters of the training set samples. The verification set validates the trained model and verifies its ability to classify fault types. The set parameters of IBA algorithm are shown in Table 3.

Table 3. Parameter setting for IBA.

Population Size	Population Dimension	Number of Iterations	Loudness Factor	Search Range of σ	Search Range of γ
50	5	100	1.5	1~100	1~100

Three hundred sets of samples are selected from each group as the training, and the rest as the prediction set. Then, all the training sets are input into the IBA-ISMO algorithm, and the values of the penalty factor ζ and kernel function parameter σ of the best fitness are obtained by IBA algorithm. The iterative process and the changing process of the evaluation function are shown in Figure 9. As can be seen from Figure 9, the evaluation function in IBA algorithm constantly calculates the fitness value produced by the matching of different penalty factor γ and kernel function parameter σ, and the fitness increases with the increase of the number of iterations until the optimal fitness value is obtained when the maximum number of iterations is close to the maximum number of iterations. At this time, the output fitness $\sigma = 87.63$, $\gamma = 5.78$.

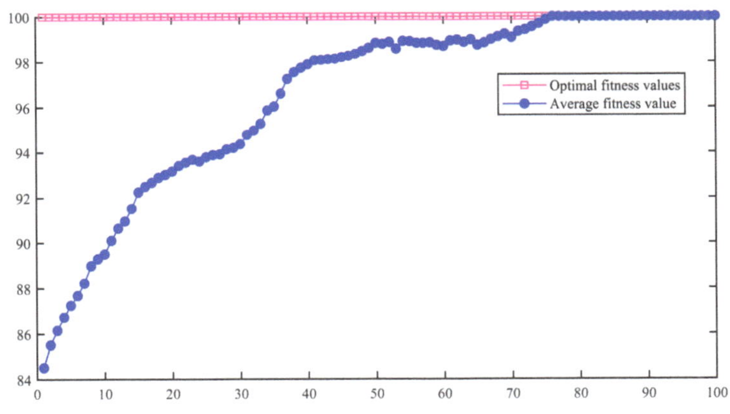

Figure 9. The optimal fitness curve of the improved bat algorithm.

In order to better prove that the improved sequence minimum optimization algorithm has a significant improvement in classification accuracy, firstly, the penalty factor and kernel function parameter obtained by IBA algorithm are input into the ISMO model as input parameters, and the optimal classification surface of the sample set is obtained. As shown in Figure 10, the optimal classification plane has completely separated different faults.

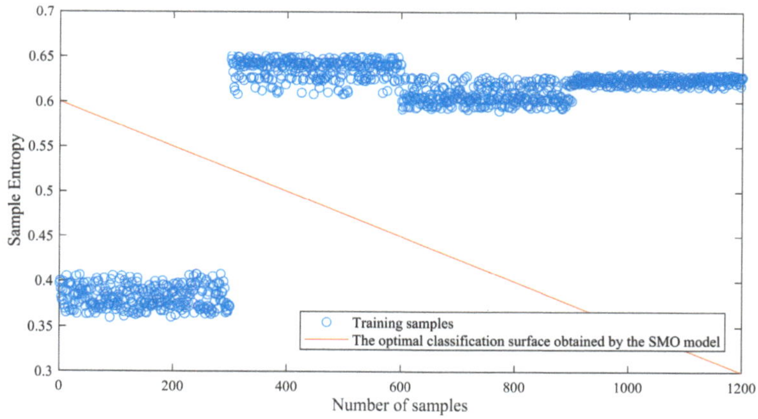

Figure 10. The optimal classification surface of the SMO model.

Figure 11 shows that IBA optimizes the fault identification accuracy of the traditional SMO model. It can be seen that there are misjudgments in some test sets, although the overall fault identification rate is 95.5%. Using the IBA-ISMO algorithm introduced in this study to re-train the training set samples and re-input the test set samples into the model derived by the IBA-ISMO algorithm, and the verification results are shown in Figure 12. Among them, class labels from 1 to 4 represent the normal state, inner ring fault, rolling fault and outer ring fault, respectively.

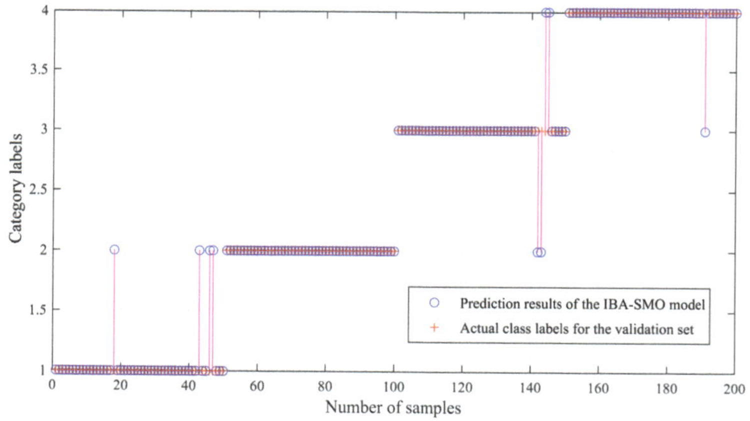

Figure 11. IBA-SMO Model verification.

The results of the validation set input into the different models are shown in Table 4. It can be seen from Table 4 that the accuracy of the IBA-ISMO model is significantly higher than that of other models except PSO-ISMO, so it shows that the IBA-ISMO model can better identify the faults and can be effectively applied to the fault diagnosis of rolling bearings.

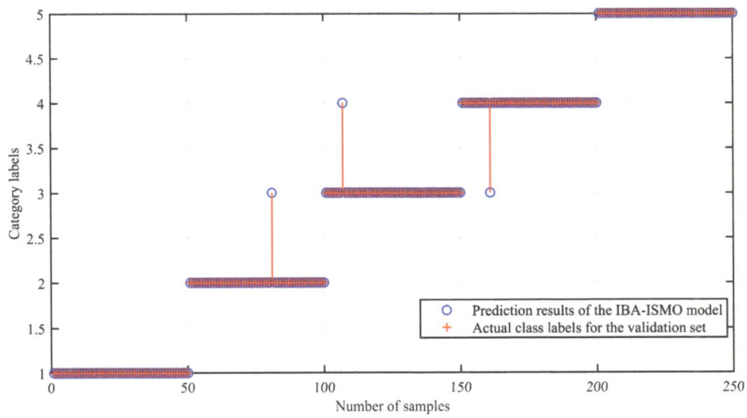

Figure 12. IBA-ISMO Model verification.

Table 4. Model recognition results.

Model Types	Inner Ring Fault	Rolling Fault	Outer Ring Fault	Normal State	Overall Accuracy Rate	Training Time (s)
BA-SMO	90%	92%	90%	96%	92%	5.94
GA-SMO	90%	96%	90%	90%	91.5%	6.65
PSO-SMO	96%	98%	96%	94%	96%	8.99
BA-ISMO	94%	100%	96%	98%	96%	3.35
GA-ISMO	92%	98%	94%	98%	95.5%	4.52
PSO-ISMO	100%	98%	96%	100%	98.5%	7.85
IBA-SMO	92%	100%	92%	98%	95.5%	6.36
IBA-ISMO	100%	98%	98%	98%	98.5%	5.58

4. Conclusions

Aiming at the fault characteristics of rolling bearings, a feature extraction algorithm based on variational modal decomposition and sample entropy has been proposed, and most importantly, an improved fault identification method, IBA-ISMO, was proposed in this study. Using the CWRU data set as a sample set to verify the IBA-ISMO, it is confirmed that the method has a higher fault recognition rate than the comparison method, while the effectiveness of feature extraction for instability vibration signals has been indirectly proven. The main work of this research is as follows:

(1) The VMD algorithm is employed to adaptively decompose the characteristic frequency of the original signal to obtain its specific frequency bandwidth, and the sample entropy is used to extract the characteristics of the IMF component, highlighting the fault information.
(2) An improved bat optimization is designed to optimize the classifier's parameters, which avoids the disadvantages of falling into local optimal solutions compared with the traditional BA algorithm.
(3) The research improves the Gaussian kernel function coefficient of the traditional SMO method, which effectively reduces the classification error rate and optimizes the algorithm's time complexity by reducing the solution vectors that meet the boundary conditions in the data samples.

It should be noted that effective features are very beneficial for fault diagnosis. In this study, only the variational mode decomposition is performed on the signal, and the sample entropy of the component is used as the fault feature. The follow-up research will focus on the fault characteristics and fault phenomena. On this basis, an in-depth analysis of the interpretability of deep learning methods will be carried out.

Author Contributions: Conceptualization, D.Z. and H.L.; methodology, H.Z.; software, Z.G.; validation, H.Z., L.X. and Z.G.; formal analysis, D.Z.; investigation, D.Z. and L.X.; resources, J.Q.; data curation, H.Z.; writing—original draft preparation, L.X.; writing—review and editing, Z.G.; visualization, Z.G.; supervision, G.C. and J.Q.; project administration, J.Q.; funding acquisition, J.Q. All authors have read and agreed to the published version of the manuscript.

Funding: This research was supported by the Science Foundation of China Coal Technology and Engineering Group Corp (China) funded by Tiandi Science & Technology Co. Ltd. (Grant No. 2022-2-TD-QN005), and the Science Foundation of China Coal Technology and Engineering Group Shanghai Company Ltd. (Grant No. 2021-TD-MS005).

Institutional Review Board Statement: Not applicable.

Informed Consent Statement: Not applicable.

Data Availability Statement: This data comes from the public data set of Case Western Reserve University, and the link is "https://engineering.case.edu/bearingdatacenter/apparatus-and-procedures (accessed on 7 August 2022)".

Conflicts of Interest: The authors declare that they have no conflict of interest to report regarding the present study.

References

1. Salahshoor, K.; Khoshro, M.S.; Kordestani, M. Fault detection and diagnosis of an industrial steam turbine using a distributed configuration of adaptive neuro-fuzzy inference systems. *Simul. Model. Pract. Theory* **2011**, *19*, 1280–1293. [CrossRef]
2. Xu, Y.; Li, S.L.; Zhang, D.Y. Identification framework for cracks on a steel structure surface by a restricted Boltzmann machines algorithm based on consumer-grade camera images. *Struct. Control. Health Monit.* **2018**, *25*, e2075. [CrossRef]
3. He, Z.; Shao, H.; Zhong, X.; Yang, Y.; Cheng, J. An intelligent fault diagnosis method for rotor-bearing system using small labeled infrared thermal images and enhanced CNN transferred from CAE. *Adv. Eng. Inform.* **2020**, *46*, 101150.
4. Song, L.Y.; Wang, H.Q.; Chen, P. Vibration-based intelligent fault diagnosis for roller bearings in low-speed rotating machinery. *IEEE Trans. Instrum. Meas.* **2018**, *67*, 1887–1899. [CrossRef]
5. Chen, X.; Cheng, G.; Li, H.; Li, Y. Research of planetary gear fault diagnosis based on multi-scale fractal box dimension of CEEMD and ELM. *Stroj. Vestn.-J. Mech. Eng.* **2017**, *63*, 45–55. [CrossRef]
6. Zhao, X.; Ye, B. Selection of effective singular values using difference spectrum and its application to fault diagnosis of headstock. *Mech. Syst. Signal Process.* **2011**, *25*, 1617–1631. [CrossRef]
7. Zhang, Q.S.; Ding, J.M.; Zhao, W.T. An adaptive boundary determination method for empirical wavelet transform and its application in wheelset-bearing fault detection in high-speed trains. *Measurement* **2021**, *171*, 108746. [CrossRef]
8. Park, C.S.; Choi, Y.C.; Kim, Y.H. Early fault detection in automotive ball bearings using the minimum variance cepstrum. *Mech. Syst. Signal Process.* **2013**, *38*, 534–548. [CrossRef]
9. Borghesani, P.; Pennacchi, P.; Randall, R.B.; Sawalhi, N.; Ricci, R. Application of cepstrum pre-whitening for the diagnosis of bearing faults under variable speed conditions. *Mech. Syst. Signal Process.* **2013**, *36*, 370–384. [CrossRef]
10. Xu, Y.G.; Zhang, K.; Ma, C.Y.; Li, X.; Zhang, J. An improved empirical wavelet transform and its applications in rolling bearing fault diagnosis. *Appl. Sci.* **2018**, *8*, 2352. [CrossRef]
11. Chen, X.H.; Cheng, G.; Shan, X.L.; Hu, X.; Guo, Q.; Liu, H.G. Research of weak fault feature information extraction of planetary gear based on ensemble empirical mode decomposition and adaptive stochastic resonance. *Measurement* **2015**, *73*, 55–67. [CrossRef]
12. Wang, C.L.; Zhang, C.L.; Zhang, P.T. Denoising algorithm based on wavelet adaptive threshold. *Phys. Procedia* **2012**, *24*, 678–685.
13. Wang, Z.P.; Jia, L.M.; Qin, Y. Adaptive diagnosis for rotating machineries using information geometrical kernel-ELM based on VMD-SVD. *Entropy* **2018**, *20*, 73. [CrossRef]
14. Miao, Y.; Zhao, M.; Yi, Y.; Lin, J. Application of sparsity-oriented VMD for gearbox fault diagnosis based on built-in encoder information. *ISA Trans.* **2020**, *99*, 496–504. [CrossRef] [PubMed]
15. Fan, Y.R.; Zhang, C.; Xue, Y.; Wang, J.; Gu, F. A bearing fault diagnosis using a support vector machine optimised by the self-regulating particle swarm. *Shock Vib.* **2020**, *2020*, 9096852. [CrossRef]
16. Runelhart, D.E.; Hinton, G.E.; Williams, R.J. Learning representations by back-propagating errors. *Nature* **1986**, *323*, 533–536. [CrossRef]
17. Lu, Q.; Yang, R.; Zhong, M.; Wang, Y. An improved fault diagnosis method of rotating machinery using sensitive features and RLS-BP neural network. *IEEE Trans. Instrum. Meas.* **2020**, *69*, 1585–1593. [CrossRef]
18. He, X.H.; Wang, D.; Li, Y.F.; Zhou, C.H. A novel bearing fault diagnosis method based on gaussian restricted Boltzmann machine. *Math. Probl. Eng.* **2016**, *2016*, 2957083. [CrossRef]

19. Wang, H.; Xu, J.W.; Yan, R.Q.; Gao, R.X. A new intelligent bearing fault diagnosis method using SDP representation and SE-CNN. *IEEE Trans. Instrum. Meas.* **2020**, *69*, 2377–2389. [CrossRef]
20. Smith, W.A.; Randall, R.B. Rolling element bearing diagnostics using the Case Western Reserve University data: A benchmark study. *Mech. Syst. Signal Process.* **2015**, *64–65*, 100–131. [CrossRef]

sensors

MDPI

Article

Degradation Feature Extraction Method for Prognostics of an Extruder Screw Using Multi-Source Monitoring Data

Jun-Kyu Park [1], Howon Lee [2,3], Woojin Kim [2], Gyu-Man Kim [3] and Dawn An [2,*]

1 Renewable Energy Solution Group, Korea Electric Power Research Institute, Naju 58277, Republic of Korea
2 Advanced Mechatronics R&D Group, Korea Institute of Industrial Technology, Daegu 42994, Republic of Korea
3 School of Mechanical Engineering, Kyungpook National University, Daegu 41566, Republic of Korea
* Correspondence: dawnan@kitech.re.kr

Abstract: Laboratory-scale data on a component level are frequently used for prognostics because acquiring them is time and cost efficient. However, they do not reflect actual field conditions. As prognostics is for an in-service system, the developed prognostic methods must be validated using real operational data obtained from an actual system. Because obtaining real operational data is much more expensive than obtaining test-level data, studies employing field data are scarce. In this study, a prognostic method for screws was presented by employing multi-source real operational data obtained from a micro-extrusion system. The analysis of real operational data is more challenging than that of test-level data because the mutual effect of each component in the system is chaotically reflected in the former. This paper presents a degradation feature extraction method for interpreting complex signals for a real extrusion system based on the physical and mechanical properties of the system as well as operational data. The data were analyzed based on general physical properties and the inferred interpretation was verified using the data. The extracted feature exhibits valid degradation behavior and is used to predict the remaining useful life of the screw in a real extrusion system.

Keywords: degradation feature; data processing; prognostics; screw; extrusion system; real operational data; multi-source data; structural health monitoring

Citation: Park, J.-K.; Lee, H.; Kim, W.; Kim, G.-M.; An, D. Degradation Feature Extraction Method for Prognostics of an Extruder Screw Using Multi-Source Monitoring Data. *Sensors* **2023**, *23*, 637. https://doi.org/10.3390/s23020637

Academic Editors: Branko Glisic and Jongmyon Kim

Received: 15 November 2022
Revised: 13 December 2022
Accepted: 30 December 2022
Published: 5 January 2023

1. Introduction

Structural health monitoring (SHM) has been employed in various fields as a cost-effective maintenance strategy based on sensing systems, such as acoustic emission [1,2], piezoelectric [3,4], vibration [5,6], and multi-source sensors [7–9]. SHM data are utilized in diagnostics and prognostics. In diagnostics, they detect, isolate, and identify the damage and/or defect of a structure; in prognostics, they predict degradation behavior and remaining useful life (RUL) of an in-service system. In recent decades, prognostics have been studied in various engineering applications such as bearings [6], aircraft engines [7], batteries [8], and fuel cell stacks [9]. Recently, Huang et al. [6] developed novel methods for bearing prognostics but employed the open-source bearing datasets that are widely used for bearing prognostics. Studies on RUL prediction of aircraft engines by Liu et al. [7] used simulation datasets of turbofan engine degradation. This type of dataset is typically generated using simulation tools such as the commercial modular aero-propulsion system Simulation [10]. Zhang and Li [8] recently provided a detailed summary of lab-scale datasets used in the field of lithium-ion batteries. Marine et al. [9] conducted a prognostic study of a fuel cell (FC) stack using datasets obtained from aging tests, determining the effect of high-frequency current ripples on FC stack durability.

As mentioned above, most existing studies used data obtained under laboratory conditions; the datasets were obtained in an easy, fast, and cost-effective manner using an accelerated test or a well-organized test plan. Before applying the developed prognostic method to a real industrial field, it is essential to employ a simple dataset for pilot testing purposes. However, the prognostic method has rarely been validated with real operational data because obtaining field data is time-consuming and expensive. Run-to-failure data are required to validate the prognostic method; however, the life span of a real operating system is several years or even decades. Considering the differences in the process and operating conditions used to obtain laboratory and field data, the importance of employing real operational data cannot be sufficiently emphasized.

In this study, real operational data were obtained from a micro-extrusion system for medical tube-grade catheters. To the best of our knowledge, in addition to using real operational data, no study has addressed the prognosis of the extrusion system or its components. It is important to maintain the quality of the product, particularly for medical catheters, as they can be applied to the human body. There have been long-standing efforts to maintain the extrusion product quality itself [11–15]; however, the maintenance strategy for the extrusion system is rather simplified, although its health condition influences the quality of products. Additional extrusion processes are performed as a maintenance strategy, which is periodically conducted under controlled conditions that restrict the process variables, raw materials, etc. [16]. Although this method is intuitive, it involves cumbersome operations that require additional time and money. Therefore, a prognostic method for an extrusion system is presented based on real operational data to monitor the health condition of the system without additional tasks.

The extrusion system consists of a variety of components, including the motor, screw, barrel, and puller, and its health condition is affected by various failure modes of the components. Among them, screw wear is the most common failure mode of the extrusion system. Further, it affects the mixing quality and temperature of the molten polymer, thereby affecting the quality of the products (catheters) [17,18]. The quality of products can be maintained under moderate wear of the screw by adjusting the process variables but cannot be maintained under severe wear of the screw. However, it is not easy to distinguish whether the level of screw wear is moderate or severe because it progresses gradually, and the quality of products also depends on other factors such as operator and environmental conditions. Inadequate control of the extrusion process due to wear can cause variations in product quality [19,20], which can lead to raw material wastage, increased energy consumption, and environmental pollution [21,22]. Therefore, this study aims to predict the RUL of the extrusion screw based on real operational data so that a timely replacement is performed before the screw under inadequate performance deteriorates the quality of the products.

The analysis of real operational data is more challenging than that of test-level data because it is obtained at the system level, wherein the mutual effect of each component on the system is reflected chaotically. Therefore, the main contribution of this study is the extraction of the degradation feature to monitor the wear of the screw used in a real extrusion system, which is based on the physical and mechanical properties of the system as well as the operational data. The extracted feature exhibits valid degradation behavior and is used to predict the RUL of the extrusion screw during its lifespan.

The remainder of this paper is organized as follows. In Section 2, the medical catheter extrusion system and the operating data are introduced. In Section 3, the process of extracting the degradation features of screw wear based on the physical interpretation of the extrusion system and experimental data is explained, and the features are applied to real operational data. In Section 4, the application of the extracted degradation features to predict the RUL of an extrusion screw is described. Finally, a brief conclusion is presented.

2. Extrusion System

An extrusion system is a continuous production system that uses polymer melts manufactured through frictional heat between a cylinder and a screw to produce a tube using a mold composed of a tip and die. An extrusion system is composed of several main components, such as a hopper, screw and barrel, tip and die, quenching system, vacuum water tank, measurement device, puller, cutter, and conveyor system. The actual micro-extrusion system (Davis-Standard Inc., Fulton, NY, USA) is shown in Figure 1. This provided the operational data used in this study. Detailed information on the extrusion process and operating data is provided in the following subsections.

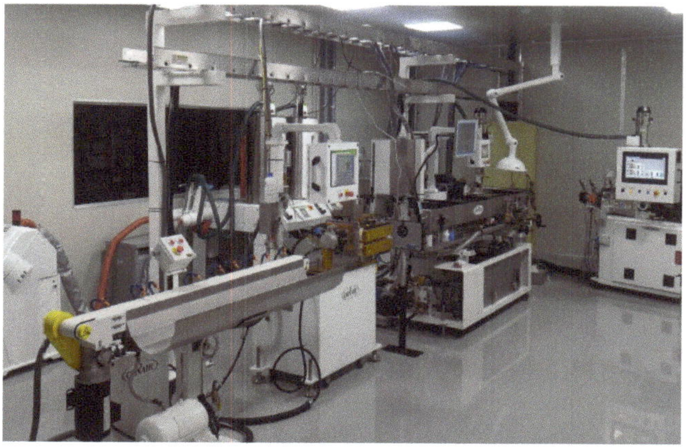

Figure 1. Medical multi-lumen tubing extrusion system.

2.1. General Extrusion Process and System Configuration

The extrusion system shown in Figure 1 is illustrated in Figure 2 to introduce the general extrusion process. After drying and dehumidifying, the polymer is injected into the barrel equipped with a screw through the hopper of the extruder and melted by frictional heat between the screw, polymer, and inner wall of the barrel. The initial shape of the tube is generated as the polymer passes through the tip and dies by adjusting the rotating speed of the screw, and the pressure of air injected into the lumen controls the ovality and shape of the tube. The lumen of the tube is stably hardened in a quenching part and vacuum tank filled with water, and the size of the tube is precisely controlled by adjusting the puller speed. The final tubes are produced on a conveyor system after the consecutive tubes are cut.

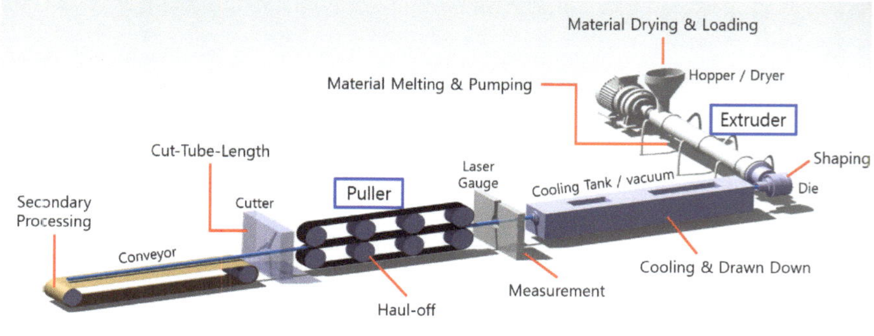

Figure 2. Illustration of the entire extrusion system.

Many factors affect the quality of products during complex extrusion processes. The screw wear-related parts are illustrated in Figure 3 with the locations of the on-board sensors, which are used to monitor the condition of each component. The components are classified into two large groups: the extruder part, which includes the motor, screw in the barrel, tip, and die; and the puller part, which includes the puller itself. Four types of sensing signals (motor load, head pressure, and screw and puller speeds) were selected from the many on-board sensors attached to the actual extrusion system by considering their relationship to screw wear. In Figure 3, the numbers represent the sequence of the operational flow: 1. The screw speed setting (called S.Spd.Set in diagrams and calculations) is a control parameter that is adjusted to make the quality of the catheter close to the final product; 2. The motor operates by following the screw-speed setting. The motor load is denoted by M.Load; 3. The motor rotates the screw in the barrel, and the actual screw speed (S.Spd.Act) is monitored; 4. The rotating screw extrudes the polymer melt in the barrel to the tip and die, and the head pressure (H.Press) at this time is monitored. The coarsely shaped catheter is then moved to puller: 1*. the puller speed setting (P.Spd.Set), another control parameter, is precisely adjusted to make the catheter of high quality; 2*. the puller motor runs the puller following the puller speed setting, and the actual puller speed (P.Spd.Act) is measured. In this study, the operational data of the system were analyzed based on the operational mechanism shown in Figure 3.

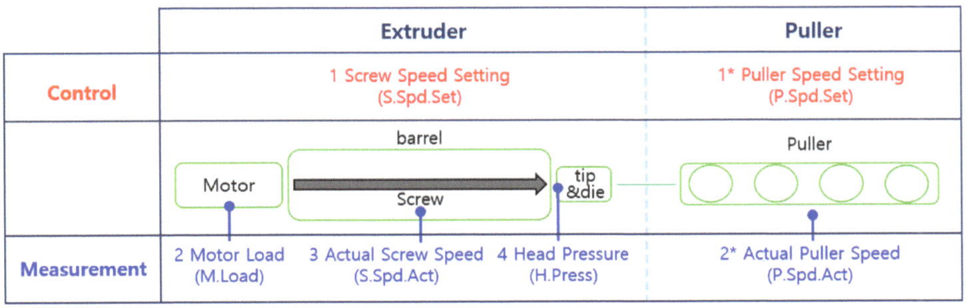

Figure 3. Screw wear-related parts and sensing location.

2.2. Data Description

The extrusion system shown in Figure 1 has been operational since March 2017, and no component has been replaced (but minimum maintenance was conducted), except for screws. Two types of screws were used in the same extrusion system according to the polymer series. The screws used for the polyurethane/polyester and polyamide series are called barrier and (spiral) Maddock screws, respectively. The barrier screw has been in use since it was replaced in July 2020, and the Maddock screw has been in use since the start of the operation and is now severely worn. The operational data used in this study were obtained from July 2020 to March 2022. In summary, the barrier screw is likely to have mild wear, whereas the Maddock screw has reached its end-of-life (EOL). Therefore, the Maddock screw was considered the target component for prognostics in this study.

The extrusion system was operated for approximately eight hours per day on weekdays. The real operational data for all the measurements are shown in Figure 4a. In the figure, the bars represent the monitoring data for each day, but these raw data are not easy to interpret. This is the reason why feature extraction is required. The operational data for one day (green box) are shown in Figure 4b. In the figure, the extrusion process data can be categorized into two types, as shown in Figure 3: control (colored red) and measurement (colored blue) data. As shown in Figure 4b, all measurement values were changed in real time by following operator-controlled speed settings.

As shown in Figure 4b, the datasets were obtained in two different ways according to the operational conditions. First, a component-level experiment was conducted to monitor the effect of screw wear on extrusion data by minimizing other effects, such as motor degradation and processing variables. Next, real operational data were collected from an actual extrusion system that produces medical catheters over a long period. A more detailed explanation of the data is provided in the following subsection.

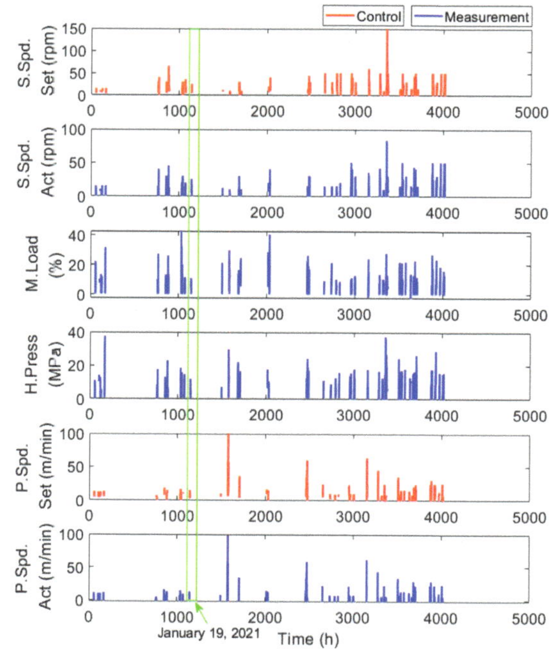

(a)

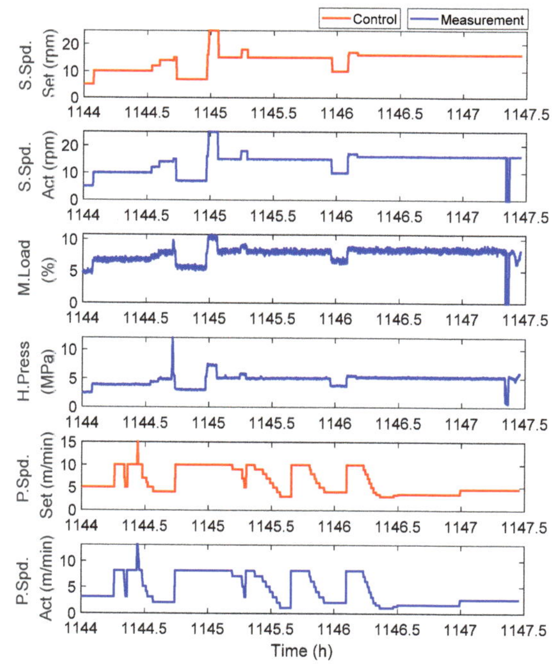

(b)

Figure 4. Real operational data: (**a**) All measurements for the Maddock screw; (**b**) Extrusion data for the Maddock screw collected on 19 January 2021.

2.2.1. Screw Experimental Data

Experimental screw wear data were obtained from three screws with different levels of wear under the same production conditions that were used to produce the same type of single-lumen catheters with an inner diameter of 2.0 mm and outer diameter of 2.5 mm using the same polymer (carbothane PC-3595A, Lubrizol, Cleveland, OH, USA) over two days. It should be noted that screw wear naturally progresses during actual use, as shown in Figure 5. In the figure, wear levels 1, 2, and 3 correspond to the intact, moderately worn, and completely worn screws, respectively. The experimental data, which are similar to the signals in Figure 4, are analyzed in Section 3.

(a)
(b)
(c)

Figure 5. Three screws used in the screw wear experiments: (**a**) wear level 1: intact; (**b**) wear level 2: moderate wear; and (**c**) wear level 3: severe wear.

2.2.2. Real Operational Data

According to the production schedule, one of the barrier and Maddock screws were used in the same extrusion system shown in Figure 1, which means that the system and two screws have different life spans. Even though the operation of the extrusion system and the use of the Maddock screw started in March 2017, the time when the monitoring started (July 2020) was assumed to be the initial cycle. Consequently, the total operating times of the extrusion system and Maddock screw were approximately 4000 h and 1200 h, respectively. The cumulative usage time of the barrier screw was approximately 2800 h (which is the actual cumulative usage time rather than the assumed life of the system and Maddock screw), but it was not considered in this study. During the extrusion process, catheters with various specifications were produced using different polymer materials. These diversities made it difficult to extract the degradation features of screw wear from monitoring signals. In the following section, the robust wear feature was extracted regardless of the type of catheter produced and the material used.

3. Degradation Feature Analysis in the System

To predict the RUL of a screw, the degradation feature must first be extracted in the form of a monotonic increase or decrease. However, it is challenging to extract degradation features from raw data, particularly from real operational data, as shown in Figure 4a. Moreover, it may take several decades to obtain sufficient data from an actual system for prognostics, including for feature extraction. Therefore, data and physical information are complementarily used; raw data are analyzed based on physical and mechanical interpretation, and the inferred interpretation is, in turn, verified by the data.

3.1. *Physical and Mechanical Properties of the Extrusion System*

Six signals were recorded, including the control and measurement signals, as shown in Figure 3. Among the six signals, the setting and actual speeds were examined first, as shown in Figure 6. In the figure, the blue solid and red dashed lines represent the setting and actual speed, respectively. They were very close to each other during the normal extrusion process between the two green vertical lines at both screw and puller speeds. Thus, the health condition of the motor, monitored until the current time, had no valid effect on the following process (refer to Figure 3). In other words, the head pressure was not affected by motor degradation because the actual screw speed followed the set value well, regardless of the motor condition.

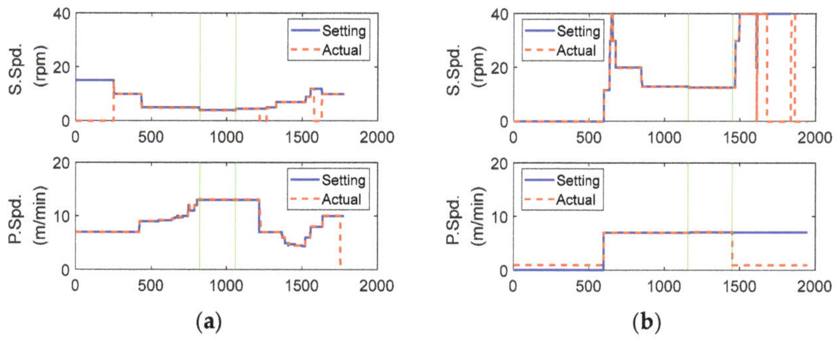

Figure 6. Difference between setting and actual speeds (Maddock): (**a**) data monitored on the first day (5 July 2020); (**b**) data monitored on the last day (25 March 2022).

However, the effect of motor degradation was reflected in the change in motor load, which usually increased as the motor degraded [23]. Screw wear was also related to the motor load, which decreased as the screw wore out because the screw speed increased owing to the reduced radius of the blade [24]. Note that the motor load decreased with increasing screw speed caused by wear but increased with an increase in the setting speed of the screw. Thus, the motor load reflected at least three aspects: the screw speed setting, screw wear, and motor degradation. This explains why the data analysis at the system level is complex and difficult.

The motor load was closely related to the mechanical aspect, whereas the head pressure and puller speed were more closely related to catheter quality. As screw wear progressed, head pressure typically decreased, and the screw and puller speeds were appropriately adjusted by the operator to maintain catheter quality. However, as shown in Figures 3 and 4, screw speed had a greater effect on the overall system control than the quality of catheters. In general, to maintain catheter quality, puller speed should increase with a decrease in head pressure, which is based on the law of conservation of energy. Maintaining the same catheter quality required the same energy; however, energy loss occurred because of a decrease in head pressure as the screw wore out over time. Thus, the puller speed must be increased to compensate for lost energy.

To summarize the above two paragraphs, screw wear over time (t) can be expressed as follows:

$$d(t) = L(t)A \times \frac{P(t)}{V(t)} \tag{1}$$

where L, P, V, and d represent the motor load, head pressure, puller speed, and screw degradation level, respectively. The motor load in Equation (1) reflects the degradation of both the screw and motor and decreases or increases depending on the predominance of the two degradations. However, motor degradation was not considered in this study because it is negligible compared with the wear level of the Maddock screw. The head pressure and puller speed decreased and increased, respectively, as the screw wore out. In conclusion, each term on the right side of Equation (1) is an indicator that can be used to monitor screw deterioration. When screw deterioration is dominant, Equation (1) clearly shows a gradual decrease. The degradation feature of the screw wear in Equation (1) was verified and improved using experimental and real operating data in the following subsections.

3.2. Experimental Data Analysis

The most common phenomenon caused by screw wear is a decrease in the head pressure, which was verified through experiments, as shown in Figure 7a. The black and red signals were obtained from the screws of wear levels 1 and 3, respectively, as shown in Figure 5. This result was obtained experimentally by letting the polymer melt flow down at the tip and die with a fixed screw speed for both the intact and worn screws. As shown in

Figure 7a, the head pressure of the intact screw (black) is approximately 2 MPa higher than that of the worn screw (red). Although it was obvious that the head pressure decreases with screw wear, the data from the real operating process for producing catheters, as shown in Figure 7b, did not support this conclusion. In the figure, the head pressures of the intact (black) and worn (red) screws are similar, and it is difficult to distinguish which dataset corresponds to the worn screw from the signals. This is because the operation settings, such as the screw and puller speeds, were adjusted to maintain product quality under different health conditions of the screws.

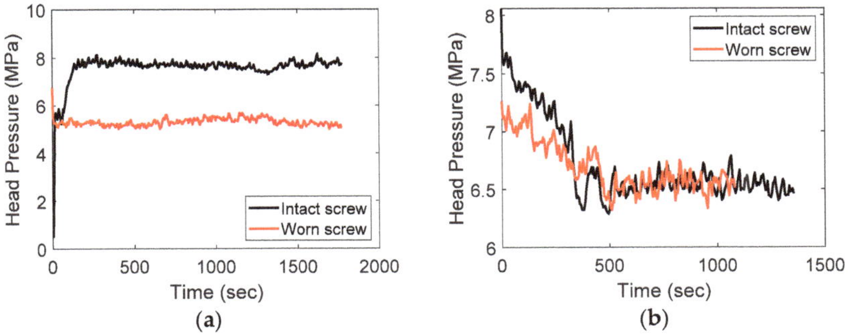

Figure 7. Head pressure change according to screw wear: (**a**) no catheter production; (**b**) during catheter production.

The experimental results using the three screws in Figure 5 are shown in Figure 8, where the black, blue, and red colors represent wear levels 1, 2, and 3, respectively. The operational data during normal extrusion are shown in Figure 8a, and the corresponding averages are shown in Figure 8b. The y-axes of the four results in Figure 8b were scaled to a 20% difference between the maximum and minimum values. In the figure, significant changes in the motor load and puller speed can be observed, whereas the head pressure did not show much change (note that the screw speed was fixed at 10.7 rpm).

As mentioned in Section 3.1, the motor load was affected by the screw speed setting and degradation of both the motor and screw. The screw speed was fixed during the experiment. In addition, the deterioration in motor performance cannot be reflected because the experiment was performed for two consecutive days. Therefore, we can conclude that the reduction in motor load in Figure 8b was caused only by screw degradation. However, Figure 8b does not show the expected result of the puller speed increasing continuously as the wear level increases. Despite the obvious screw wear, the puller speed increased at wear level 2, but decreased at wear level 3. To understand this phenomenon, an unexpected inflow of energy, such as fluctuation energy, was investigated by considering the severity of wear level 3.

The production quality of the catheters could not be maintained when a screw with a wear level of 3 was used. This was demonstrated by a flow-rate test and the quality of the produced catheter. The flow rate was measured for one minute at a fixed screw speed of 10.7 rpm according to the wear level listed in Table 1. The tests were performed three times each, and the average and standard deviation were listed with the rate of increase relative to wear level of 1. The flow rate was expected to increase as the physical space increased owing to an increase in screw wear; however, the flow rate at wear level 3 was lower than the flow rate at wear level 2. This is because the melted polymer adheres to the screw surface and prevents it from falling. The severity of wear level 3 led to a large variation in not only the standard deviation of the flow rate (a 652% increase over wear level 1), but also in the quality of the catheter, as shown in Figure 9. In the figure, the yellow circles are the catheters produced with the three wear levels of the screws, and the red circles that are the same for each figure represent the ideal shape and size of the catheter. As shown in the figure, the catheter produced using the screw of wear level 1 was very close to the red circle, and the catheter of wear level 2 was slightly distorted in the upper part (between two green lines). On the other hand, the catheter using the screw with wear level 3 showed a large discrepancy between the product and the right half of the red circle (between the two green lines). In conclusion, the wear level 3 screw was not suitable for producing a fair quality catheter. Therefore, the useful life of the screw should be considered between wear levels 2 and 3.

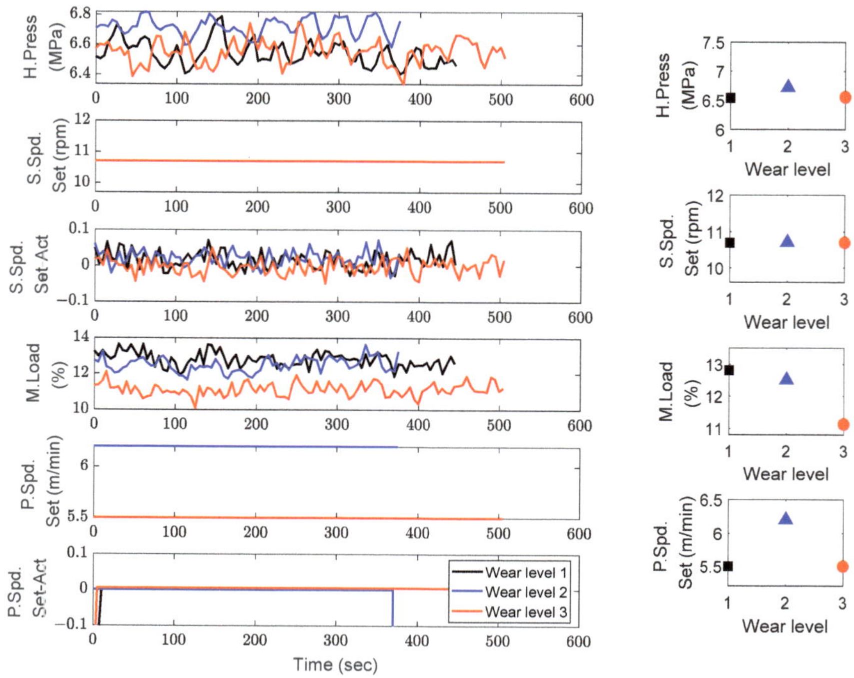

(**a**) (**b**)

Figure 8. Experimental data according to the level of screw wear, keeping catheter production and materials the same: (**a**) operational data; (**b**) mean of the operational data.

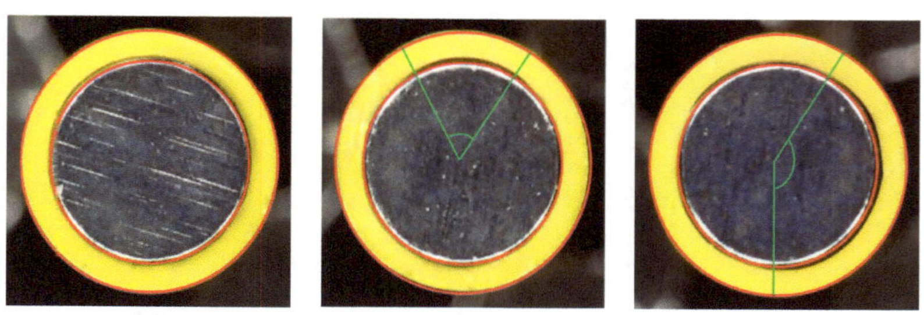

(**a**) (**b**) (**c**)

Figure 9. Quality of catheters according to the level of screw wear: (**a**) wear level 1; (**b**) wear level 2; (**c**) wear level 3.

Table 1. Results of the flow rate measurement test according to screw wear level.

							Unit: [g/min]
Screw Wear Level	Test 1	Test 2	Test 3	Average		Standard Deviation	
				Amount	Ratio to Wear L. 1	Amount	Ratio to Wear L. 1
1	13.88	13.85	13.89	13.87	-	0.021	-
2	15.67	15.64	15.66	15.66	12.9%	0.015	−28.6%
3	14.17	14.27	14.48	14.30	3.1%	0.158	652%

Before employing real operational data, an additional aspect should be considered. To date, experimental data have been obtained under conditions that produce the same types of catheters at a fixed screw speed using the same polymer. Because various types of catheters and polymers are used during actual operation, the degradation feature in Equation (1) must be modified to reflect these variabilities. The simplest method is to employ the screw speed setting corresponding to the overall system control as a correction factor as follows:

$$d_{\mathrm{M}}(t) = \frac{1}{R(t)} A \times L(t) \tag{2}$$

$$d_{\mathrm{E}}(t) = \frac{1}{R(t)} A \times \frac{P(t)}{V(t)} \tag{3}$$

$$d_{\mathrm{T}}(t) = \frac{L(t) A \times P(t)}{R(t) A \times V(t)} \tag{4}$$

where d_{M}, d_{E}, d_{T}, and R represent the mechanical, energy, total degradations, and screw speed, respectively. The degradation feature in Equation (1) is divided into the mechanical aspect (d_{M}) in Equation (2) and the energy aspect (d_{E}) in Equation (3) because they show different behaviors at the very late stage of screw life. The total degradation (d_{T}) in Equation (4), which becomes the final degradation feature, is obtained by multiplying the two aspects and avoiding the duplication of the screw speed. The denominators (screw and puller speeds) and numerators (motor load and head pressure) in Equations (2)–(4) increase and decrease, respectively, as the screw wear progresses. To be precise, the head pressure can be maintained during the actual operating process but is not expected to increase.

Consequently, all three aspects of the degradation feature in Equations (2)–(4) are expected to decrease, provided the screw functions properly, even with gradual degradation, as shown in Figure 10. In the figure, the black, blue, and red markers indicate wear levels 1, 2, and 3, respectively. Note that the x-axis ranges from zero to five years. For these plots, the average life span of the screws was assumed to be five years. The time index on the x-axis can vary because the three screws may have different useful lives. The gray curves in Figure 10 depict the expected behavior based on the analysis thus far. The gray star marks the EOL of the screws because screws should be replaced before they deteriorate the catheter quality.

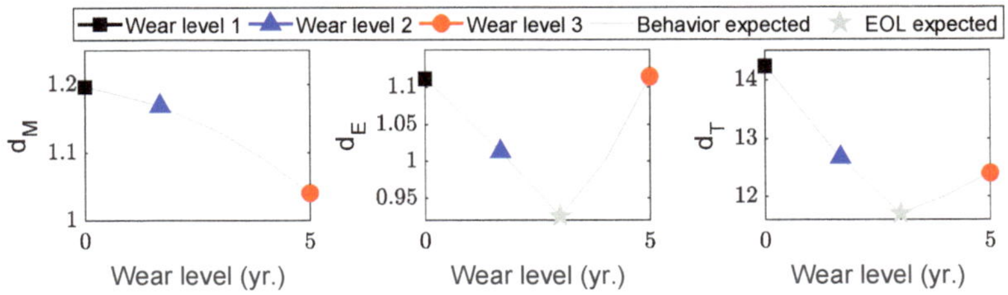

Figure 10. Degradation feature analysis; the material and product were kept the same.

One screw for each wear level was insufficient for validating the degradation feature. Thus, real operational data were applied to the same feature extraction methods, as detailed in the following section.

3.3. Real Operational Data Analysis

The corrected degradation features of the screw wear using Equations (2)–(4) are shown in Figure 11. In the figure, each marker represents the mean of the real operational data for a day, and the linear fitting results are shown with black lines. The barrier and Maddock screws represent the control and test groups, respectively. In addition, the cumulative usage time of the system is used for both the barrier and Maddock screws to compare the changes in the monitoring signals according to the degree of screw wear rather than to consider the usage time of screws. The slope values in Figure 11 show that the amount of variation in the degradation features of the Maddock screw is more than twice that of the barrier screw. This is because the cumulative usage time of the barrier screw corresponds to wear level 2 in Figure 10, which shows little degradation of the screw. In Figure 11a, the positive slope indicates that motor degradation is more dominant than screw wear. However, the negative slope of the Maddock screw in Figure 11b clearly shows the screw wear.

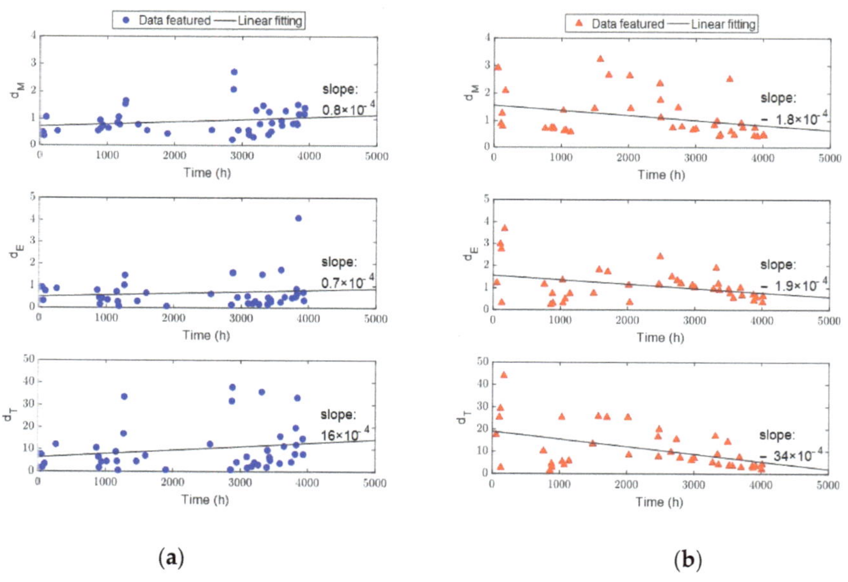

Figure 11. Degradation feature with correction factor using Equations (2)–(4): (**a**) barrier screw; (**b**) Maddock screw.

However, the distribution of the degradation features fluctuated and was scattered, which was not sufficient to predict the degradation behavior and RUL. Therefore, a weighted cumulative average (WCA) was applied to the degradation features shown in Figure 11b to further highlight the degradation characteristics. The weight in the WCA was defined as the value uniformly divided between 1 and 1/n, where n is the number of data obtained up to the current time. For example, when the current time was 3000 h, n = 26 (refer to Figure 11b), and the weights decreased from 1 to 0.0385 (1/n) with a uniform interval of 0.0385 (1/n). Each value of the degradation features in Figure 11b was then multiplied by the calculated weight. Finally, the mean of the weighted feature cumulated from zero to the current time was obtained as the WCA at 3000 h. This process was repeated each time, and the results are shown in Figure 12. As shown in Figure 12a, the WCA of the degradation feature over the entire range contained unstable data with large fluctuations in the early stages of the lifespan. When the full time was reduced after 1000 h, as shown in Figure 12b, the characteristics of each feature could be observed more clearly. The WCA of d_M did not show a monotonic trend, which is a basic condition for degradation behavior. However, the WCA of d_E and d_T decreased continuously, and either of the two could be considered the final degradation feature for the prediction. However, whereas the WCA of d_E decreased almost linearly after 1500 h, the rate of decrease of the WCA of d_T increased. The behavior of the WCA of d_T was an expected characteristic of the general degradation feature. This demonstrated that the degradation feature proposed in Equation (4) was reasonable for describing the wear behavior of the extruder screw. Consequently, the WCA of d_T in Figure 12 was extracted from the raw data in Figure 4a and used as the final degradation feature to predict the RUL of the Maddock screw.

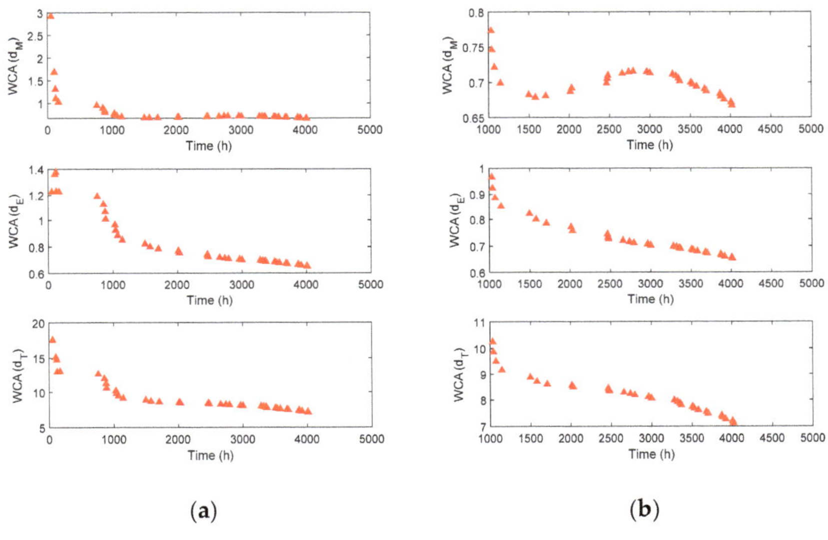

Figure 12. Degradation feature analysis based on proposed methods: (**a**) entire time scale; (**b**) after 1000 h.

4. Prediction Result

The RUL of a component measures the time remaining before repair or replacement. In this study, the component of interest was the Maddock screw, and thus the time label in Figure 12 is important. Therefore, the time label of the WCA of d_T in Figure 12a was converted to the time scale of the Maddock screw, as shown in Figure 13a. During 4000 h of system operation, Maddock screws were used for approximately 1200 h (more accurately, 1128 h). In the figure, only the solid black dots are considered for RUL prediction because the circle markers up to 400 h indicate the data points where the degradation feature differed from the expected behavior. The final data are depicted by dotted markers in Figure 13b.

Once the degradation data are obtained, a degradation model is assumed for the RUL prediction. Paris and Erdogan [25] proved that crack growth behaves exponentially and developed the Paris model, which is a physical degradation model that describes crack growth behavior under different loading conditions. Goebel et al. [26] used the exponential function as an empirical model to describe battery degradation behavior. Because physical models such as the Paris model are rare, exponential functions are generally employed as empirical degradation models in most prognostic studies. Therefore, in this study, it was assumed that the degradation behavior (z) is described by the following equation:

$$z = a + b \times \exp\left(\frac{t}{1000}\right) \tag{5}$$

where a, b, and t are the model parameters and the time, respectively. The model parameters were estimated using the data obtained up to the current time by minimizing the error between the data and model output, z. There are several methods for parameter estimation, and the Bayesian method [27] was used in this study. In the Bayesian method, the model parameters are estimated in the form of a probability density function (PDF). Subsequently, the Markov chain Monte Carlo sampling method was employed to draw samples from the PDF of the model parameters. Consequently, the final model output at time t was obtained by substituting the estimated values (drawn samples) for a and b in Equation (5). For example, the gray curve in Figure 13b is the model output at EOL obtained using all data and the Bayesian method. Because all monitoring data were used, the curve was considered to be the true degradation behavior, excluding the noise in the data. More details on the Bayesian method and implementation code can be found in the book by Kim et al. [27]. Next, a degradation threshold should be determined by considering the trade-off between risk and cost. However, determining the threshold is another specialized field; thus, in this study, the threshold was assumed to be the true wear level at EOL. In Figure 13b, the green horizontal line represents the threshold of 7.24, where the true model (gray curve) reaches an EOL of 1128 h (red vertical line).

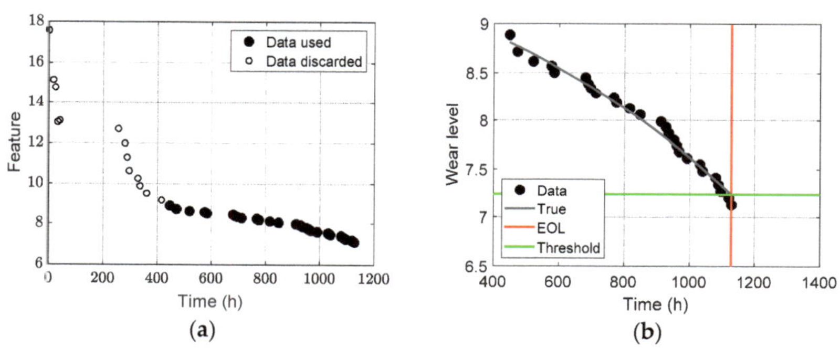

Figure 13. Given information and assumption for prognostics: (**a**) data for RUL prediction; (**b**) assumed true model and threshold.

The prognostics can be performed using the above information and assumptions, that is, the data, degradation model, and threshold. The prediction results of the degradation behavior at 768 h are shown in Figure 14a. In the figure, the dotted markers represent the data obtained up to the current 768 h, which were used to estimate the model parameters in Equation (5). The dashed and dotted red curves represent the median and 90% interval of the degradation prediction results, respectively, with the estimated model parameters. The EOL prediction is distributed between 1050 h and 1200 h, which is when the red curves reach the threshold (green horizontal line). The RUL was predicted by subtracting the current time from the predicted EOL, as indicated by the magenta vertical line in Figure 14b. The RUL prediction results were obtained by repeating the process for the degradation prediction in Figure 14a every time data was obtained. The results in Figure 14b show that the RUL prediction results (red lines) are very close to the true results (gray diagonal line) with narrow distributions after approximately 750 h. In other words, an accurate prediction is possible approximately 380 h before the EOL. The extrusion system is operated for approximately eight hours a day on weekdays; thus, the maintenance and/or replacement of Maddock screws can be scheduled approximately two months before EOL (it is a twofold improvement in the prediction of the RUL for the system using both barrier and Maddock screws).

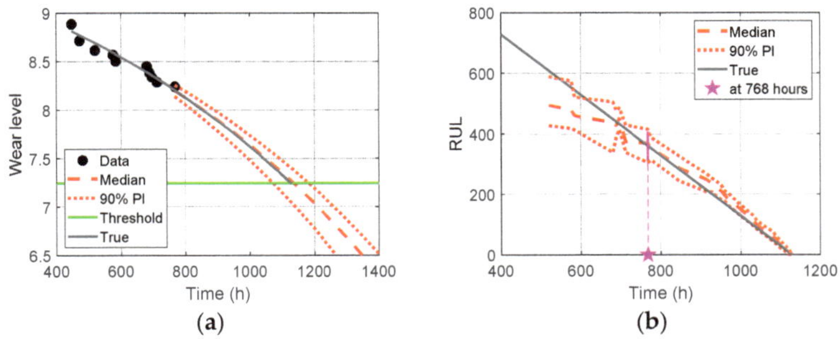

Figure 14. Prognostics results for the Maddock screw: (**a**) degradation prediction at 768 h; (**b**) RUL prediction for entire time.

5. Conclusions

In this study, a novel method for degradation feature extraction is proposed to predict the RUL of an extrusion screw using real operational data. The micro-extrusion system has been operational for more than five years, and in the last two years real operational data were obtained while the system produced various specifications of medical catheters. During this period, the Maddock screw reached its EOL owing to severe wear and was targeted for RUL prediction based on the proposed degradation feature. The proposed degradation feature exhibited typical characteristics of degradation behavior, which monotonically decreased with an increase in the degradation rate over time. The degradation feature was used to predict the RUL of the Maddock screw, and accurate results for the RUL were obtained approximately 380 h before EOL.

The main objective of this study was to accomplish the prognostics process using real operational data from a micro-extrusion system in service. First, real operational data are invaluable compared with test-level data, which are usually obtained at the component level under severe load conditions and well-planned operation conditions. Next, a valid degradation feature was extracted from the system-level data based on the fusion of the physical interpretation by the authors and the data information. Notably, the target object for prognostics was at the component level; however, the data used in this study were obtained at the system level, which reflects multiple complex aspects of the system within one type of signal. Finally, a prognostic study for the extrusion system was addressed, not by common objects such as bearings and batteries.

This study had some limitations. First, the proposed method for degradation feature extraction has not been fully validated owing to a lack of operational data. However, real operational data are still being collected and will be used in further studies. Next, the degradation of other components, including the motor, was ignored because screw wear is currently predominant. The degradation features of other components will become distinct over time, and they will be applied to improve the degradation feature of the screw. Lastly, the prognostics study for other components will be conducted aiming for system-level prognostics.

Author Contributions: Conceptualization, J.-K.P. and D.A.; methodology, J.-K.P. and D.A.; software, J.-K.P. and D.A.; validation, J.-K.P. and D.A.; formal analysis, W.K.; investigation, J.-K.P. and D.A.; resources, W.K. and G.-M.K.; data curation, H.L.; writing—original draft preparation, J.-K.P.; writing—review and editing, D.A.; visualization, J.-K.P. and D.A.; supervision, D.A. and W.K.; project administration, D.A.; funding acquisition, D.A. All authors have read and agreed to the published version of the manuscript.

Funding: This work was supported by a National Research Foundation of Korea (NRF) grant funded by the Korean government (MSIT) (No. 2020R1A4A4079904).

Institutional Review Board Statement: Not applicable.

Informed Consent Statement: Not applicable.

Data Availability Statement: Not applicable.

Conflicts of Interest: The authors declare no conflict of interest.

References

1. Hidle, E.L.; Hestmo, R.H.; Adsen, O.S.; Lange, H.; Vinogradov, A. Early Detection of Subsurface Fatigue Cracks in Rolling Element Bearings by the Knowledge-Based Analysis of Acoustic Emission. *Sensors* **2022**, *22*, 5187. [CrossRef] [PubMed]
2. Gao, Y.; Piltan, F.; Kim, J.-M. A Hybrid Leak Localization Approach Using Acoustic Emission for Industrial Pipelines. *Sensors* **2022**, *22*, 3963. [CrossRef] [PubMed]
3. Karayannis, C.G.; Golias, E.; Naoum, M.C.; Chalioris, C.E. Efficacy and Damage Diagnosis of Reinforced Concrete Columns and Joints Strengthened with FRP Ropes Using Piezoelectric Transducers. *Sensors* **2022**, *22*, 8294. [CrossRef] [PubMed]
4. Wu, Y.; Liu, K.; Li, D.; Shen, X.; Lu, P. Numerical and Experimental Research on Non-Reference Damage Localization Based on the Improved Two-Arrival-Time Difference Method. *Sensors* **2022**, *22*, 8432. [CrossRef] [PubMed]
5. Ahang, M.; Jalayer, M.; Shojaeinasab, A.; Ogunfowora, O.; Charter, T.; Najjaran, H. Synthesizing Rolling Bearing Fault Samples in New Conditions: A Framework Based on a Modified CGAN. *Sensors* **2022**, *22*, 5413. [CrossRef]
6. Huang, G.; Zhang, Y.; Ou, J. Transfer remaining useful life estimation of bearing using depth-wise separable convolution recurrent network. *Measurement* **2021**, *176*, 109090. [CrossRef]
7. Liu, L.; Song, X.; Zhou, Z. Aircraft engine remaining useful life estimation via a double attention-based data-driven architecture. *Reliab. Eng. Syst. Saf.* **2022**, *221*, 108330. [CrossRef]
8. Zhang, Y.; Li, Y.-F. Prognostics and health management of Lithium-ion battery using deep learning methods: A review. *Renew. Sustain. Energy Rev.* **2022**, *161*, 112282. [CrossRef]
9. Jouin, M.; Gouriveau, R.; Hissel, D.; Péra, M.-C.; Zerhouni, N. Degradations analysis and aging modeling for health assessment and prognostics of PEMFC. *Reliab. Eng. Syst. Saf.* **2016**, *148*, 78–95. [CrossRef]
10. Frederick, D.K.; DeCastro, J.A.; Litt, J.S. *User's Guide for the Commercial Modular Aero-Propulsion System Simulation (C-MAPSS)*; NASA Technical Manuscript 2007–215026; NASA Technical Reports Server: Washington, DC, USA, 2007.
11. Deng, J.; Li, K.; Harkin-Jones, E.; Price, M.; Karnachi, N.; Kelly, A.; Vera-Sorroche, J.; Coates, P.; Brown, E.; Fei, M. Energy monitoring and quality control of a single screw extruder. *Appl. Energy* **2014**, *113*, 1775–1785. [CrossRef]
12. Rauwendaal, C. *SPC: Statistical Process Control in Injection Molding and Extrusion*; Hanser Verlag: Munich, Germany, 2008.
13. Ge, Z.; Song, Z. Process monitoring based on independent component analysis—principal component analysis (ICA-PCA) and similarity factors. *Ind. Eng. Chem. Res.* **2007**, *46*, 2054–2063. [CrossRef]
14. Weighell, M.; Martin, E.B.; Morris, A.J. Fault Diagnosis in Industrial Process Manufacturing Using MSPC. In Proceedings of the IEE Colloquium (Digest), London, UK, 21 April 1997.
15. Liu, X.; Xie, L.; Kruger, U.; Littler, T.; Wang, S. Statistical-based monitoring of multivariate non-Gaussian systems. *AIChE J.* **2008**, *54*, 2379–2391. [CrossRef]
16. Plastic Technology. What Is Your Extruder Tring to Tell You? 2018. Available online: https://www.ptonline.com/articles/what-is-your-extruder-trying-to-tell-you (accessed on 15 August 2022).
17. Klein, I. Predicting the effect of screw wear on the performance of plasticating extruders. *Polym. Eng. Sci.* **1975**, *15*, 444–450. [CrossRef]
18. Plastic Technology. Screw Wear: Understanding Causes, Effects, and Solutions. Available online: https://www.ptonline.com/articles/screw-wear-understanding-causes-effects-and-solutions (accessed on 15 August 2022).
19. Jiang, Z.; Yang, Y.; Mo, S.; Yao, K.; Gao, F. Polymer extrusion: From control system design to product quality. *Ind. Eng. Chem. Res.* **2012**, *51*, 14759–14770. [CrossRef]
20. García, V.; Sánchez, J.S.; Rodríguez-Picón, L.A.; Méndez-González, L.C.; Ochoa-Domínguez, H.D.J. Using regression models for predicting the product quality in a tubing extrusion process. *J. Intell. Manuf.* **2019**, *30*, 2535–2544. [CrossRef]
21. Vera-Sorroche, J.; Kelly, A.; Brown, E.; Coates, P.; Karnachi, N.; Harkin-Jones, E.; Li, K.; Deng, J. Thermal optimisation of polymer extrusion using in-process monitoring techniques. *Appl. Therm. Eng.* **2013**, *53*, 405–413. [CrossRef]
22. Abeykoon, C.; McMillan, A.; Nguyen, B.K. Energy efficiency in extrusion-related polymer processing: A review of state of the art and potential efficiency improvements. *Renew. Sustain. Energy Rev.* **2021**, *147*, 111219. [CrossRef]
23. Kim, W.-H.; Kim, K.-C.; Kim, S.-J.; Kang, D.-W.; Go, S.-C.; Lee, H.-W.; Chun, Y.-D.; Lee, J. A Study on the Optimal Rotor Design of LSPM Considering the Starting Torque and Efficiency. *IEEE Trans. Magn.* **2009**, *45*, 1808–1811. [CrossRef]
24. Rahim, M.K.; Jidin, A.; Sutikno, T. Enhanced Torque Control and Reduced Switching Frequency in Direct Torque Control Utilizing Optimal Switching Strategy for Dual-Inverter Supplied Drive. *Int. J. Power Electron. Drive Syst. (IJPEDS)* **2016**, *7*, 328. [CrossRef]
25. Paris, P.; Erdogan, F. A critical analysis of crack propagation laws. *J. Basic Eng.* **1963**, *85*, 528–533. [CrossRef]

26. Goebel, K.; Saha, B.; Saxena, A.; Celaya, J.R.; Christophersen, J.P. Prognostics in battery health management. *IEEE Instrum. Meas. Mag.* **2008**, *11*, 33–40. [CrossRef]
27. Kim, N.H.; An, D.; Choi, J.H. Chapter 4.3. Bayesian Method (BM). In *Prognostics and Health Management of Engineering Systems*; Springer International Publishing: Cham, Switzerland, 2017.

sensors

MDPI

Article

High Precision Feature Fast Extraction Strategy for Aircraft Attitude Sensor Fault Based on RepVGG and SENet Attention Mechanism

Zhen Jia [1,*], Kai Wang [2], Yang Li [2], Zhenbao Liu [2], Jian Qin [1] and Qiqi Yang [1]

1 School of Mechanical and Electrical Engineering, Xi'an University of Architecture and Technology, Xi'an 710055, China
2 School of Civil Aviation, Northwestern Polytechnical University, Xi'an 710072, China
* Correspondence: jiazhen@xauat.edu.cn

Abstract: The attitude sensor of the aircraft can give feedback on the perceived flight attitude information to the input of the flight controller to realize the closed-loop control of the flight attitude. Therefore, the fault diagnosis of attitude sensors is crucial for the flight safety of aircraft, in view of the situation that the existing diagnosis methods fail to give consideration to both the diagnosis rate and the diagnosis accuracy. In this paper, a fast and high-precision fault diagnosis strategy for aircraft sensor is proposed. Specifically, the aircraft's dynamics model and the attitude sensor's fault model are built. The SENet attention mechanism is used to allocate weights for the collected time-domain fault signals and transformed time-frequency signals, and then inject the fused feature signals with weights into the RepVGG based on the convolutional neural network structure for deep feature mining and classification. Experimental results show that the proposed method can achieve good precision speed tradeoff.

Keywords: attitude sensor; fault diagnosis; attention mechanism; time-frequency signal; RepVGG

Citation: Jia, Z.; Wang, K.; Li, Y.; Liu, Z.; Qin, J.; Yang, Q. High Precision Feature Fast Extraction Strategy for Aircraft Attitude Sensor Fault Based on RepVGG and SENet Attention Mechanism. *Sensors* **2022**, *22*, 9662. https://doi.org/10.3390/s22249662

Academic Editor: Sandra Verhagen

Received: 17 November 2022
Accepted: 8 December 2022
Published: 9 December 2022

1. Introduction

The complex structure, numerous equipment, system cross-linking and diverse flight environment of aircraft make it easy to have faults. The flight control system is the core system of the aircraft, in which the sensor is used to transmit the real-time measured aircraft flight state parameters to the flight control system. Therefore, the state of the sensor will directly affect the flight state [1,2]. Once a fault occurs, it will cause the sensor to transmit the wrong information to the flight control system, which may cause greater economic losses and even endanger people's lives. Therefore, the diagnosis of aircraft sensors is essential to ensure aircraft flight safety [3,4].

The traditional fault diagnosis methods consist of mainly two types, one is based on signal analysis or artificial feature extraction [5,6], and the other is based on models [7,8]. Among them, a common application of the first method is to determine whether a fault occurs by designing a threshold value and comparing whether the signal reaches the threshold value [9]. Other methods based on manual feature extraction have also been widely studied. A fault diagnosis method based on signal decomposition and two-dimensional feature clustering is designed to diagnose battery status [10]. The data processing method of high-speed railway fault signal diagnosis based on MapReduce algorithm was designed [11]. Statistical method and wavelet packet decomposition method are used for feature extraction of vibration signal to identify the fault type of rotor [12].

The model-based method refers to establishing the model of the object to be diagnosed, and analyzing the situation when various faults occur by setting different types of faults in the model. Fault diagnosis is realized through the corresponding relationship between the output difference of the model in different faults and the fault type. A review of model-based fault diagnosis methods was published, focusing on fault detection and fault estimation [13]. N. Valceschini et al. proposed a model-based fault detection and isolation scheme for the transmission components of electro-mechanical actuators, which was applied to the drive of sliding doors [14]. In addition, a model-based battery fault diagnosis method is proposed, which is based on multiple equivalent circuit models [15]. In addition, Wang et al. established the equivalent circuit model of battery pack insulation fault diagnosis using the high fidelity unit model [16].

However, the two traditional methods mentioned above have some limitations, specifically , the method based on signal analysis diagnoses by manually selecting feature types, which is difficult to avoid the problem of insufficient representation of selected features [17]. The main problem of model-based method is that it requires high accuracy of the model, and it is no longer applicable when the object changes a little. With the development of machine learning and artificial intelligence technology, data-driven fault diagnosis method is very popular in recent years because of its advantages of automatically exploring the characteristics of signals and high applicability. More and more data driven diagnostic methods with higher accuracy have emerged [18–23]. A data driven method based on improved Elman neural network was proposed to realize the fast diagnosis of open circuit fault of IGBT [18]. Nicholas et al. [19] proposed a general robust data-driven scheme for fault detection, isolation and estimation of multiple sensor faults, and verified it with multiple flight data records. A fault diagnosis method based on Deep belief network (DBN) to generate local random graph to intuitively explain the fault action mechanism was proposed, which realized the diagnosis of different faults of air conditioner [20] Guo et al. [21] established a predictive model for photovoltaic power generation under normal conditions through clustering algorithm and long short-term memory neural network (LSTM), and used the predictive model to conduct quantitative fault diagnosis through transfer learning. In addition, some work related to fault diagnosis combines signal based and data-driven, or uses the transfer learning strategy [24–27] to achieve high-precision fault diagnosis results.

Different types of machine learning models have been targeted for development and used in data-driven diagnosis. However, there are still two problems in the processing of input data: the type of input signal data is single, which has no advantage in ensuring the integrity of data information; in a small amount of work considering multiple input signals, the importance of different signals is rarely considered, which is not conducive to subsequent feature extraction and fault classification. In addition, in view of the fact that most of the deep learning diagnosis models cannot give consideration to both time cost and computational efficiency, this paper also proposes a targeted scheme.

Specifically, the innovation points of this paper are as follows:

(i) The time sequence signal of aircraft attitude sensor is transformed into time-frequency domain, and the time-domain signal and time-frequency domain signal are taken as the feature mining object.
(ii) A signal representation weight analysis and allocation strategy is designed, and the representativeness of each channel of time-domain signal and time-frequency signal is analyzed by using Squeeze-and-excitation networks (SENet) attention mechanism.
(iii) A fast and high-precision diagnostic technology based on Re-parameterization visual geometry group (RepVGG) is proposed, which achieves a good diagnostic accuracy speed tradeoff.

The following text is arranged as follows: the relevant theories and methods are given in Section 2. Section 3 describes the experimental setup and the preparation process of the fault data set, including fault model building and data collection, experimental parameter settings, etc. The Section 4 presents the experimental results and discussions. Section 5 summarizes the full text.

2. Relevant Theories and Proposed Methods

2.1. RepVGG

The maturity of the "convolutional" neural network has made it a solution to many mainstream tasks. The commonly used convolutional neural network models in image classification include VGG-16 and ResNet. The performance of VGG network model will increase with its depth, which may lead to over fitting and gradient disappearance, and the accuracy will decline. ResNet model's residual element can solve the gradient disappearance phenomenon well, but it is powerless for the common over fitting phenomenon of deep network. The multi-level branches in the residual structure in ResNet make the model difficult to implement. RepVGG network is a single path convolutional network architecture, which integrates the ideas of VGG and ResNet, and only adds 3 times. The 3-volume integration layer can also achieve simple and more efficient performance [28].

2.1.1. RepVGG Block

As shown in Figure 1, RepVGG uses a multi branch model similar to ResNet style during training, and converts it into a single path model of VGG style during reasoning. Figure 1a shows the network structure used in RepVGG training, while Figure 1b is used in reasoning. Figure 1b shows the RepVGG network in the reasoning stage. The structure of the network is very simple. The whole network is composed of convolution with kernel size 3 × 3 and ReLu activation function, which is easy for model reasoning and acceleration.

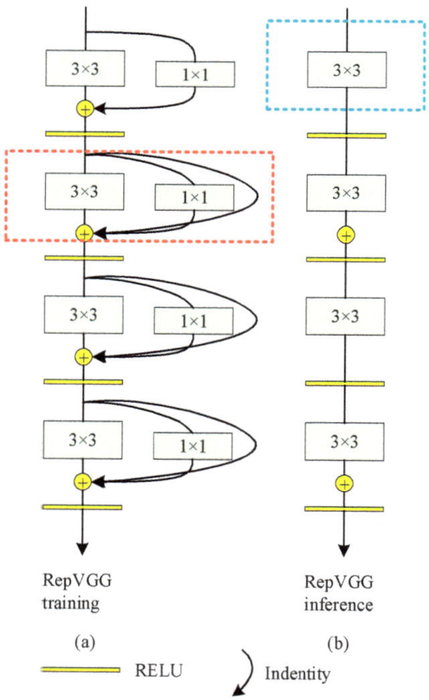

Figure 1. Schematic diagram of partial structure of RepVGG. (**a**) RepVGG training. (**b**) RepVGG inference.

RepVGG is formed by continuously stacking RepVGG Blocks. During the training, RepVGG Block paralleled three branches: a main branch with a convolution core size of 3 × 3, a shortcut branch with a convolution core size of 1 × 1, and a shortcut branch with only BN connected. Since the residual structure has multiple branches, it is equivalent to adding multiple gradient flow paths to the network. Such a network is trained, which is similar to training multiple networks and integrating multiple networks into one network.

It is similar to the idea of model integration, which can improve the training effect of the network.

2.1.2. Structural Reparameterization

RepVGG reparameterization transforms the multi branch structure in the training process into 3 × 3 convolution with deviation, which improves the reasoning speed of the network, reduces the network parameters and reduces the memory occupation. The process of reparameterization is shown in Figure 2, which includes four processes: merging Conv2d and BN; Convert 1 × 1 convolution to 3 × 3 convolution and BN to 3 × 3 convolution, and fuse multiple branches.

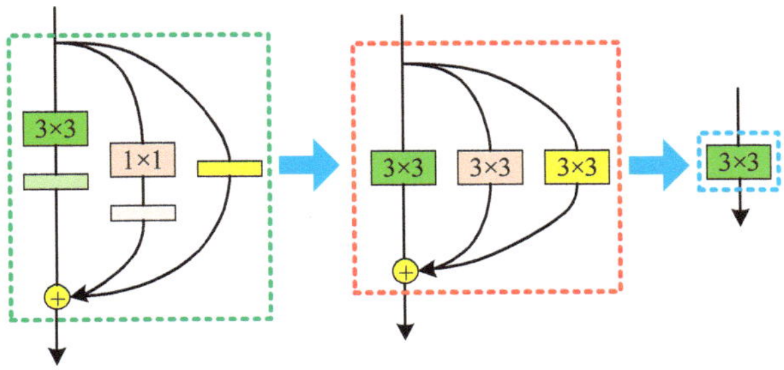

Figure 2. Schematic Diagram of Reparameterization.

At the stage of merging 3 × 3 convolution layer and *BN* layer, the formula of convolution layer and *BN* layer is as follows:

$$\begin{aligned} &\mathrm{Conv}(x) = W(x) \\ &BN(x) = \gamma \times \tfrac{(x-\mu)}{\sqrt{\sigma_{\mathrm{i}}^2+\epsilon}} + \beta \end{aligned} \quad (1)$$

where the input is x, the fusion of Conv into *BN* can be expressed as:

$$BN(\mathrm{conv}(x)) = \gamma \times \frac{W(x)-\mu}{\sqrt{\sigma_{\mathrm{i}}^2+\epsilon}} + \beta = \left(\frac{\gamma \times W(x)}{\sqrt{\sigma_{\mathrm{i}}^2+\epsilon}}\right) + \left(\frac{\gamma \times W(x)}{\sqrt{\sigma_{\mathrm{i}}^2+\epsilon}} + \beta\right) \quad (2)$$

The above formula can be regarded as the convolution layer incorporating *BN* operation, where $\sqrt{\sigma_{\mathrm{i}}^2}$ is the variance of *BN* layer, γ Is the scale factor of *BN* layer, β Indicates the offset factor of *BN* layer. If the content in the first bracket of the above formula is regarded as W', and the content in the second bracket is regarded as B', then:

$$W' = \frac{\gamma \times W}{\sigma_{\mathrm{i}}} \quad (3)$$

$$B' = \beta - \frac{\gamma \times \mu}{\sigma_{\mathrm{i}}} \quad (4)$$

Finally, it can be rewritten as:

$$BN(\mathrm{Conv}(x)) = W'(x) + B'(x) \quad (5)$$

When converting 1 × 1 convolution to 3 × 3 convolution form, take a convolution core in 1 × 1 convolution layer as an example, just add a circle of zeros around the original convolution core weight, which becomes a 3 × 3 convolution layer. Note that in order to ensure that the height and width of the input/output feature map remain unchanged, the padding is usually set to 1. Finally, the above convolution layer and BN layer can be fused.

When converting BN to 3 × 3 convolution, as there is no convolution layer for branches with only BN, a 3 × 3 convolution layer needs to be constructed first, and the convolution layer only carries out identity mapping, that is, the input and output characteristic maps remain unchanged. With this convolution layer, BN layer can be converted into 3×3 convolution.

Finally, multi branch fusion is carried out. The process of merging is relatively simple. The parameters of the three convolution layers are added together. In this step, the weight W and offset B of all branches are superposed to obtain a fused 3×3 Convolutional network layer.

2.2. SENet Attention Mechanism

Convolution often focuses on the fusion of scale information in space. Through the introduction of an attention mechanism, SENet focuses on the connection between different channels, so that it can learn the importance of each channel feature [29]. For the fault classification task in this paper, an attention mechanism is introduced to improve the attention of different characteristic channels of input signals. The SE module contains two operations: Squeeze and Exception; the global characteristics of each trace in the feature map can be obtained by the Squeeze operation. The relationship between channels can be learned through the Exception, and the weights between different channels can be obtained.

Its implementation is shown in Figure 3. The input feature layer is pooled globally. Then two full connections are made. The number of fully connected neurons in the first time is less, and the number of fully connected neurons in the second time is the same as the input feature layer. A ReLU layer is set between two full connections. Then, another sigmoid is performed to fix the value between 0–1. At this time, the weight value of each channel of the input feature layer is obtained. Finally, the weighted feature layer can be obtained by multiplying this weight value by the original input feature layer.

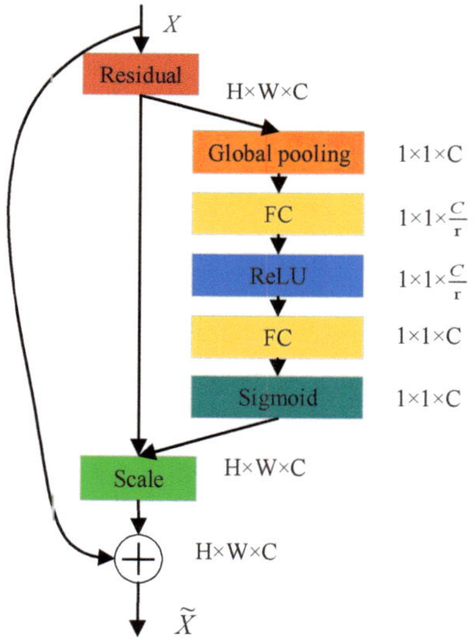

Figure 3. Structure of SENet.

2.3. The Proposed Strategy

In order to fully exploit the characteristics of sensor fault signals, the fault diagnosis strategy shown in Figure 4 is proposed in this paper. First, different faults are injected into the flight control system model of the aircraft, and then the signals under the fault state are collected. The time-frequency characteristic diagram is obtained by processing one-dimensional time-domain residual signal through S-transform. The one-dimensional time-domain residual signal is sliced and stacked into a $50 \times 50 \times 1$ format. As the first channel data, the data size of the RGB three channels of time-frequency characteristic map is $50 \times 50 \times 3$. The data of these four channels $50 \times 50 \times 4$ are used as the processing data of the subsequent SENet attention mechanism and RepVGG. Because RepVGG only obtains features in spatial dimension, SENet module is integrated into RepVGG to obtain feature association between different channels. The proposed diagnostic algorithm consists of two parts: training and testing. The training part is to learn the model's parameters by using the training data set, and the testing part is to test the effect of the proposed model by using the testing data set.

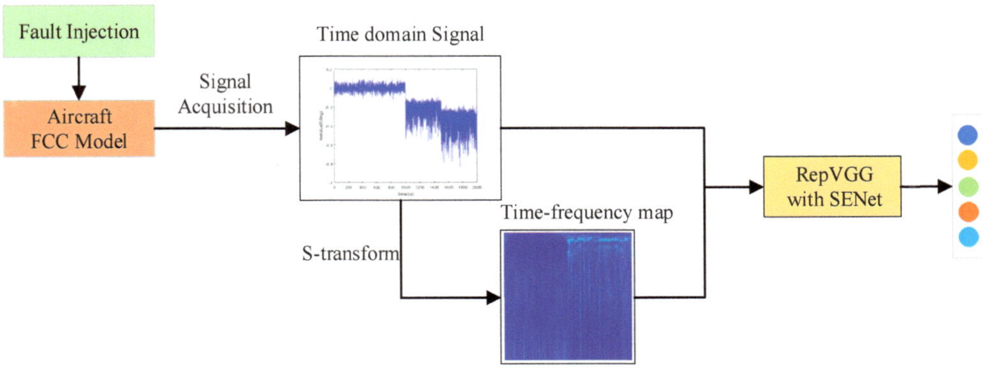

Figure 4. Flowchart of the algorithm proposed in this paper.

Add SE attention module to the 3 × 3 channel of RepVGG, and the feature map with size H × W × C is obtained after the convolution kernel operation (in the fault diagnosis task of this paper, the value of parameter C is 4). At this time, the convolution is only the characteristic diagram obtained by spatial operation, and there is no relationship between each channel; The 2D feature p_c of each channel is mapped to the global feature f_c through a global average pooling, and the formula is as follows:

$$f_c = F_{sq}(p_c) = \frac{1}{h \times h}\sum_{i=1}^{h}\sum_{j=1}^{h} p_c(i,j), f \in R^C \tag{6}$$

Then, two Fully connected (FC) layers are used, one to reduce the dimension characteristics, and the other to upgrade back to the original dimension. Finally, the normalized weight is obtained through Sigmoid, and the formula is as follows:

$$\mathrm{s} = F_{ex}(f, W) = \sigma(g(f, W)) = \sigma(W_2 \mathrm{ReLU}(W_1 z)) \tag{7}$$

$$W_1 \in R^{\frac{C}{r} \times C}, W_2 \in R^{\frac{C}{r} \times C} \tag{8}$$

Finally, the weight s obtained is weighted to each characteristic channel f_c. This allows important channels to gain greater attention and ensure the accuracy of classification.

3. Experiment Setup and Data Set Preparation

3.1. Establishment of Fault Model

Navion aircraft model is built in this paper, and different fault types of its attitude sensor are set. Navion aircraft model is a navigation aircraft model. By the end of 1947, more than 1100 aircraft of this type had been produced in the United States. The aircraft has a total length of 8.38 m, a wingspan of 10.19 m, a height of 2.65 m, a maximum takeoff weight of 1338 kg, a maximum flight speed of 260 km per hour, and a maximum range of 1120 km. The aircraft was once designated as a training aircraft of the US Air Force. Today, there are still a large number of such aircraft in civilian use. In this paper, according to the published aerodynamic parameters of Navion aircraft, the first order Taylor expansion method is used to linearize the small disturbance equation of fixed wing aircraft at a certain equilibrium point, and the linearized model of Navion aircraft is obtained. As shown in

Figure 5, a fault signal generation model is designed, including the normal sensor sensing part and the fault sensor sensing part. The input control signal is input into the control model of the aircraft. The attitude sensor in the normal attitude frame can correctly perceive the attitude information of the Unmanned aerial vehicles (UAV), while the attitude sensor in the fault attitude frame cannot correctly perceive the attitude information.

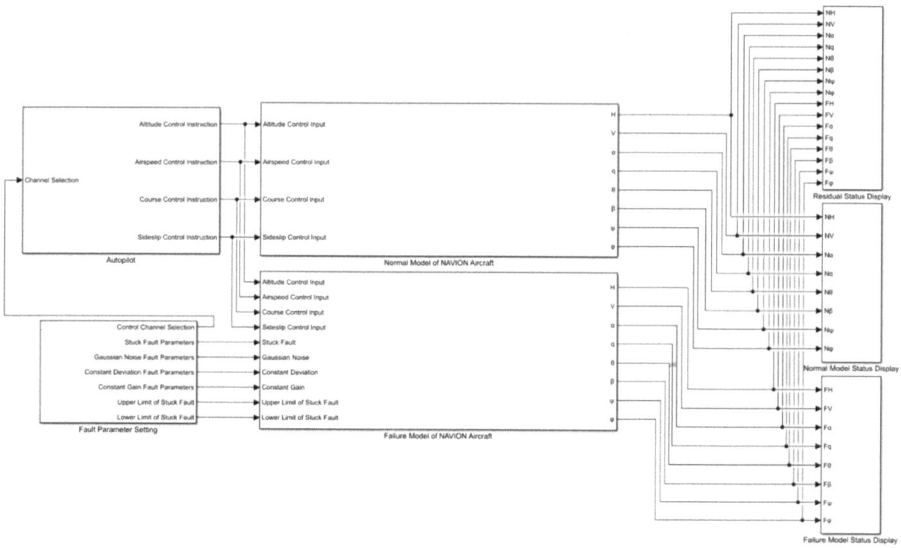

Figure 5. Simulation diagram of attitude sensor fault.

Among them, the attitude sensor in the normal attitude frame can correctly perceive the attitude information of the UAV, while the attitude sensor in the fault attitude frame cannot correctly perceive the attitude information. After the sensor fault model is established, the attitude information is measured and the fault output is obtained. Then calculate the residual of normal sensor data and fault sensor data, and use the residual time series signal as the data processed by the subsequent fault diagnosis model. The fault type settings are shown in Table 1, which contains the fault manifestations and corresponding labels. Taking the pitch angle sensor of an aircraft as an example, four common faults are set, including jamming, lateral gain, lateral deviation and excessive noise. In addition, if the fault free state is regarded as a special fault state, there are five fault types in total.

Table 1. Setting of fault types.

Fault Type	Fault Manifestation	Fault Label
No fault	The fault free state represents the health state and is regarded as a special fault	F0
Stuck	The measured value of the sensor output deviates from the normal value and reaches a stuck position	F1
Constant gain	The measured value of the sensor output maintains a constant proportion to the normal output value	F2
Constant deviation	The measured value of the sensor output deviates from the normal value and keeps the deviation constant	F3
Excessive noise	The measured value of the sensor output contains large noise	F4

3.2. Acquisition of Fault Data

Set the corresponding input control signal to change within a certain angle range, collect the residual signals of various faults, the length of each residual signal is 2500, and rearrange them into a 50 × 50 format. In addition, the time-frequency diagram obtained by

S-transform is also cut to 50 × 50 size, and 50 × 50 × 3 data obtained by extracting RGB three channels of color time-frequency diagram is used as the input of SENet attention mechanism model. Figure 6 shows the time-frequency diagram of different fault types after S transformation.

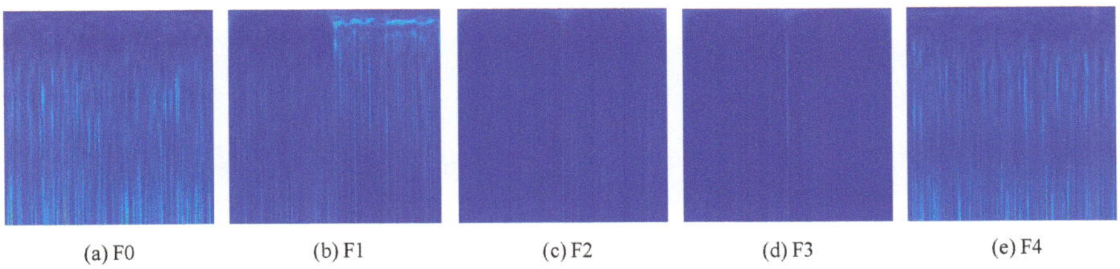

Figure 6. Signal diagrams for five different types of faults. (**a**) F0. (**b**) F1. (**c**) F2. (**d**) F3. (**e**) F4.

In order to explore the influence of different data sets on the fault diagnosis accuracy of the algorithm proposed in this paper, five data sets with different sizes are set. Table 2 lists the size information of training set, verification set and test set for each fault type.

Table 2. Different data set size for each fault.

Data Set	Number of Training Sets	Number of Validation Sets	Number of Test Sets
1	140	20	40
2	280	40	80
3	420	60	120
4	700	100	200
5	980	140	280

3.3. Basic Parameter Settings of the Proposed Method

The parameters of the proposed method are shown in Table 3, where the kernel size is the size of the convolution kernel; Padding is the matrix filling value, that is, the filling is added to all four sides input, and the default value is 0; padding_ Mode is the matrix filling mode, and the default value is 'zero'; num_ Blocks is the number of modules, that is, the number of sub modules in different stages; num_ Classes is the number of fault classifications, which is set as five fault states; width_ Multiplier is the stage multiplication coefficient, that is, the different coefficients multiplied at different stages; Groups is the number of input channel groups, that is, the number of blocked connections from the input channel to the output channel. The default value is 1; Street is the convolution step, and the default value is 1; The division is the expansion flag bit, that is, the spacing between kernel elements. The default value is 1; Bias adds a learnable deviation to the output. This parameter is a Boolean value. If it is true, a learnable deviation is added to the output, indicating that the parameters learned in the backward feedback are applied.

Table 3. Size of the dataset.

Parameter	Value
kernel size	3
padding	0
padding_mode	'zeros'
num_blocks	[2, 4, 14, 1]
num_classes	5
width_multiplier	[0.75, 0.75, 0.75, 2.5]
groups	1
stride	1
dilation	1
bias	True

4. Experimental Results and Discussion

The experiments are conducted with Python 3.9, CUDA 11.6 and Pytorch 1.12.1 libraries on Windows11 operation system. The key experimental hardware configurations are NVIDIA GeForce RTX 3060 Laptop GPU with 6 GB memory and 12th Gen Intel(R) Core(TM) i9-12900H 2.50 GHz CPU with 16GB memory.

Accuracy is used to evaluate the diagnostic performance. Two indexes are as follows.

$$Accuracy = \frac{N_{cp}}{N_{cp} + N_{wp}} \tag{9}$$

where N_{cp} represents the number of cases whose label is correctly predicted, N_{wp} refers the number of cases whose label is wrongly predicted.

4.1. Effect of Data Set on Diagnosis Results

The average precision and average training time are selected as the evaluation indicators of this paper to study the fault diagnosis performance of the method proposed in this paper under different data sets. The results are shown in Table 4.

Table 4. Diagnostic performance of the proposed method under different data sets.

Different Datasets	Dataset 1	Dataset 2	Dataset 3	Dataset 4	Dataset 5
Average accuracy (%)	93.45%	96.10%	99.28%	99.37%	99.42%
Average training time (s)	356.08	693.76	1019.73	1754.73	2390.49

Table 4 shows that: (i) As the size of the dataset increases (from dataset 1 to dataset 5), the model training time of the proposed algorithm becomes longer and longer. (ii) As far as the average precision is concerned, the precision has reached more than 99% in the case of data volume shown in dataset 3. Later, as the dataset continues to grow, for example, when it changes to dataset 5, compared with dataset 3, the average accuracy of the algorithm is only improved by 0.14%. According to Table 4, for the smallest dataset 1, its accuracy is the lowest, while for the medium dataset 3, the proposed algorithm has reached a satisfied accuracy. Therefore, we have added a Table 5 to list the accuracy of each fault type of a test in detail under the conditions of datasets 1 and 3.

Table 5. Accuracy detail presentation of data sets 1 and 3 in a test.

Dataset 1		Dataset 3	
Fault code	Accuracy (%)	Fault code	Accuracy (%)
F0	92.50	F0	98.33
F1	100.00	F1	99.17
F2	90.00	F2	100.00
F3	85.00	F3	98.33
F4	90.00	F4	99.17
Average accuracy	91.50	Average accuracy	99.00

As can be seen from Table 5, the corresponding classification accuracy of each fault label in dataset 3 is generally higher than that in dataset 1. In addition, in dataset 3, the number of tests for each fault type is 120, and the number of the wrong classification is less than 2.

In order to better show the classification of the algorithm proposed in this paper under different data. The confusion matrix of dataset 1 and dataset 3 in an experiment is shown in Figure 7.

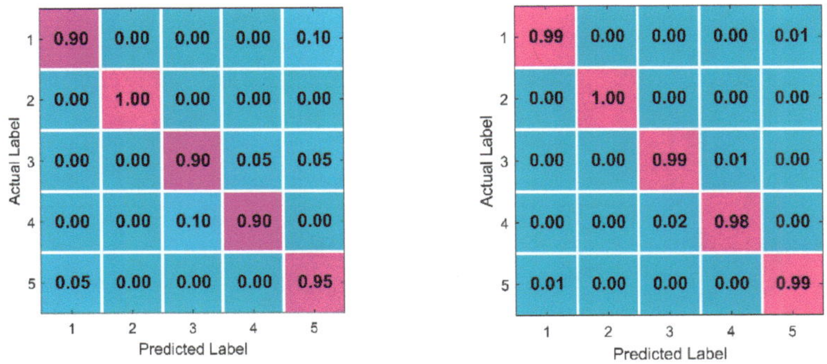

(**a**) Dataset 1 (**b**) Dataset 3

Figure 7. Confusion matrix under different data sets.

It can be seen from Figure 7 that the misclassification of the proposed algorithm in dataset 3 is obviously better than that in dataset 1. Further, the model training process in the case of dataset 3 is shown in Figure 8.

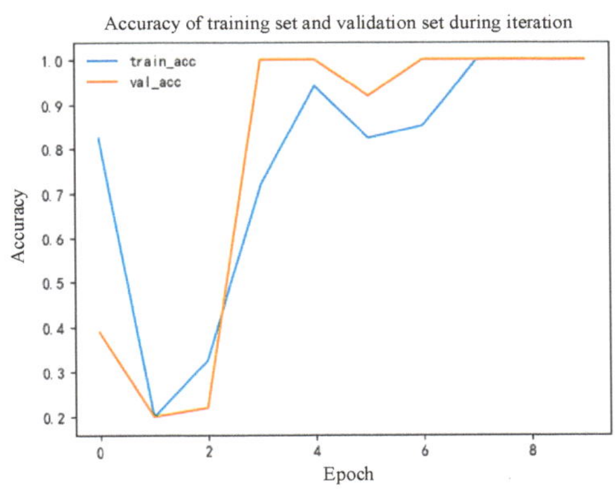

Figure 8. Iteration diagram of the accuracy of the algorithm proposed in this paper in the case of dataset 3.

It can be seen from Figure 8 that the training and verification stages of the model of the proposed diagnostic algorithm achieve the best accuracy at the 7th Epoch. In addition, the convergence speed of the model is relatively fast.

4.2. Ablation Experiment

In order to explore the role of each module of the method proposed in this paper. Ablation experiment was set up to conduct ablation research by deleting each module from the proposed method. (we conduct ablation studies by removing the identity and/or 1×1 branch from every block of RepVGG-B0.) Specifically, as shown in Table 6, respectively cancel the S transform module to use only the time domain signal (No. 1), cancel the time domain signal to use only the time-frequency domain signal (No. 2), cancel the SENet module (No. 3), and simultaneously cancel the S transform module and the SENet module (No. 4). The fault diagnosis accuracy under each condition is listed in this table.

Table 6. Fault diagnosis accuracy is tested in ablation study.

No.	Time Domain Signal	S-Transform	SENet	Average Accuracy
1	√		√	80.68%
2		√	√	86.35%
3	√	√		92.37%
4	√			62.75%

Note: The modules marked with √ in the table are reserved in the model.

Table 6 shows that the performance is the best when the SENet module is canceled, reaching 92.37%. Secondly, the performance is the second best when only time-frequency signals are used, reaching 86.35%. When the S-transform module is cancelled and only the time-domain signal is used, the corresponding diagnostic accuracy ranks third, only 80.68%. The performance is the worst when S-transform module and SENet module are canceled at the same time, which is only 62.75%. It can be seen from the results that the signal processed by the diagnosis strategy has the greatest impact on the diagnosis performance, and it is very important to obtain time-frequency diagram signal through S-transform. Secondly, the SENet module shows some advantages in data preprocessing, which provides a better basis for RepVGG to mine effective features. In addition, by comparing No. 1 and No. 2, it can be seen that compared with the timing signal, the time-frequency map obtained by using S-transform can better reflect the characteristics of the object.

5. Conclusions

In this paper, a fault diagnosis method for aircraft attitude sensor is proposed. Research shows that sensor residual signals can reflect the difference of various faults and can be used as a diagnostic signal. One dimension time-domain signals and two-dimensional time-frequency domain features are processed by SENet attention mechanism, and key feature categories are enhanced by high weight. Subsequently, the depth RepVGG is used to conduct in-depth feature mining and achieve a fast and high accuracy diagnosis effect. Therefore, the SENet attention mechanism is an effective feature importance ranking scheme, which realizes the weight division of signal categories while ensuring the integrity of diagnostic signals. In addition, RepVGG has also been proven to be a potential fault diagnosis algorithm, with obvious advantages in diagnosis speed while ensuring accuracy.

Author Contributions: Conceptualization, Z.J. and K.W.; methodology, Z.J.; software, K.W.; validation, K.W., Y.L. and Z.J.; formal analysis, Z.J.; investigation, J.Q.; resources, Q.Y.; data curation, K.W.; writing—original draft preparation, Z.J.; writing—review and editing, Z.J.; visualization, K.W.; supervision, Z.J.; project administration, Z.L.; funding acquisition, Z.L. All authors have read and agreed to the published version of the manuscript.

Funding: This research was funded by National Natural Science Foundation, grant number 52072309.

Institutional Review Board Statement: Not applicable.

Informed Consent Statement: Not applicable.

Data Availability Statement: Data available on request due to restrictions, e.g., privacy or ethical.

Conflicts of Interest: The authors declare no conflict of interest.

References

1. Qi, X.; Theilliol, D.; Qi, J.; Zhang, Y.; Han, J. A literature review on Fault Diagnosis methods for manned and unmanned helicopters. *Int. Conf. Unmanned Aircr. Syst.* **2013**, 1114–1118. [CrossRef]
2. Fuggetti, G.; Zanzi, M.; Ghetti, A. Safety improvement of fixed wing mini-UAV based on handy FDI current sensor and a FailSafe configuration of control surface actuators. *Metrol. Aerosp.* **2015**, 356–361. [CrossRef]
3. Ai, S.; Song, J.; Cai, G. A real-time fault diagnosis method for hypersonic air vehicle with sensor fault based on the auto temporal convolutional network. *Aerosp. Sci. Technol.* **2021**, *119*, 107220. [CrossRef]
4. Guo, K.; Liu, L.; Shi, S.; Liu, D.; Peng, X. Uav sensor fault detection using a classifier without negative samples: A local density regulated optimization algorithm. *Sensors* **2019**, *19*, 771. [PubMed]

5. Gao, Z.; Cecati, C.; Ding, S. A survey of fault diagnosis and fault-tolerant techniques-part i: Fault diagnosis with model-based and signal-based approaches. *IEEE Trans. Ind. Electron.* **2015**, *62*, 3757–3767. [CrossRef]
6. Gao, L.; Li, D.; Yao, L.; Gao, Y. Sensor drift fault diagnosis for chiller system using deep recurrent canonical correlation analysis and k-nearest neighbor classifier. *ISA Trans.* **2021**, *122*, 232–246. [CrossRef]
7. He, Z.; Shao, H.; Ding, Z.; Jiang, H.; Cheng, J. Modified deep auto-encoder driven by multi-source parameters for fault transfer prognosis of aero-engine. *IEEE Trans. Ind. Electron.* **2022**, *69*, 845–855.
8. Liu, Z.; Jia, Z.; Vong, C.; Bu, S.; Han, J.; Tang, X. Capturing high-discriminative fault features for electronics-rich analog system via deep learning. *IEEE Trans. Ind. Inform.* **2017**, *13*, 1213–1226. [CrossRef]
9. Garramiola, F.; Poza, J.; Madina, P.; Olmo, J.; Ugalde, G. A Hybrid Sensor Fault Diagnosis for Maintenance in Railway Traction Drives. *Sensors* **2020**, *20*, 962. [CrossRef]
10. Li, S.; Zhang, C.; Du, J.; Cong, X.; Zhang, L.; Jiang, Y.; Wang, L. Fault diagnosis for lithium-ion batteries in electric vehicles based on signal decomposition and two-dimensional feature clustering. *Green Energy Intell. Transp.* **2022**, *1*, 100009. [CrossRef]
11. Cao, Y.; Li, P.; Zhang, Y. Parallel processing algorithm for railway signal fault diagnosis data based on cloud computing. *Future Gener. Comput. Syst.* **2018**, *88*, 279–283. [CrossRef]
12. Wang, S.; Wang, Q.; Xiao, Y.; Liu, W.; Shang, M. Research on rotor system fault diagnosis method based on vibration signal feature vector transfer learning. *Eng. Fail. Anal.* **2022**, *139*, 106424. [CrossRef]
13. Zhong, M.; Xue, T.; Steven, X.D. A survey on model-based fault diagnosis for linear discrete time-varying systems. *Neurocomputing* **2018**, *306*, 51–60. [CrossRef]
14. Valceschini, N.; Mazzoleni, M.; Previdi, F. Model-based fault diagnosis of sliding gates electro-mechanical actuators transmission components with motor-side measurements. *IFAC-PapersOnLine* **2022**, *55*, 784–789. [CrossRef]
15. Wei, J.; Dong, G.; Chen, Z. Model-based fault diagnosis of Lithium-ion battery using strong tracking Extended Kalman Filter. *Energy Procedia* **2019**, *158*, 2500–2505. [CrossRef]
16. Wang, Y.; Tian, J.; Chen, Z.; Liu, X. Model based insulation fault diagnosis for lithium-ion battery pack in electric vehicles. *Measurement* **2019**, *131*, 443–451. [CrossRef]
17. Li, X.; Shao, H.; Lu, S.; Xiang, J.; Cai, B. Highly-efficient fault diagnosis of rotating machinery under time-varying speeds using LSISMM and small infrared thermal images. *IEEE Trans. Syst.* **2022**, *52*, 7328–7340. [CrossRef]
18. An, Y.; Sun, X.; Ren, B.; Li, H.; Zhang, M. A data-driven method for IGBT open-circuit fault diagnosis for the modular multilevel converter based on a modified Elman neural network. *Energy Rep.* **2022**, *8*, 80–88. [CrossRef]
19. Nicholas, C.; Marcello, R.; Napolitano, G.C.; Paolo, V.; Mario, L. Fravolini, Aircraft robust data-driven multiple sensor fault diagnosis based on optimality criteria. *Mech. Syst. Signal Process.* **2022**, *170*, 108668. [CrossRef]
20. Li, T.; Zhao, Y.; Zhang, C.; Luo, J.; Zhang, X. A knowledge-guided and data-driven method for building HVAC systems fault diagnosis. *Build. Environ.* **2021**, *198*, 107850. [CrossRef]
21. Guo, H.; Hu, S.; Wang, F.; Zhang, L. A novel method for quantitative fault diagnosis of photovoltaic systems based on data-driven. *Electr. Power Syst. Res.* **2022**, *210*, 108121. [CrossRef]
22. Andreas, L.; Daniel, J. Data-driven fault diagnosis analysis and open-set classification of time-series data. *Control. Eng. Pract.* **2022**, *121*, 105006. [CrossRef]
23. Wang, Z.; Li, G.; Yao, L.; Qi, X.; Zhang, J. Data-driven fault diagnosis for wind turbines using modified multiscale fluctuation dispersion entropy and cosine pairwise-constrained supervised manifold mapping. *Knowl.-Based Syst.* **2021**, *228*, 107276. [CrossRef]
24. Jia, Z.; Liu, Z.; Vong, C.M.; Wang, S.; Cai, Y. DC-DC Buck circuit fault diagnosis with insufficient state data based on deep model and transfer strategy. *Expert Syst. Appl.* **2023**, *213*, 118918. [CrossRef]
25. Wang, S.; Liu, Z.; Jia, Z.; Li, Z. Composite fault diagnosis of analog circuit system using chaotic game optimization-assisted deep ELM-AE. *Measurement* **2022**, *202*, 111826.
26. Zhao, K.; Jiang, H.; Wang, K.; Pei, Z. Joint distribution adaptation network with adversarial learning for rolling bearing fault diagnosis. *Knowl.-Based Syst.* **2021**, *222*, 106974.
27. Saeed, R.; Mehdi S, A.; Stefania, S.; Francesco, F. Fault diagnosis in industrial rotating equipment based on permutation entropy, signal processing and multi-output neuro-fuzzy classifier. *Expert Syst. Appl.* **2022**, *206*, 117754. [CrossRef]
28. Ding, X.; Zhang, X.; Ma, N.; Han, J.; Ding, G.; Sun, J. RepVGG: Making VGG-style ConvNets Great Again. *IEEE CVPR* **2021**, 13733–13742.
29. Hu, J.; Shen, L.; Albanie, S.; Sun, G.; Wu, E. Squeeze-and-excitation networks. *IEEE Trans. Pattern Anal. Mach. Intell.* **2020**, *42*, 7132–7141. [CrossRef]

sensors

MDPI

Article

A TCP Acceleration Algorithm for Aerospace-Ground Service Networks

Canyou Liu [1,*], Jimin Zhao [2], Feilong Mao [3], Shuang Chen [1], Na Fu [1], Xin Wang [1] and Yani Cao [1]

1 State Key Laboratory of Astronautic Dynamics, Xi'an Satellite Control Center, Xi'an 710043, China
2 Space Star Technology Co., Ltd., Beijing 100086, China
3 Department of Electronic and Optical Engineering, Space Engineering University, Beijing 101416, China
* Correspondence: liu_yusi1@163.com

Abstract: The transmission of satellite payload data is critical for services provided by aerospace ground networks. To ensure the correctness of data transmission, the TCP data transmission protocol has been used typically. However, the standard TCP congestion control algorithm is incompatible with networks with a long time delay and a large bandwidth, resulting in low throughput and resource waste. This article compares recent studies on TCP-based acceleration algorithms and proposes an acceleration algorithm based on the learning of historical characteristics, such as end-to-end delay and its variation characteristics, the arrival interval of feedback packets (ACK) at the receiving end and its variation characteristics, the degree of data packet reversal and its variation characteristics, delay and jitter caused by the security equipment's deep data inspection, and random packet loss caused by various factors. The proposed algorithm is evaluated and compared with the TCP congestion control algorithms under both laboratory and ground network conditions. Experimental results indicate that the proposed acceleration algorithm is efficient and can significantly increase throughput. Therefore, it has a promising application prospect in high-speed data transmission in aerospace-ground service networks.

Keywords: network delay; packet loss rate; aerospace-ground service network; BoostTCP acceleration algorithm; bottleneck bandwidth and round-trip propagation time congestion control algorithm; cubic congestion control algorithm

Citation: Liu, C.; Zhao, J.; Mao, F.; Chen, S.; Fu, N.; Wang, X.; Cao, Y. A TCP Acceleration Algorithm for Aerospace-Ground Service Networks. *Sensors* **2022**, *22*, 9187. https://doi.org/10.3390/s22239187

Academic Editor: Hamed Kalhori

Received: 22 September 2022
Accepted: 23 November 2022
Published: 26 November 2022

1. Introduction

The distance between the sending and receiving ends of an aerospace-ground service network can exceed several thousand kilometers. Therefore, data transmission between the sending and receiving ends represents ultra-long-distance optical fiber transmission through a special line. It should be noted that without using a relay, the maximum effective transmission distance of an optical fiber is tens of kilometers, as optical signals attenuate to a certain extent to meet transmission bandwidth requirements. Accordingly, relay stations must be added to the transmission route to compensate for optical signal attenuation to realize ultra-long-distance transmission. However, the bit error rate (BER) increases with the number of used relay stations, which can result in packet loss and cause a packet error during data transmission. Namely, in ultra-long-distance optical fiber transmission over a special line, packet loss and an error in data transmission are caused by the attenuation of optical signals and BER, not by congestion on a physical link. However, in the standard TCP protocol, packet loss is treated as link congestion, thus reducing the transmission rate. Furthermore, this processing mechanism contradicts the reality of ultra-long-distance optical fiber transmission through a dedicated line, which results in bandwidth waste. The TCP protocol ensures data flow reliability using sequence confirmation and packet retransmission mechanisms. In addition, it achieves excellent adaptability under various network conditions and, thus, has significantly contributed to the rapid development and popularization of the Internet. However, the TCP protocol was designed more than two decades ago; consequently, it is unsuitable to model high-bandwidth, long-delay services in current ground networks. When packets are lost or delayed along the network path, the throughput of a TCP connection is significantly reduced. As a result, bandwidth is frequently underutilized, causing idle and unexploited bandwidth. Therefore, using the TCP will significantly increase long-distance data transmission and slow application response time, and it can even cause failure in data transmission. The literature [1,2] proposed some quantum logic gates and proved the success of the operations in implementing these gates. The literature [3–6] proposed a multi-qubit system consisting of two trapped ions coupled in a laser field. These devices may provide the next-generation design for quantum computers. To adapt to the current network characteristics of wide bandwidth and long delay, it is necessary to modify the TCP design to increase the transmission rate.

This article discusses the application of lightweight learning-based congestion control. The term "lightweight" refers to a type of congestion control algorithm that does not include deep learning, such as heuristic algorithms, utility functions, or gradient descent. A lightweight algorithm requires short training time and has a low cost, which makes it "light." In addition, it accelerates TCP transmission and improves TCP connection stability by improving the standard TCP protocol and its handling of congestion, and the algorithm can detect and compensate for packet loss accurately and in a timely manner.

2. Materials and Methods

The TCP protocol was developed based on the RFC793 standard document published by the Internet Engineering Task Force (IETF) in 1981. The early development of the TCP protocol considered the effects of a transmission environment on the transmission rate, and both sender and receiver employed a sliding window strategy to control the data flow dynamically. However, as network services became more complex, it has been found that a simple flow control considers only the receiver's accepting capacity. Nevertheless, from a macro perspective, the entire network contains a large number of routers and other network devices, and their storage and forwarding functions can affect the network's congestion. Still, relying only on the receiver's information cannot mitigate the effect of congestion by other network devices. This shortcoming caused a TCP collapse in 1986, resulting in a reduced link throughput between the LBL and UC Berkeley from 32 kbps to 40 kbps. Since then, researchers have recognized the critical nature of congestion control protocols, and pertinent research results have rapidly emerged [7,8].

The first congestion control algorithm was proposed by Van Jacobson et al. [9,10], which introduced mechanisms, such as slow start and congestion avoidance, for the first time. However, this type of algorithm immediately executes the slow-start strategy after judging the link as congested. This is because the link frequently reduces the size of the windows sent, impacting bandwidth utilization. In [11,12], a TCP-Reno algorithm was proposed to solve the bandwidth utilization problem by adding a fast recovery mechanism based on TCP Tahoe. Since the Reno algorithm can ensure network stability but not optimal resource utilization [13], in [14], a BIC algorithm, which consists of the binary searching and linear growth stages, was proposed. The Reno algorithm was modified in [15], and a modified TCP-Reno algorithm was developed. Further, in [16], the TCP-BIC algorithm was enhanced, and the TCP cubic algorithm, which improves the TCP-BIC algorithm's window adjustment method, was developed. The TCP cubic algorithm is a default congestion control algorithm of the current Linux and Android kernels.

The conventional TCP protocol is incapable of correctly distinguishing the causes of packet loss, and it performs only indiscriminate window-reduction operations, thus limiting future network transmission efficiency enhancements [17–19]. In [20], a performance-based congestion control (PCC) protocol and a rate control mechanism were proposed to address the two mentioned issues. The proposed algorithm could increase the network's transmission bandwidth, but the convergence rate was extremely slow. However, research on the PCC algorithm provided a large amount of information for subsequent analyses.

Further, Google proposed an innovative congestion control scheme in 2016 named the dubbed BBR (Bottleneck Bandwidth and Round-trip propagation time) [21,22]. Certain concepts in the BBR are consistent with the PCC algorithm. Nonetheless, in [21,22], it was demonstrated that the BBR algorithm performed excellently in environments with a high bandwidth, long delay, and high packet loss rate.

Unlike traditional congestion control algorithms, the BBR algorithm uses the bandwidth-delay product (BDP) [12] as an identification indicator rather than the packet loss or long transmission delay to identify network congestion. When the total number of data packets in the network exceeds the BDP value, the BBR algorithm considers the network congested. Therefore, the BBR algorithm can be referred to as a congestion-based control algorithm. It should be noted that it is impossible for network data flow to achieve both an enormous link bandwidth and a very small network delay simultaneously. Accordingly, the BBR algorithm detects network capacity regularly, measures maximum link bandwidth and minimum network delay alternately, and then uses their product to determine the congestion window size. The congestion window can be used to characterize network capacity, providing a more accurate identification of congestion. Because of the BBR algorithm's unique mechanism for measuring congestion window size, it neither increases the number of congestion windows indefinitely like ordinary congestion control algorithms nor uses the buffer of the switch node, thus avoiding the emergence of buffer bloat (buffer overflow) [13], which shortens the transmission delay significantly. Another advantage of the BBR algorithm is that it measures network capacity actively, adjusting the congestion window. In addition, the autonomous adjustment mechanism enables the BBR algorithm to control the data flow sending rate independently. In contrast, ordinary congestion control algorithms only calculate the congestion window, whereas the TCP protocol completely determines the data flow sending rate. As a result, when the data flow sending rate is close to the link's bottleneck bandwidth, there is data packet queuing or data packet loss due to the rapid increase in the sending rate.

Initially, the BBR drew great attention from researchers and was considered a paradigm-shifting achievement in the field of congestion control. However, with research progress, it has been discovered that the BBR protocol has several shortcomings, including a slow convergence speed in the bandwidth detection stage, a low sensitivity, and a lack of consideration for delay and jitter.

3. Transmission Acceleration Using BoostTCP

This paper proposes the BoostTCP, which represents a learning-based TCP transmission acceleration method based on transmission history learning. By improving the judgment and handling of congestion, the BoostTCP can judge and recover packet loss more accurately and rapidly, thus accelerating TCP transmission and increasing TCP connection stability.

3.1. Improved Congestion Judging and Handling Mechanism

Many congestion estimation and recovery strategies were developed for standard TCP over the last two decades to meet network requirements under different conditions. The fundamental premise was that a packet loss represented a result of congestion. However, this assumption does not hold for a transmission network with ultra-long-distance special-line optical fiber. In such a network, packet loss is typically caused by the BER of long-distance transmission, not by congestion-related factors. Therefore, standard TCP can frequently enter an excessively conservative transmission state. Meanwhile, when a network path contains deep-queue network devices, packet loss does not occur for a long period after the congestion occurs. The standard TCP is insensitive to congestion, resulting in excessive transmission, which not only affects network congestion but can also cause significant packet losses. As a result, the TCP enters a lengthy recovery phase for packet loss, resulting in transmission stagnation. All of these factors contribute to the poor performance of the standard TCP protocol for an ultra-long-distance optical fiber transmission network with a special line.

Considering both packet loss and delay variation, the proposed BoostTCP algorithm can dynamically learn the network path characteristics of each specific connection during data transmission, including end-to-end delay and its variation characteristics, the arrival interval of feedback packets (ACK) at the receiving end and its variation characteristics, the degree of data packet reversal and its variation characteristics, delay and jitter caused by the security equipment's deep data inspection, and random packet loss caused by various factors. These characteristics are monitored in real time and analyzed holistically to derive precursor signals and available bandwidth that reflect congestion and packet loss along the TCP connection network path. They also determine the degree of congestion and the transmission rate, and show whether the congestion recovery mechanism is compatible with the available bandwidth on the current path and can achieve accurate and timely packet loss judgment and recovery.

Based on network characteristics, the congestion degree and available bandwidth can be estimated accurately and in a timely manner. When congestion occurs, the transmission is realized based on the mentioned result. The unnecessarily slow data transmission rates caused by BER-induced packets can be avoided in an ultra-long-distance optical fiber transmission network with a special line. Specifically, the advanced congestion judgment and control algorithm of BoostTCP mainly uses the two following mechanisms: Prevent excessive conservative transmission and Prevent congestion deterioration.

3.1.1. Prevent Excessive Conservative Transmission

Because the current TCP protocol stack has difficulty determining the cause of a packet loss (caused by network congestion) and the actual bandwidth available on the connection path following the packet loss, restoration has been typically performed to reduce the transmission rate significantly. This mechanism results in an idle path bandwidth, which is one of the primary reasons for TCP's inefficient transmission performance.

The ultra-long-distance optical fiber transmission network with a special line often has sufficient bandwidth but a relatively long delay. In such a case, both the initial sending window and the current TCP protocol stack's sending window increase at a relatively conservative rate. BoostTCP begins sending data with a large initial sending window and rapidly increases the sending window size to reach the upper limit of available bandwidth in the shortest time.

In an ultra-long-distance optical fiber transmission network with a special line, the BoostTCP congestion judge algorithm considers network characteristics, determining whether packet loss results from network congestion or not. As packet loss occurs as a result of random errors in an optical fiber network, and the transmission rate increases and is maintained at a higher rate, the rate is adjusted instantly when real congestion occurs. Namely, the bandwidth closest to the available bandwidth on the current path is used to perform transmission, and a slightly lower transmission rate is used to clear the queue on the path, which contributes to the recovery of nodes in a congested network. Moreover, transmission behavior is maintained to be consistent with and related to the network state. The judge algorithm eliminates idle bandwidth, resulting in a faster, more consistent transmission rate.

3.1.2. Prevent Congestion Deterioration

Congestion may also occur in an ultra-long-distance special-line optical fiber transmission network due to a large number of networks and relays. In addition, congestion may worsen if not handled properly, causing two problems. First, the time required for retransmission and hole filling will be extremely long due to the high packet loss rate, and as a result, the TCP transmission window will become stuck for an extended period, and transmission will become slower or even fail. Second, retransmission is required due to increased packet loss; the retransmission rate increases while the effective data rate declines.

Therefore, users will notice that although online traffic increases, the actual application rate does not change.

BoostTCP determines the congestion degree in real time and slows it down to prevent congestion from deterioration and reduce the number of lost packets, resulting in faster and smoother transmissions with an effective data rate.

In summary, the BoostTCP congestion judgment algorithm is an automatic state machine that considers various network characteristics along the transmission path. Its function is to learn and improve congestion judgment skills intelligently in a connection-by-connection manner. Since learning the network characteristics requires data accumulation, and the BoostTCP algorithm allocates resources to each TCP connection, the optimal application scenario for the BoostTCP algorithm is a long-connection scenario rather than a high-concurrency scenario, which is satellite payload data in this article.

3.2. Fast Prediction-Based Packet Loss Judgment and Recovery Mechanism

The standard TCP protocol stack determines packet loss in two ways, based on the number of Dup-ACKs received at the receiving end and based on the ACK timeout. When a large number of packets are lost, the ACK timeout has been frequently used to determine the timeout condition and initiate a retransmission. It should be noted that packet loss is frequently sporadic in modern networks, and it is not uncommon for multiple data packets to be lost concurrently on a connection. As a result, the standard TCP protocol frequently relies on timeouts to retransmit data to fill gaps, resulting in a waiting state of several seconds, which can even last up to ten seconds. As a result, the transmission may pause for an extended period or even disconnect entirely, which can affect the standard TCP efficiency significantly.

In addition to the two methods used by standard TCP, the BoostTCP's packet loss judgment mechanism uses a dynamic self-learning algorithm to predict packet loss based on the network characteristics of the TCP connection path. The prediction algorithm considers network characteristic factors similar to those considered by the self-learning algorithm for BoostTCP congestion detection. The BoostTCP packet loss detection algorithm calculates a probability of loss for each packet sent but not confirmed by the other party's ACK. The probability changes as the transmission process continues. When the probability reaches a certain value, the algorithm considers the data packet lost and initiates retransmission immediately. This mechanism significantly reduces the likelihood of the TCP transmission relying on timeout and determining the packet loss, allowing it to fill holes faster, transmit data more smoothly, and achieve a higher average transmission rate. This packet loss-to-retransmission mechanism, which is faster than the standard TCP, is beneficial for maintaining faster and smoother data transmission in ultra-long-distance special-line optical fiber transmission networks.

Due to the untimely packet loss detection of standard TCP, its transmission efficiency is frequently very low, and transmission quality is unstable, which is difficult to predict and impacts user experience. BoostTCP acceleration can predict packet loss in real time and recover the lost packets on time. The transmission is smoother and faster, significantly improving the user experience.

3.3. Congestion Control Algorithm

The flowchart of the BoostTCP congestion control algorithm is presented in Figure 1. It defines the smoothed throughput rate, which can reflect the actual throughput rate and roundtrip time, and controls the growth mode of the congestion window (CWND) based on different factors, such as the actual throughput rate and roundtrip time. The smoothed throughput rate variation is used to determine the most suitable CWND growth mode, which follows the principle of maximum throughput rate. As long as increasing the CWND value improves the smoothed throughput rate, the CWND value will be continuously increased. However, BoostTCP does not use the smoothed throughput rate to determine the necessary CWND reduction, and the CWND value to reduce is determined based on packet loss.

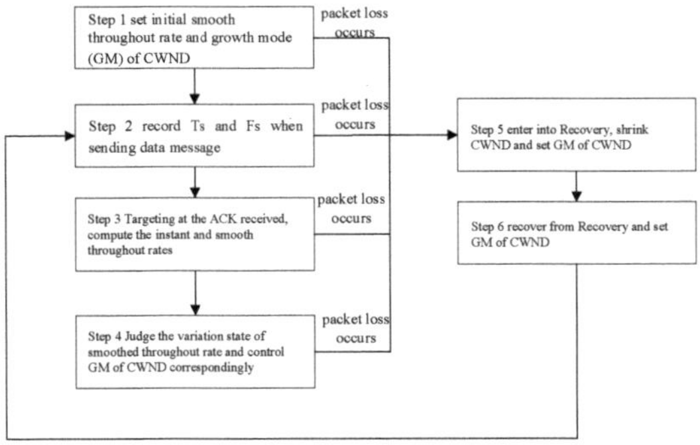

Figure 1. Flowchart of the BoostTCP congestion control algorithm.

CWND growth can be classified into three types: exponential growth, linear growth, and termination. The exponential growth mode assumes that the current CWND value is one. After the first, second, and third increases, the CWND value is two, four, and eight, and further changes follow the exponential trend. Each time the CWND value increases linearly, it is increased by a fixed value. It should be noted that the CWND value does not increase during the termination stage and remains constant.

The specific steps of the BoostTCP congestion control algorithm are as follows:

Step 1. In the initial state, set the smoothed throughput rate to B = 0 and the growth mode (GM) of CWND to exponential growth;

Step 2. Every time a new data package is sent, record the sending time T_S of the package and the total amount of data F_S that has been sent and has not been acknowledged (ACKed) yet;

Step 3. When the ACK response is received, if the ACK corresponds to one or more data packages that have been sent and there are no retransmitted data packages, the data package in the acknowledged messages with the highest sequence number (SEQ) is selected, and the following parameters called instant throughput rate and smoothed throughput rate are calculated:

The instant throughput rate is calculated according to $B_C = F_S/(T - T_S)$, where T denotes the current time; T_S denotes the sending time of the data package with the highest SEQ; and F_S denotes the total data amount that has been sent at the time T_S but has not been subject to ACK yet. As mentioned above, T_S and F_S are recorded when the data package with the highest SEQ is sent.

The smoothed throughput rate is obtained by $B = (1 - \alpha)B' + \alpha B_C$, where α denotes a constant parameter and B' denotes the previous smoothed throughput rate set in the initial state or obtained in the previous calculation iteration. BoostTCP uses a first-order exponential smoothing formula to compute the smoothed throughput rate. This is because network delay often fluctuates constantly due to various reasons, causing the real-time throughput rate to fluctuate accordingly. After smoothing, some high-frequency noise can be eliminated, and the network throughput can be estimated more accurately;

Step 4. Determine the variation state of the smoothed throughput rate B and control the CWND growth mode GM accordingly. Particularly, the two following situations are possible:

- If B is higher than the previous smoothed throughput rate set in the initial state or obtained in the last calculation and exceeds the set value γ, then the GM is an exponential GM;
- If B decreases three times in a row and the total amount of the three reductions is not less than the preset value of Δ, then judge the SRTT value: if SRTT $\leq \eta \cdot$ RTTMIN,

then the GM is a linear GM; otherwise, the GM is a termination GM. SRTT denotes the smooth roundtrip time, and RTTMIN denotes the minimum roundtrip time.

Step 5. If packet loss occurs at any time, set CWND = $\beta \cdot$ CWND and the GM to a termination GM when entering the recovery mode.

Step 6. Set GM to the exponential GM when exiting the recovery state as being recovered from the congestion state and perform operations similar to the above initial state. After exiting from the recovery mode, the smoothed throughput rate B is not cleared but performs operations similar to the initial state based on the original smoothed throughput rate.

The definitions and values of the above α, β, γ and other parameters are shown in Appendix A.

3.4. Implementation Architecture

The BoostTCP consists of several modules, as shown in Figure 2. The modules are explained in the following:

- Learning state machine: This is an information and control hub of BoostTCP, which accumulates knowledge about network paths and makes real-time decisions about the transmission of specific connections, such as the rate at which data are transmitted and the timing of data retransmission;
- Traffic monitor: This module extracts and learns the external features of each TCP flow and records and maintains the learning state machine;
- Packet loss monitor: This module monitors packet loss and determines the most probable cause of data loss using a learning state machine, for instance, whether the loss is caused by simple random packet drops or network congestion;
- Congestion controller: This controller executes the core congestion control logic based on a learning state machine;
- Exception handler: This module leverages knowledge of the learning state machines to identify flaws in peer TCP stacks or certain devices along the data transmission path, such as security detection devices. This module is used to detect specific characteristics of TCP to ensure maximum acceleration. Exception handlers also contribute to the knowledge accumulation of learning state machines;
- Window controller: This controller calculates the size of the TCP broadcast window and balances incoming packets from the LAN and WAN sides;
- Resource manager: This module tracks and controls system resources, including memory and computing power, and dynamically balances system resource consumption across all active TCP flows. The knowledge of learning state machines is the input to resource management.

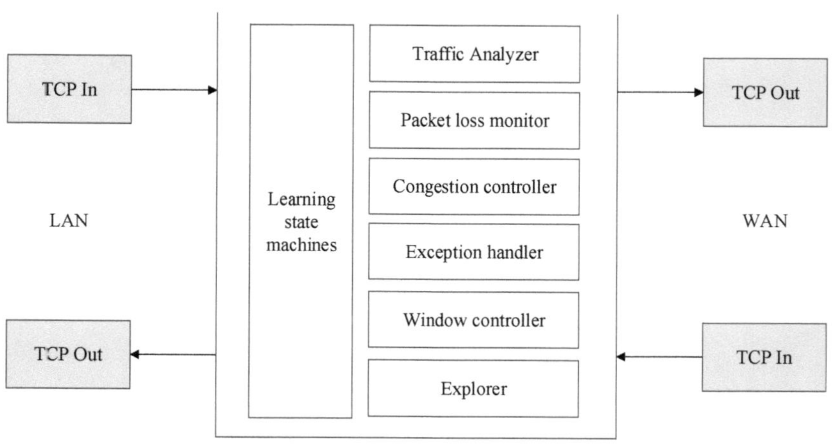

Figure 2. Schematic diagram of the BoostTCP modules.

3.5. Deployment Location

As an acceleration engine, BoostTCP follows the network driver interface specification and is located between the protocol stack and the hardware network interface card (NIC). It is fully compatible with the standard TCP protocol and does not attempt to replace the original TCP protocol stack in the operating system. When an application continues to interact with the TCP stack of the operating system in which it resides, BoostTCP is completely transparent to the application. When traffic is routed through the BoostTCP module, BoostTCP accelerates it by changing the timing of data packet transmission and retransmission without changing the data content or TCP encapsulation format. The position of BoostTCP in a multi-layer network architecture is presented in Figure 3.

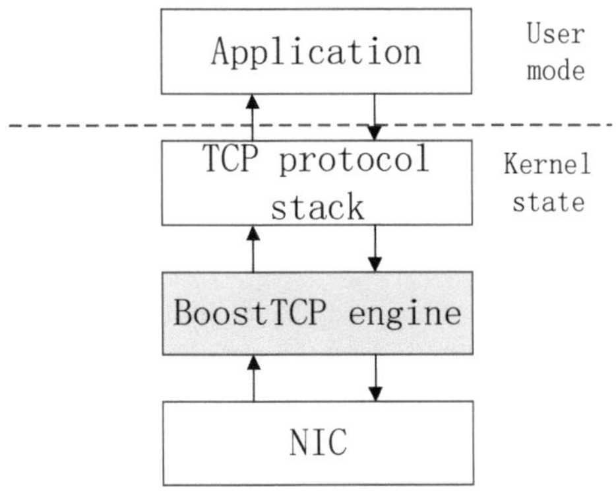

Figure 3. Illustration of the BoostTCP position in the multi-layer network transmission architecture.

4. Experimental Results

4.1. Experimental Environment

The study simulated the network environment with the TC and used the Reno, BBR, BoostTCP, and standard TCP algorithm in the simulation tests. The performance indicators of throughput, fairness, and preemption were compared for scenarios with varying bandwidth, delay, and random packet loss rate. The experiment was conducted on five Inspur NF5280M5 servers equipped with two Intel Xeon-GoXD 6136 (3.0 GHz/12-core) processors and 64 GB memory. The simulation network structure is presented in Figure 4. In the presented structure, Server 1 acted as a sender, employing the Cubic and Reno congestion control algorithms; Server 2 acted as a sender, employing the BoostTCP algorithm; Server 3 acted as a sender, employing the BBR algorithm; Server 4 acted as a receiver; and, lastly, Server 5 acted as a simulated controller for the designed network environment. The two network ports on Server 5 were connected to the switch and Server 4, forming a network bridge. TC managed the delay and packet loss and simulated the environment of a wide-area network.

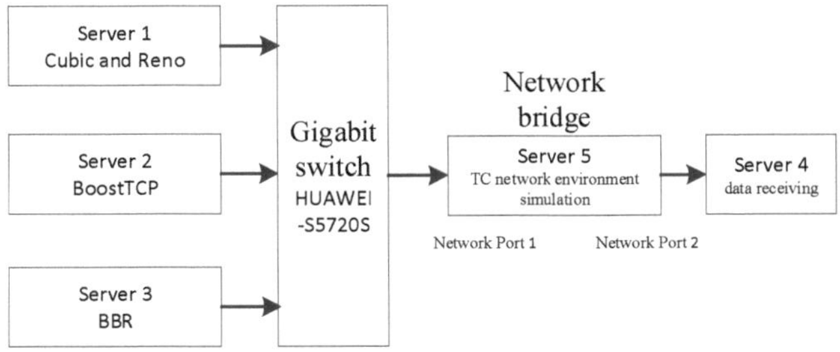

Figure 4. The experimental network structure.

The experimental topology was an end-to-end configuration with 1 Gbps network bandwidth. Multi data flows were sent from the sender to the receiver at the specified network delay and packet loss rate, with a default data packet size of 500 MB.

Four well-known congestion control methods were used in the experiment: three lightweight learning-based methods (Reno, BBR, and BoostTCP) and the conventional Cubic algorithm as a contrast method. To simulate the characteristics of a long-distance, long-delay, and low-packet-loss-rate data transmission in a real aerospace business network, the network delay in the experiment was set to 5 ms, 10 ms, 20 ms, 30 ms, 40 ms, 50 ms, 80 ms, and 100 ms, and five different random packet loss rates were used: zero, 0.01%, 0.05%, 0.1%, and 0.5%.

4.2. Results Analysis

To analyze the performance of the proposed method, it was tested and compared with the other algorithms regarding throughput, fairness, and preemptibility.

4.2.1. Average Throughput

The average throughput curves of the four congestion control algorithms at a bandwidth of 1 Gbps and network delays of 20 ms, 30 ms, 50 ms, and 80 ms are presented in Figures 5–8, respectively. The throughput was tested three times and averaged at each packet loss rate. The average throughput rate of BoostTCP was always the highest among all algorithms, and its predominant position became more apparent with packet loss, achieving a 26–41% enhancement over the BBR. The average throughput rate of the Reno algorithm was always the lowest among all algorithms. The average throughput rate of the Cubic algorithm decreased the most rapidly with the packet loss rate among all algorithms.

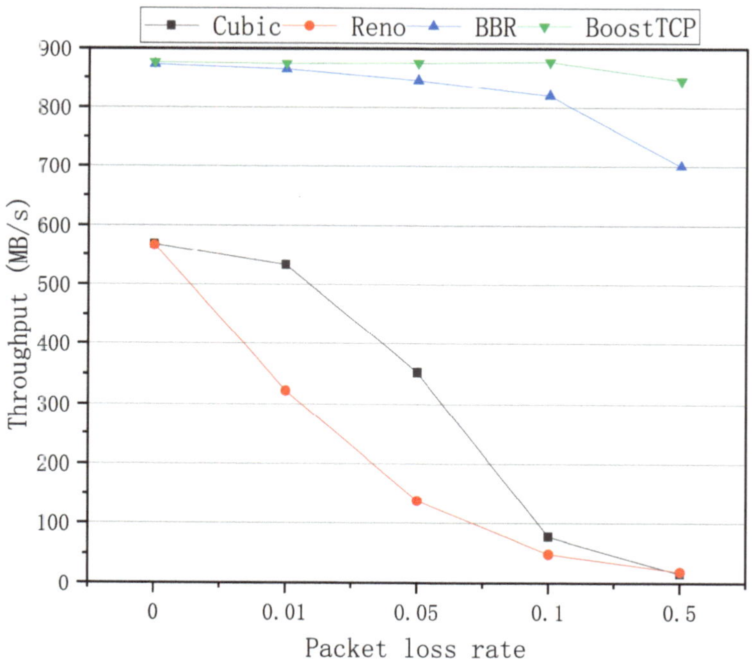

Figure 5. The throughput curves of the four congestion control algorithms as a function of the packet loss rate at a network delay of 20 ms.

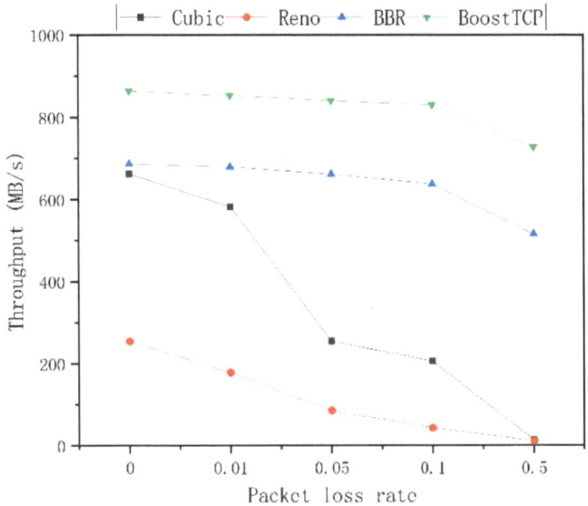

Figure 6. The throughput curves of the four congestion control algorithms as a function of the packet loss rate at a network delay of 30 ms.

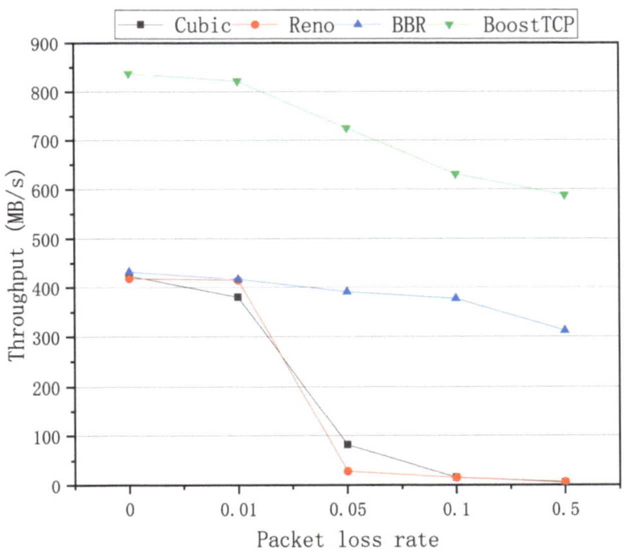

Figure 7. The throughput curves of the four congestion control algorithms as a function of the packet loss rate at a network delay of 50 ms.

The average throughput results of the four congestion control algorithms at a packet loss rate of 0.05% and a bandwidth of 1 Gbps is shown in Figure 9. The results presented in Figure 9 were averaged for each time delay. The average throughput rate of the BoostTCP algorithm was always the highest among all algorithms, and its advantage over the other algorithms became even more obvious at a longer time delay. Compared to the BBR algorithm, the average throughput of the BoostTCP algorithm was nearly identical in the early stages and increased to 2.3 times that of the BBR algorithm at a network delay of 100 ms. Among all algorithms, the Reno algorithm had the lowest average throughput. The average throughput rates of the Cubic and BBR algorithms decreased more rapidly than that of the BoostTCP algorithm with a time delay.

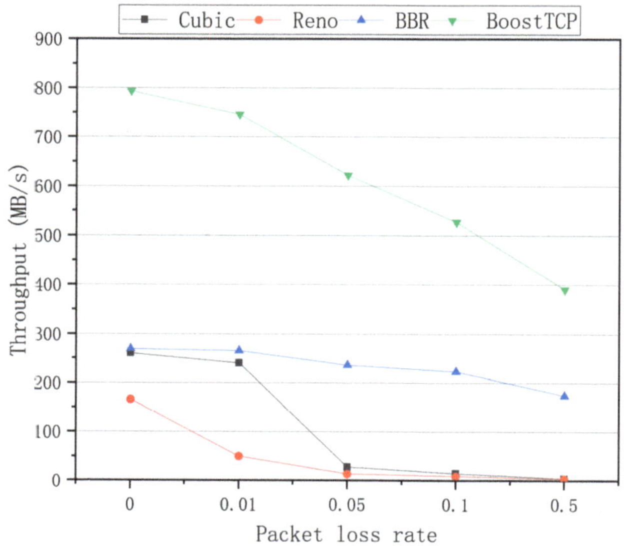

Figure 8. The throughput curves of the four congestion control algorithms as a function of the packet loss rate at a network delay of 80 ms.

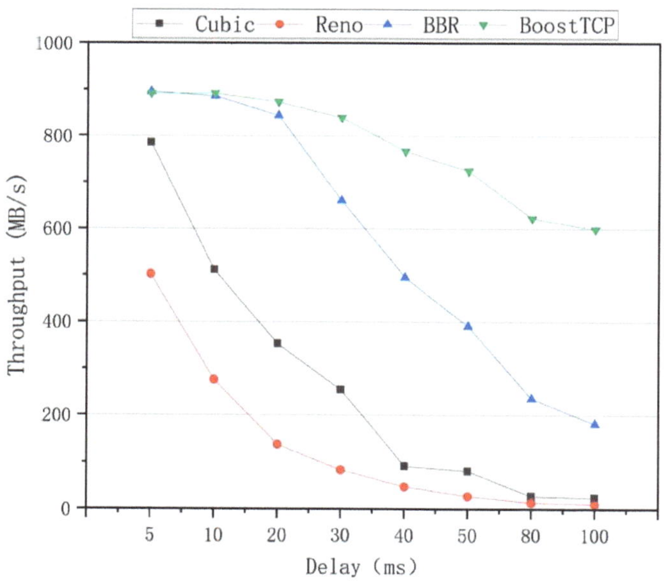

Figure 9. The throughput curves of the four congestion control algorithms versus the network delay at a packet loss rate of 0.05%.

According to the experimental results, BoostTCP had the highest average throughput under most conditions among all algorithms. This was because BoostTCP's bandwidth detection mechanism was based on learning transmission history and considered actual throughput and roundtrip time factors, which could fully use the link's excess bandwidth. Compared to the BoostTCP algorithm, the BBR algorithm's throughput was less, and the rate dropped more rapidly with a longer time delay. The reasons for this were that the convergence speed of the BBR algorithm was too slow, the sensitivity of the bandwidth detection stage was insufficient, and issues, such as delay and jitter, were ignored. The

Reno algorithm had the lowest average throughput for various random packet loss rates among all algorithms. Further, the Cubic algorithm had a higher throughput than the Reno algorithm, but the throughput rapidly decreased as the rate of random packet loss increased. Since the Cubic congestion control was based on packet loss, this was necessary. Random packet loss could significantly impact its judgment of network conditions, resulting in performance degradation.

The main idea of the BBR algorithm is to detect the maximum bandwidth and minimum roundtrip time continuously and alternately and then estimate overall network congestion using the two extreme values. Thus, the minimum roundtrip time accuracy is critical in determining the BBR algorithm's impact on network congestion. The ground network environment's primary characteristics are a high delay and sufficient bandwidth. In this case, the minimum roundtrip time is no longer capable of responding to network congestion accurately. Therefore, if the BBR algorithm continues to estimate the congestion window using the detected minimum roundtrip time, the estimated CWND value of the congestion window will be less than the link's actual ideal capacity. Further, reduced CWND limits the sender's sending rate, causing the bandwidth value measured by the BBR algorithm in detecting the link's maximum bandwidth to be less than the link's best achievable bandwidth. For instance, a lower maximum bandwidth results in a lower CWND value. As a result, the BBR algorithm can operate only at a reduced rate, thus causing significant network resource waste.

BoostTCP dynamically learns the network path characteristics of each TCP connection during transmissions, such as the end-to-end delay and its variation characteristics, the arrival interval and variation characteristics of the receiver's feedback data packet (ACK), the degree of data packet reversal and its variation characteristics, delay and jitter caused by deep data inspection by security equipment, and random packet loss caused by various factors. While tracking these characteristics in real time, BoostTCP analyzes them holistically and derives precursor signals that reflect congestion and packet loss along the TCP connection network path. Further, it determines the congestion degree based on the results of the dynamic, intelligent learning processes; determines the transmission rate and the congestion recovery mechanism that are compatible with the available bandwidth on the current path; and then performs the packet loss judgment and recovery accurately and in a timely manner. The BoostTCP algorithm can detect congestion in real time, automatically slow down, avoid mechanism congestion caused by excessively aggressive transmission, and accurately identify packet loss caused by random error codes. Thus, high-speed transmission is maintained and transmission behavior is smoother, which indirectly increases the effective data transmission rate.

4.2.2. The Fairness of Single Algorithm for Multiple Flows

To investigate the fairness of sharing link bandwidth when multiple flows coexist in the same scheme, this study sent one data flow at 0 s, 10 s, and 20 s in the test to determine whether three data flows can finally share the link bandwidth evenly, as well as the time required to evenly share bandwidth and reach convergence. Using a 1 Gbps link bandwidth, 100 ms network delay, and zero random packet loss rate as an example, the tested fairness of each scheme is summarized as follows.

The fairness results of the algorithms are presented in Figure 10, where it can be seen that the BBR had a higher throughput rate than the other algorithms when only one data flow was used. However, when two data flows of 10 s and 20 s are added, the throughput rates of the two data flows significantly differed. After 30 s, the throughput rates of the three data flows fluctuated, indicating that the BBR algorithm was unable to achieve an effective link bandwidth share. The results of the Cubic congestion algorithm for a network delay of 100 ms and a packet loss rate of zero are presented in Figure 11. After adding data flows at 10 s and 20 s intervals, the Cubic algorithm could average the throughput of three flows and ensure efficient link bandwidth sharing. The results of the BoostTCP congestion algorithm at a network delay of 100 ms and a packet loss rate of zero are presented in Figure 12. As demonstrated in Figure 12, the BoostTCP algorithm could maintain fairness between three data flows in a steady-state. However, the BoostTCP algorithm had a larger bandwidth-sharing fluctuation range than the Cubic algorithm, achieving an average value of 6.6 Mbps. BoostTCP could achieve a transmission rate of 320 Mbps after stabilization, which was faster than those of BBR and Cubic, indicating a more efficient use of network resources.

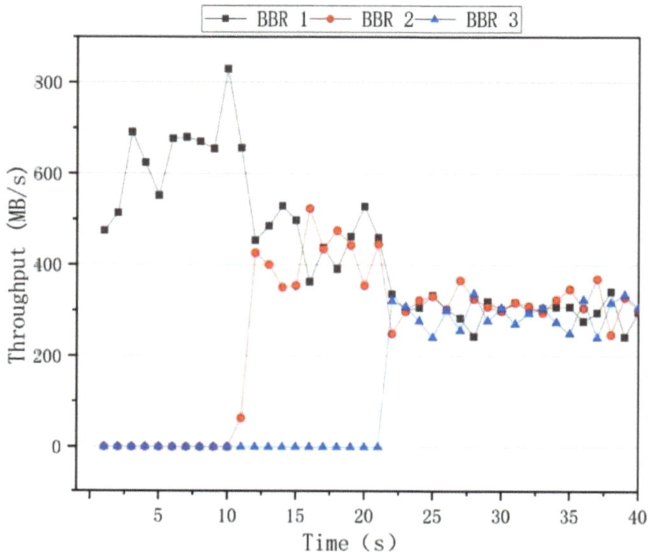

Figure 10. The fairness curves of the BBR congestion control algorithm at a network delay of 100 ms and a packet loss rate of zero.

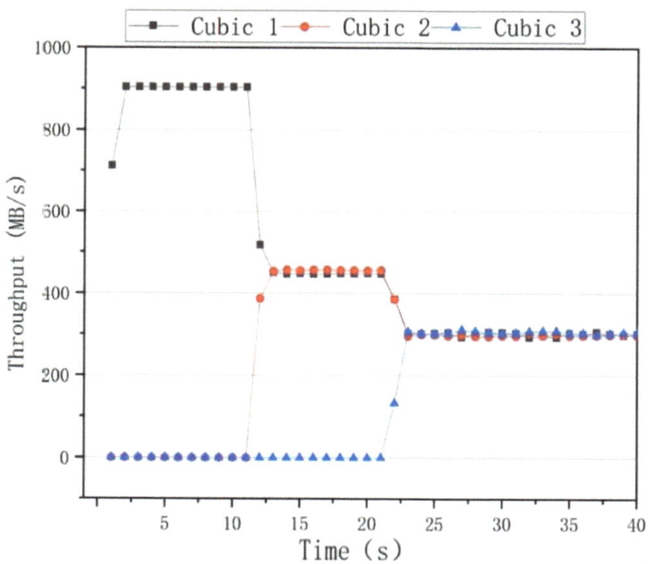

Figure 11. The fairness test curves of the Cubic congestion control algorithm at a network delay of 100 ms and a packet loss rate of zero.

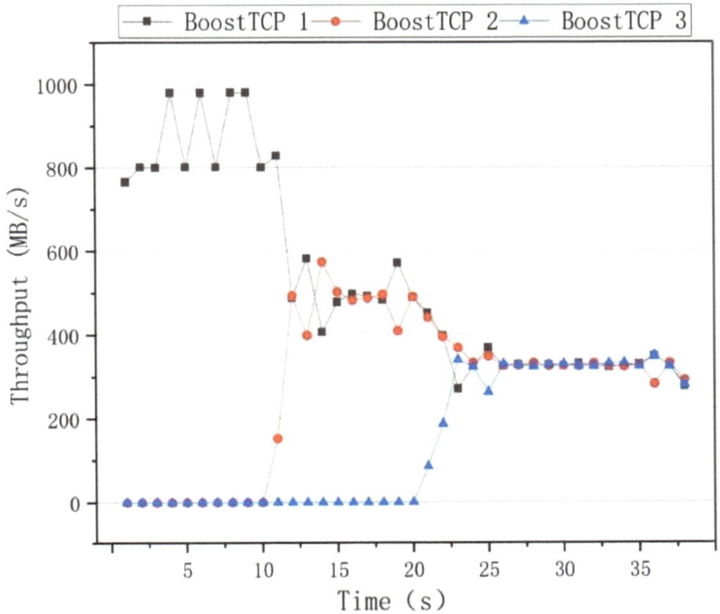

Figure 12. The fairness test curves of the BoostTCP congestion control algorithm at a network delay of 100 ms and a packet loss rate of zero.

4.2.3. Analysis Results of Preemption Ability

Different TCP connections have different bandwidth preemption levels in a real-world transmission network because they use different congestion control protocols. The bandwidth preemption level shows the ability to preempt bandwidth in terms of transmission performance. The greater the preemption capability is, the more efficiently network resources are used. The preemptive results of the BBR, Cubic, and BoostTCP algorithms are presented in the following figures.

Figures 13 and 14 illustrate the preemptive test curves of the congestion algorithms for the zero network delay and packet loss rate. The first test was conducted with the Cubic algorithm, followed by the BBR algorithm 10 s later. At the moment, the Cubic and BBR algorithms coexisted, and the Cubic algorithm severely preempted the BBR's bandwidth, resulting in no significant increase in the BBR algorithm's throughput. BoostTCP was restarted after 20 s, after which congestion occurred. The BoostTCP algorithm's throughput rate reached a stable value quickly, within 3 s, while the BBR algorithm's throughput rate gradually increased. After 35 s, the three algorithms' throughput rates converged to a steady-state. The BoostTCP algorithm had a higher throughput rate than the Cubic and BBR algorithms. The second test started with the BBR algorithm, and was followed by the Cubic algorithm 10 s later. Thus, the Cubic and BBR algorithms coexisted 10 s after the test began. The Cubic algorithm severely restricted the BBR's bandwidth, resulting in a throughput rate of nearly zero. After 20 s, the BoostTCP algorithm was invoked and congestion occurred. The BoostTCP algorithm's throughput rate reached a stable value quickly, within 3 s, while the BBR algorithm's throughput rate gradually increased. After 35 s, the three algorithms' throughput rates stabilized. The BoostTCP algorithm had a higher throughput rate than the Cubic and BBR algorithms.

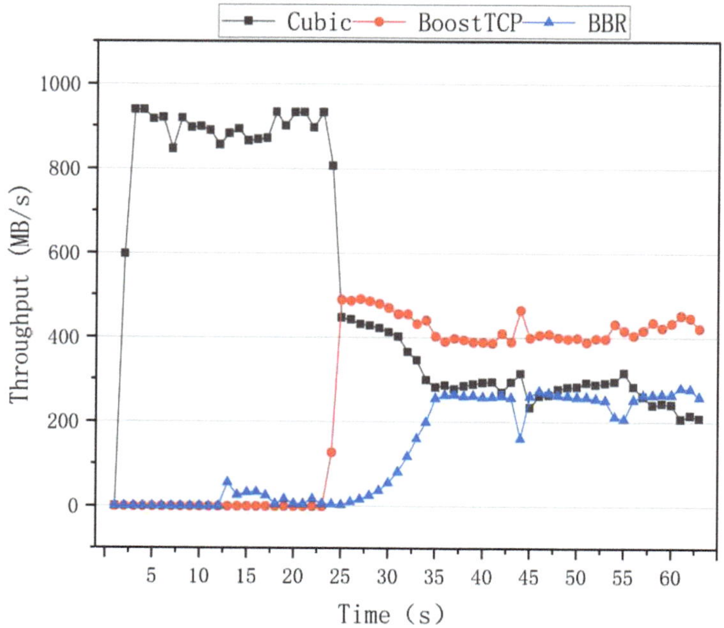

Figure 13. The preemptive test curves of the three congestion control algorithms at the zero network delay and packet loss rate.

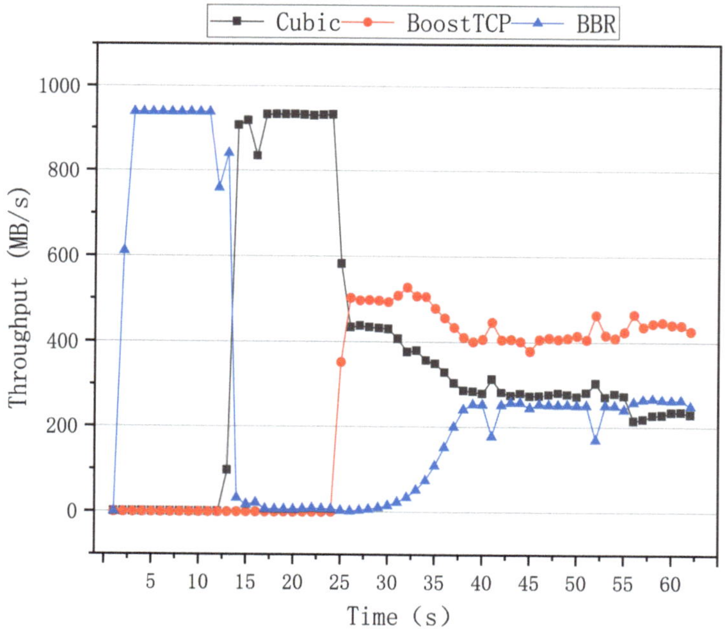

Figure 14. The preemptive test curves of the three congestion control algorithms at a network delay of zero and a packet loss rate of zero.

The preemptive test curves of the congestion algorithm for a network delay of 80 ms and a packet loss rate of zero are shown in Figure 15. The third test began with the Cubic algorithm, and the BBR algorithm was run after a 10 s delay. After that moment, the Cubic and BBR algorithms coexisted, and their throughput rates were essentially identical. After 20 s, the BoostTCP algorithm was started. Congestion occurred during this period. The BoostTCP algorithm's throughput rate rapidly stabilized after 3 s, whereas the BBR and Cubic algorithms' throughput rates decreased slightly. Finally, the three algorithms' throughput rates reached their steady states. The BoostTCP algorithm had a higher smoothed throughput rate than the Cubic and BBR algorithms.

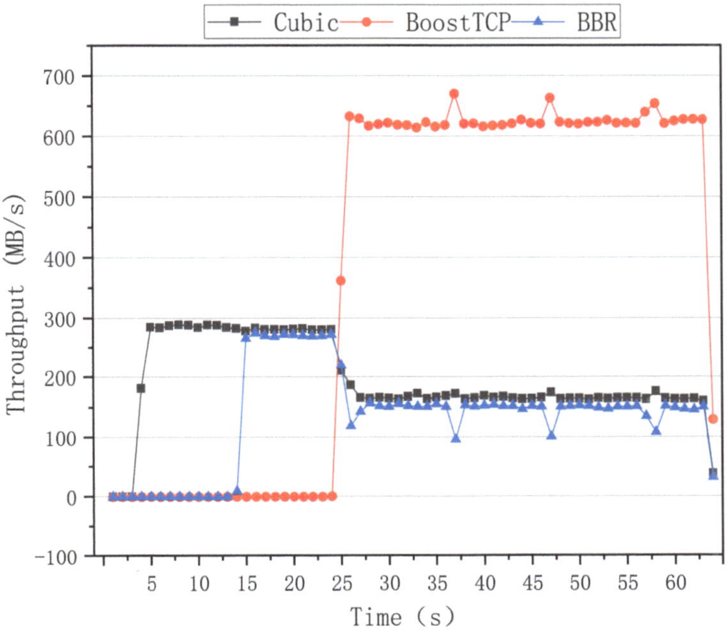

Figure 15. The preemptive test curves of the three congestion control algorithms at a network delay of 80 ms and a packet loss rate of zero.

The results of the three tests indicated that the BoostTCP algorithm had a better ability to preempt bandwidth than the BBR and Cubic algorithms under different delay conditions. Additionally, the results demonstrated that the BoostTCP algorithm was beneficial to the BBR and Cubic algorithms by assisting suppressed algorithms in resuming their normal throughput rates, demonstrating the BoostTCP algorithm's correctness.

4.3. Test in Actual Environment

To validate the BoostTCP algorithm's performance in real-world network transmission, a real-world network test was conducted analyzing the data transmission throughput rate between ground stations and satellite user centers. The real-world network test results of BoostTCP and standard TCP are presented in Figure 16, where the data transmission performances of the two algorithms were compared for a network consisting of seven ground stations and a satellite user center. The relationship between the increase in data transmission throughput rate and the network's maximum bandwidth is depicted in Figure 17. The relationship between the speed-up ratio of data transmission throughput rate and network delay is shown in Figure 18.

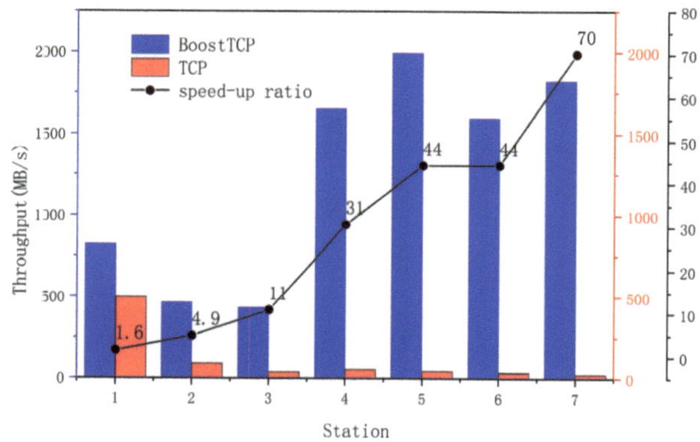

Figure 16. The actual network test results of the BoostTCP and ordinary TCP.

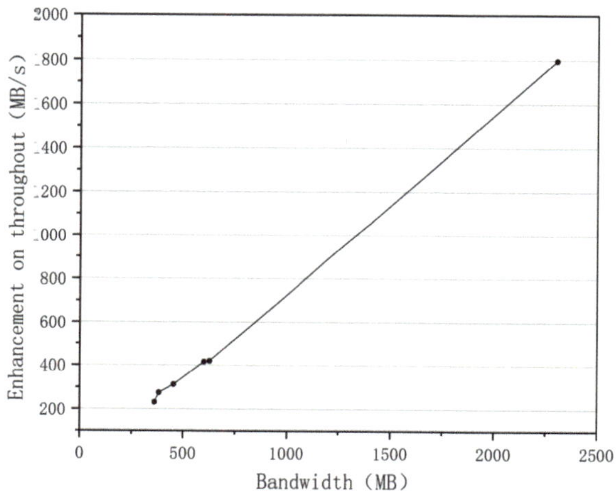

Figure 17. The throughout enhancement versus the maximum network bandwidth.

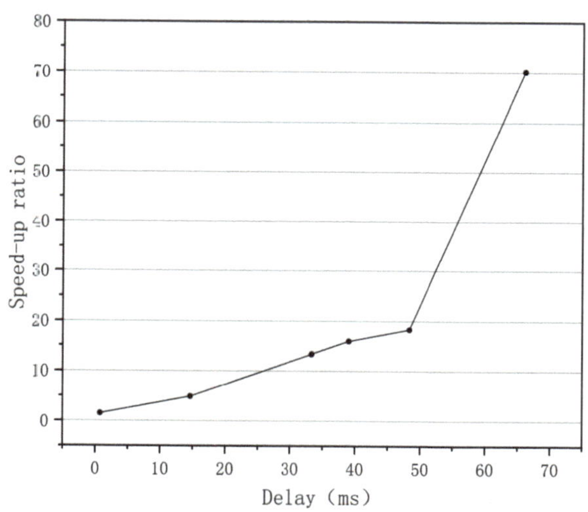

Figure 18. The speed-up ratio of data transmission versus the network delay.

As presented in Figure 16, the BoostTCP algorithm performed significantly better than the standard TCP algorithm. Data transmission rates between the satellite user center and seven ground stations were significantly increased. For instance, the speed-up ratio was typically tenfold and could reach seventyfold. The result indicates that the BoostTCP algorithm significantly increased data transmission throughput and effectively increased network resource utilization.

The relationship between the increase in data transmission throughput and the maximum network bandwidth is presented in Figure 17. As shown in Figure 17, when the network bandwidth increased from 300 MB to 2300 MB, the data transmission throughput rate increased significantly. The relationship between the speed-up ratio of the data transmission throughput rate and the network delay is displayed in Figure 18. When the network delay increased from 1 ms to 70 ms, the data transmission throughput rate's speed-up ratio increased proportionately. As a result, the greater the network bandwidth and delay were, the greater the performance advantage of the BoostTCP algorithm was. The measured data have conclusively demonstrated that the BoostTCP algorithm is more suitable for networks with high bandwidth and a long delay than the conventional TCP algorithm.

5. Conclusions

Due to the high precision requirements for satellite payload data transmission via a ground network, the TCP protocol can be considered competitive. However, the limitations of the standard TCP protocol on bandwidth utilization for networks with a long delay and large bandwidth reduce data transmission efficiency. This article uses TC to design a WAN simulation environment. Four congestion control algorithms—Reno, BBR, BoostTCP, and Cubic—are tested and compared in terms of throughput, fairness, and preemptibility. The results indicate that BoostTCP is more adaptable to network conditions and has a significantly higher throughput than the other three algorithms. In addition, it is fairly distributed across multiple data flows and has relatively strong preemption capability when multiple protocols are used. Finally, the throughput of BoostTCP is verified and tested in a real-world environment, and the results indicate that the real-world performance is identical to that in a simulated environment. Therefore, the proposed TCP acceleration algorithm can be used to improve the performance of ground networks when transmitting satellite payload data. In recent years, as a new field of quantum technology, the implementation of a quantum algorithm and quantum network has received increasing attention from scholars. In the next step, we will discuss the application of quantum algorithms and quantum networks in aerospace-ground service networks.

Author Contributions: Conceptualization, C.L.; methodology, C.L.; software, C.L. and J.Z.; validation, C.L., F.M. and S.C.; formal analysis, C.L. and J.Z.; investigation, N.F.; resources, X.W. and Y.C.; data curation, C.L.; writing—original draft preparation, C.L.; writing—review and editing, C.L. and S.C.; visualization, F.M.; supervision, N.F.; project administration, C.L.; funding acquisition, X.W. and Y.C. All authors have read and agreed to the published version of the manuscript.

Funding: This research received no external funding.

Institutional Review Board Statement: Not applicable.

Informed Consent Statement: Not applicable.

Data Availability Statement: Not applicable.

Conflicts of Interest: The authors declare no conflict of interest.

Appendix A

Algorithm A1. Pseudo-Code of Congestion Control Algorithm.

```
While(1)
{
Recv ACK //Compute window's smoothing rate
      B_C = Fs / (T − Ts);
      B = (1 − α) ∗ B′ + α ∗ B_C
If(B > B′) //window's smoothing rate increases
      {
            If((B − B′) > γ)
            CWND = CWND′ + ε^j
            Else
            CWND = CWND′ + ζ
            j++
            }
Else //window's smoothing rate decreases
      {
            i++
            if(i ≥ 3)
                  If((Σ_{k=1}^{3} (B′ − B)) < δ)
                        {
                        If(SRTT ≤ η ∗ RTTMIN)
                        CWND = CWND′ + ζ
                        Else
                        CWND = CWND′
                        }
                  Else
                        CWND = β ∗ CWND′
            else
                  CWND = CWND′ + ζ
            }
Recv NACK //packet loss occurs
      CWND = β ∗ CWND′
}
```

The meaning of the parameters in the preceding pseudo-code are as follows:

α: The value range is from zero to one, where the value of one refers to the computed throughput of the standard TCP; the closer the value is to zero, the longer the history tracking will be, which is actually a fixed constant that has been tuned and optimized;
β: The value range is from 0.5 to one, where the value of 0.5 indicates standard TCP congestion flow control. The closer the value is to one, the more aggressive it is. Three critical values have been tuned for the acceleration engine: maximum mode (β = 0.9), normal mode (β = 0.8), and conservative mode (β = 0.6). The conservative mode is nearly identical to the congestion control provided by standard TCP;
γ: A constant calculated based on the bandwidth of the acceleration engine, and it has basically the same meaning as the threshold parameter in the standard TCP. This threshold prevents congestion. If the increase in throughput exceeds the threshold value, the throughput increases rapidly. If the throughput continues to increase but remains below the threshold, the bandwidth threshold is reached and the throughput continues to increase linearly;
δ: A dynamically computed value based on the bandwidth setting in the acceleration engine. This value is used to judge whether the congestion is avoided or the bandwidth threshold is exceeded;
ε: A constant larger than two. The value for the standard TCP is two, and a larger value indicates stronger aggressiveness. The ε value of the current BoostTCP has a value of three.

ζ: An integer with a value range greater than one; one represents the constant value after tuning and fixing in the linear increase of standard TCP.

η: A constant that indicates the network's jitter tolerance and has a value range from one to two. The larger the constant value is, the less perceptible the response to the network jitter will be, which means that the occurrence of jitter cannot be easily detected. If η equals one, the constant has been tuned out and fixed when the network jitter is judged to be congested.

References

1. Obada, A.-S.F.; Hessian, H.A.; Mohamed, A.A.; Homid, A.H. Quantum logic gates generated by SC-charge qubits coupled to a resonator. *J. Phys. A Math. Theor.* **2012**, *45*, 485305–485310. [CrossRef]
2. Obada, A.-S.; Hessian, H.; Mohamed, A.-B.; Homid, A.H. A proposal for the realization of universal quantum gates via superconducting qubits inside a cavity. *Ann. Phys.* **2013**, *334*, 47–57. [CrossRef]
3. Abdel-Aty, M.; Bouchene, M.; Mcgurn, A.R. Entanglement rebirth of multi-trapped ions with trap phonon modes: Entanglement sudden death with recovery. *Quantum Inf. Process.* **2014**, *13*, 1937–1950. [CrossRef]
4. Obada, A.-S.F.; Hessian, H.A.; Mohamed, A.A.; Homid, A.H. Implementing discrete quantum Fourier transform via superconducting qubits coupled to a superconducting cavity. *J. Opt. Soc. Am. B Opt. Phys.* **2013**, *30*, 1178–1185. [CrossRef]
5. Buluta, I.; Ashhab, S.; Nori, F. Natural and artificial atoms for quantum computation. *Rep. Prog. Phys.* **2011**, *74*, 104401–104416. [CrossRef]
6. Homid, A.H.; Sakr, M.R.; Mohamed, A.-B.A.; Abdel-Aty, M.; Obada, A.-S.F. Rashba control to minimize circuit cost of quantum Fourier algorithm in ballistic nanowires. *Phys. Lett. A* **2019**, *383*, 1247–1254. [CrossRef]
7. Carofiglio, G.; Gallo, M.; Muscariello, L. Optimal multipath congestion control and request forwarding in information-centric networks: Protocol design and experimentation. *Comput. Netw.* **2016**, *110*, 104–117. [CrossRef]
8. Langley, A.; Riddoch, A.; Wilk, A.; Vicente, A.; Krasic, C.; Zhang, D.; Yang, F.; Kouranov, F.; Swett, I.; Iyengar, J.; et al. The QUIC Transport Protocol: Design and Internet-Scale Deployment. In Proceedings of the Conference of the ACM Special Interest Group on Data Communication, Los Angeles, CA, USA, 21–25 August 2017; pp. 183–196.
9. Jacobson, V. Congestion avoidance and control. *ACM SIGCOMM Comput. Commun. Rev. ACM* **1988**, *18*, 314–329. [CrossRef]
10. Stevens, W. TCP Slow Start, Congestion Avoidance, Fast Retransmit, and Fast Recovery Algorithms. *RFC 2001* **1997**. [CrossRef]
11. Bostic, K. V1.90:4.3 BSD-Reno/Fourth Berkeley Software Distribution. 1990. Available online: http://gunkies.org/wiki/4.3_BSD_Reno (accessed on 21 September 2022).
12. Padhyel, J.; Firoiu, V.; Towsley, D.F.; Kurose, J.F. Modeling TCP Reno performance: A simple model and its empirical validation. *IEEE/ACM Trans. Netw. (TON)* **2000**, *8*, 133–145. [CrossRef]
13. Li, L.; Xu, K.; Wang, D.; Peng, C.; Zheng, K.; Mijumbi, R.; Xiao, Q. A Longitudinal Measurement Study of TCP Performance and Behavior in 3G/4G Networks Over High Speed Rails. *IEEE/ACM Trans. Netw.* **2017**, *25*, 2195–2208. [CrossRef]
14. Xu, L.; Harfoush, K.; Rhee, I. Binary increase congestion control (BIC) for fast long-distance networks. In Proceedings of the INFOCOM 2004. Twenty-Third Annual Joint Conference of the IEEE Computer and Communications Societies, Hong Kong, China, 7–11 March 2004; Volume 4, pp. 2514–2524.
15. Floyd, S.; Henderson, T.; Gurtov, A. The New Reno Modification to TCP's Fast Recovery Algorithm. 2004. Available online: https://www.rfc-editor.org/rfc/rfc6582.txt (accessed on 21 September 2022).
16. Ha, S.; Rhee, I.; Xu, L. CUBIC: A new TCP-friendly high-speed TCP variant. *ACM SIGOPS Oper. Syst. Rev.* **2008**, *42*, 64–74. [CrossRef]
17. Li, Z. Research on TCP Congestion Control in Wireless Networks. Ph.D. Thesis, University of Electronic Science and Technology of China, Chengdu, China, 2013.
18. Huang, Y. Research on TCP Congestion Control in Wireless Networks Environment. Ph.D. Thesis, Hefei University of Technology, Hefei, China, 2009.
19. Raimagia, D.G.; Chanda, C.N. A novel approach to enhance performance of Linux-TCP Westwood on wireless link. In Proceedings of the Nirma University International Conference on Engineering, Ahmedabad, India, 28–30 November 2013; IEEE: New York, NY, USA, 2013; pp. 1–6.
20. Dong, M.; Li, Q.; Zarchy, D.; Zarchy, D.; Godfrey, P.B.; Schapira, M. PCC: Re-architecting Congestion Control for Consistent High Performance. *NSDI* **2015**, 395–408.
21. Cardwell, N.; Cheng, Y.; Gunn, C.S.; Yeganeh, S.H.; Jacobson, V. BBR: Congestion-Based Congestion Control. *ACM Queue* **2016**, *14*, 20–53. [CrossRef]
22. Shi, R.; Long, T.; Ye, N.; Wu, Y.; Wei, Z.; Liu, Z. Metamodel-based multidisciplinary design optimization methods for aerospace system. *Astrodynamics* **2021**, *5*, 185–215. [CrossRef]

sensors

MDPI

Review

Current Status and Applications for Hydraulic Pump Fault Diagnosis: A Review

Yanfang Yang [1], Lei Ding [1], Jinhua Xiao [1,*], Guinan Fang [1] and Jia Li [2]

[1] School of Transportation and Logistics Engineering, Wuhan University of Technology, Wuhan 430063, China
[2] Naval Submarine Academy, Qingdao 266071, China
* Correspondence: xiaojinh@whut.edu.cn

Abstract: To implement Prognostics Health Management (PHM) for hydraulic pumps, it is very important to study the faults of hydraulic pumps to ensure the stability and reliability of the whole life cycle. The research on fault diagnosis has been very active, but there is a lack of systematic analysis and summary of the developed methods. To make up for this gap, this paper systematically summarizes the relevant methods from the two aspects of fault diagnosis and health management. In addition, in order to further facilitate researchers and practitioners, statistical and comparative analysis of the reviewed methods is carried out, and a future development direction is prospected.

Keywords: hydraulic pump; fault diagnosis; fault prediction; remaining service life prediction; health status monitoring

Citation: Yang, Y.; Ding, L.; Xiao, J.; Fang, G.; Li, J. Current Status and Applications for Hydraulic Pump Fault Diagnosis: A Review. *Sensors* **2022**, *22*, 9714. https://doi.org/10.3390/s22249714

Academic Editors: Yongbo Li and Bing Li

Received: 16 November 2022
Accepted: 8 December 2022
Published: 11 December 2022

1. Introduction

Hydraulic systems are applied to all crucial mechanical equipment and play an irreplaceable role in the field of industrial production and manufacturing [1]. As the "heart" of the hydraulic system, the hydraulic pump is responsible for converting mechanical energy into hydraulic energy and providing pressure oil for the system [2]. With the development of the hydraulic industry, the structure of hydraulic pumps becomes more and more complex, and the probability of failure also increases; When it breaks down, it may cause the equipment controlled by the system to shut down for a long time, thus reducing the efficiency of the production process, bringing economic and safety problems, and even causing casualties in serious cases [3]. Therefore, it is of great practical significance to make reasonable and accurate fault diagnoses for hydraulic pumps; Under the premise of fault diagnosis, fault prediction, remaining service life prediction and health state detection can further master the safety of the hydraulic pump in operation, which is more conducive to improving the flexibility of the system, so as to prevent the occurrence and development of catastrophic faults in industrial systems, resulting in major losses.

The fault diagnosis method of hydraulic pumps mainly uses different sensors to collect different kinds of state monitoring signals of the hydraulic pump to analyze and reflect the change in the operating state of a hydraulic pump [4]. These state monitoring signals mainly include vibration signals [5], temperature signals [6], flow signals [7], and pressure signals [8], but other signals that can characterize the change of the operating state of the hydraulic pump also belong to the state monitoring signals [9]. Hydraulic pump fault diagnosis methods mainly include signal processing methods [10] and artificial intelligence methods [11], as well as mechanism analysis-based diagnosis methods [12]. The structural composition and operation mechanism of the hydraulic pump is complex, so it is difficult to quantitatively diagnose the fault under the mechanism analysis method. In different operating states of the hydraulic pump, the state monitoring signals present different information, and it is feasible to diagnose faults according to the information presented by the monitoring signals. With the development of artificial intelligence, fault diagnosis can be carried out by analyzing the signal data information when the operating mechanism of the hydraulic pump is fuzzy. In the process of fault diagnosis, there are two crucial problems: one is which state monitoring signals are selected as characteristic signals. The second is how to build a fault diagnosis model. On the premise of fault diagnosis, the "fault threshold" of various faults is extracted, and early fault prediction, remaining service life prediction, and health state detection can be carried out for the hydraulic pump. In view of the above problems, more and more research and investigations have been conducted in recent years, but there is a lack of a timely summary of the developed methods. The purpose of this paper is to provide the latest research progress and application.

This paper takes the hydraulic pump as the research object and analyzes the application and development of hydraulic pump fault diagnoses in recent years. Collate the articles on fault diagnosis and health management of various hydraulic pumps, and analyze and summarize the articles; Summarize the main causes of hydraulic pump failure; The methods used for fault diagnosis of hydraulic pumps are classified, and the paper evaluation index is proposed to evaluate the selected articles; The methods used for fault prediction, remaining service life prediction and health state detection of hydraulic pumps are described; Finally, the selected articles are statistically analyzed, and the research prospect of hydraulic pump fault diagnosis is given. The research flow of this paper is shown in Figure 1.

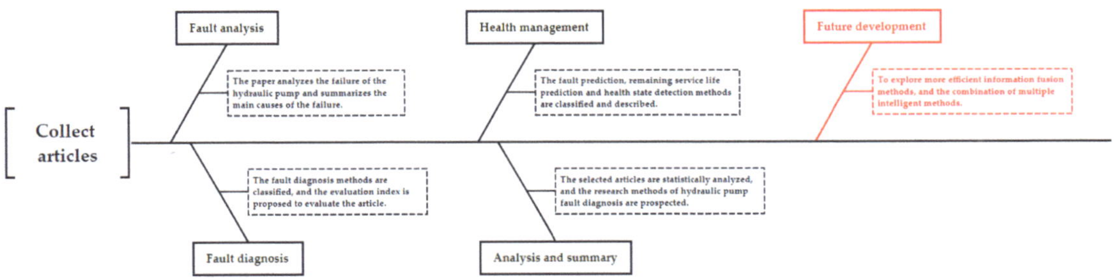

Figure 1. The research process of the paper.

This paper is structured as follows. Section 1 explains the importance and challenges of hydraulic pump fault diagnosis for application. Section 2 introduces the research on hydraulic pump faults in published papers and summarizes the fault types. Section 3 proposes the classification scheme of hydraulic pump diagnosis methods and summarizes the application of these methods. Section 4 briefly mentions the research and application of health management of hydraulic pumps. Section 5 makes a statistical analysis of the published papers and outlines future research trends. Section 6 gives a summary of this paper.

2. Fault Analysis of Hydraulic Pump

According to the different structures, hydraulic pumps can be divided into gear-type hydraulic pumps, vane-type hydraulic pumps, plunger-type hydraulic pumps, and screw-type hydraulic pumps. Although the components of various hydraulic pumps are different, their oil supply principle is the same, and they all belong to positive displacement hydraulic pumps. Its working principle is essentially the change of the sealing volume, that is, the oil is sucked by the local vacuum formed by the gradual increase of the sealing volume on the side of the oil inlet of the hydraulic pump, and the oil is squeezed into the hydraulic system by the gradual decrease of the sealing volume on the side of the oil outlet.

After a certain period of normal operation of the hydraulic pump, its parts and components will be gradually worn and damaged, or when the hydraulic pump operates under abnormal conditions, various fault phenomena such as increased noise, increased vibration, and decreased flow will occur. The failure of the hydraulic pump may be caused by excessive wear or damage to certain parts in the structure of the hydraulic pump, so the failure of the whole hydraulic pump can be studied from the study of certain parts.

In the hydraulic pump, the rotation of the shaft drives the operation of the whole medium, so the shaft of the hydraulic pump is one of the research contents. Xu et al. [13] analyzed the cause of the driving shaft fracture by calculating the radial force of the driving shaft of the hydraulic gear pump and used the finite element analysis software Ansys to simulate and verify the correctness of the fault cause. Xiao et al. [14] used the life acceleration experiment to analyze the deterioration and failure of the shaft of the hydraulic gear pump, checked the static strength of the broken part, and analyzed the main reasons for the failure of the shaft. Shawkis et al. [15] analyzed the annular crack on the drive shaft of the high-pressure hydraulic screw pump and concluded that one of the reasons for the shaft fracture was fatigue caused by misalignment during the rotation bending process. Xu et al. [16] believed that the main reason for the fracture was the increase of rotation and bending load caused by low viscosity medium through the analysis of macro morphology and microstructure, chemical composition, fracture metallography, and pump operation. Through the metallographic and fracture analysis of different parts of the hydraulic pump shaft, Yordanov B. et al. [17] can see the mixed characteristics in the morphology of the damaged surface, and conclude that the oxidation of the shaft surface and the intergranular corrosion at the grain boundary are one of the reasons for the crack generation and fracture propagation.

In the hydraulic pump, there are faults caused by other parts and hydraulic oil. Li et al. [18] analyzed the mechanics and microstructure of the broken pump housing of the hydraulic gear and found the main reason for the failure of the pump housing. Sekercioglu T. et al. [19] used hardness, chemical analysis, and metallographic examination to analyze the broken gear of the hydraulic gear pump, carried out geometric analysis of the gear of the hydraulic gear pump, and obtained the reason for the fracture of the gear of the hydraulic gear pump. Pflum et al. [20] used the pressure sensor to detect the detection signal in the narrow band frequency domain to analyze the spalling of the mechanical bearing of the hydraulic pump and the failure of the hydraulic gear pump. Hemati et al. [21] used signal processing technologies such as mechanical spectrum, envelope spectrum, and acceleration spectrum to conduct vibration analysis and signal processing of the hydraulic gear pump, and studied the failure of the hydraulic gear pump caused by the looseness of the bearing bush. Lee et al. [22] analyzed the characteristics of hydraulic oil, calculated the friction heat value, and analyzed the phenomenon that caused the failure to study the cause of the failure of the pilot check valve of the hydraulic pump caused by hydraulic oil pollution and leakage. Wang et al. [23] conducted the vibration fatigue test of the flameproof housing of the hydraulic pump regulator and analyzed the factors that caused the housing failure.

In addition to single-component failures, there are also some combined failures. By analyzing the structure and working principle of the external gear hydraulic pump, Zhang et al. [24] analyzed the failure of the external gear pump and proposed corresponding failure solutions. Das et al. [25] analyzed the microscopic cause of rapid wear of hydraulic pumps from the influence of the microstructure of hydraulic gear pump on the corrosion wear behavior of materials. Jiang et al. [26] carried out detailed statistics on various failures of screw pumps to analyze the failure modes of hydraulic screw pumps. Milović et al. [27] took the damage of the high-pressure three-screw oil pump in the regulating oil of the hydropower station as an example to analyze the failure of screw pump wear, thread tear, and filter blockage. Shang et al. [28] analyzed the failure and main causes of hydraulic pump damage and proposed corresponding effective solutions. Hidayath et al. [29] comprehensively considered the hydraulic pump failure caused by hardware and hydraulic oil. UłAnowicz et al. [30] established a simplified three-dimensional solid model of the cylinder piston assembly and gave the piston cylinder block, the inclination adjustment mechanism of the axial-flow hydraulic pump, and the fracture load model of the selected components of the pump, and discussed the actual damage of the axial piston pump.

When studying hydraulic pump faults, there are methods based on mechanism analysis and modeling, software simulation, signal fusion, and artificial intelligence. Fabiś-Domagała et al. [31] proposed the method of combining FMEA matrix analysis and Error Diagram to analyze the fault of the hydraulic gear pump and find out the factors causing the fault. Major et al. [32] proposed the fatigue failure finite element model of screw pump for the most serious fatigue fracture failure of a reciprocating screw of screw pump and carried out model simulation in Ansys. Ma et al. [33] established a simulation model of the hydraulic system by using AMESim software and analyzed the failure modes and mechanisms of key components in the system and their failure effects. Lee et al. [34] proposed to use FMECA to carry out extensive fault analysis of hydraulic gear pumps and proposed to use MFCC combined with a random forest classifier (RFC) to extract features and identify faults of vibration signals.

For the hydraulic pump failures studied in the above literature, it is concluded that the main reason for the failure of the hydraulic pump is the wear of the hydraulic pump. The wear of the hydraulic pump is divided into the situations shown in Table 1.

Table 1. Wear classification.

Wear Type	Form Factor
Friction wear	The surface of the parts after manufacturing is always uneven when carefully observed with a magnifying glass. After the operation wear of the hydraulic pump, the metal particles fall off from the surface of the parts, and the uneven parts on the surface of the parts are relatively smoothed. If friction is continued later, deep marks or small-size wear will be produced. This kind of wear is normal natural friction wear.
Abrasive wear	According to the analysis of oil pollutants used in hydraulic pumps, more than 20% of the pollution particles are silica and metal oxides. These abrasive particles are the most serious components of pump parts wear. They are sandwiched between the surfaces of moving pair parts. When moving, they act as grinding sand, resulting in severe abrasive wear.
Pit wear	This is a kind of fatigue damage to hydraulic components. Under the action of alternating load, due to periodic compression and deformation, residual stress and metal fatigue will occur, resulting in tiny cracks on the parts, which will gradually cause small pieces of parts to peel off.
Corrosive wear	The surface of the hydraulic pump components is subjected to corrosive substances such as acids and moisture in the oil, and the metal surface is gradually damaged.

3. Failure Diagnosis Method

The idea of hydraulic pump fault diagnosis based on condition monitoring signals is to collect the condition monitoring signals by sensors, then use signal processing methods to pre-process the collected status monitoring parameters, and then combine the fault diagnosis model to diagnose faults. In this investigation, based on the correct signal acquisition process, the hydraulic pump fault diagnosis methods are divided into the following three categories:

(1) Fault diagnosis based on a single signal;
(2) Fault diagnosis based on multi-signal;
(3) Other diagnostic methods.

3.1. Fault Diagnosis Based on Single Signal

At present, among the fault diagnosis methods based on a single signal, the vibration signal is the most widely used condition monitoring signal as the feature input of the fault diagnosis model. This is because once the internal parts of the pump fail, it usually causes changes in the characteristics of the load state structure and other characteristics, so the vibration response of the pump structure will change. Through the measurement of structural vibration signals, and relying on the principle of signal analysis, specific fault information is extracted, and the fault diagnosis is realized by artificial intelligence or signal analysis. Additionally, a few are based on other types of state monitoring signals such as sound signals. In the methods of hydraulic pump fault diagnosis, there are two main categories: the method of hydraulic pump fault diagnosis based on signal processing and the method of hydraulic pump fault diagnosis based on artificial intelligence.

3.1.1. Fault Diagnosis Based on Vibration Signal

(1) Method based on signal processing

The vibration signal has been proven to be useful for fault diagnosis of hydraulic pumps, but it contains noise, interference, and other information without fault characteristics. Therefore, it is necessary to use effective signal processing methods to extract available fault information from vibration signals. The following article has conducted some research on noise removal of vibration signals.

Yu et al. [35] proposed an EWT-VCR fusion method based on EWT and VCR to deal with the nonlinear, multi-frequency, and noise data of vibration signals. Jiang et al. [36] used the method of combining EEMD and PCC to denoise the collected hydraulic pump vibration signals, converted the denoised data into snowflake images by using the symmetric polar coordinate method, and converted the obtained images into gray level co-occurrence matrix, and used the fuzzy c-means algorithm for fault diagnosis. In view of the problem that the vibration signal of the hydraulic pump will be polluted by stronger Gaussian and non-Gaussian noise, Zheng et al. [37] proposed using PSE to extract fault information, effectively highlighting fault features and suppressing noise pollution. Wang et al. [38] studied the DCT denoising method and the CNC denoising method in view of the serious noise problem in the vibration signal of the hydraulic pump. Finally, CNC denoising was adopted, and then HHT was used to extract the fault information of the signal. In order to reduce noise and other interference, Sun et al. [39] carried out local feature scale decomposition for high-frequency harmonic correction of vibration signals and proposed discrete cosine transform high-order spectrum analysis algorithm to extract singular entropy as the degradation feature of hydraulic pumps. Liu et al. [40]. proposed a new rough set fault diagnosis algorithm for hydraulic pumps guided by PCA, aiming at the characteristics of fuzzy fault features and low signal-to-noise ratio of hydraulic pumps, using WA for noise reduction processing, extracting effective fault features, using PCA method for dimensionality reduction and decoupling correlation analysis of these features, using rough set theory to establish a knowledge base of diagnosis rules. Hou et al. [41] proposed a WPD-based denoising method for hydraulic pump fault feature extraction to solve the problem that the feature signal is weak and covered by noise. Wang et al. [42] introduced the idea of WNC denoising in view of the problems of the DCT denoising method, proposed a CNC denoising method, and extracted fault features from the output signal by HHT, effectively solving the problem of missing vibration signal components.

Under the actual conditions, the fault information of hydraulic pumps is still relatively poor, so it is necessary to solve the problem of fault diagnosis under the condition of poor information. Jia et al. [43] proposed a fault diagnosis method based on SPIP and HMM in order to realize fault diagnosis in the case of poor information. This method converts vibration signals into symbol sequences as feature sequences of hidden Markov models, uses genetic algorithms to optimize the symbol space division scheme, and then uses hidden Markov models for fault diagnosis. In view of the shortage of single-scale arrangement entropy when measuring the complexity of vibration signals on a single scale,

Wang et al. [44] proposed an MPE entropy value and MMPE. The analysis results of the measured vibration signals of hydraulic pumps verified the effectiveness and superiority of this index as a fault feature of hydraulic pumps. Aiming at the problem of poor detection of fault signals of the hydraulic pump in the early stage, Yu et al. [45] proposed a method of using EWT to decompose the vibration signals of three channels, then defining VCR to divide the weights of components to form a single signal, and using HT to demodulate the characteristic frequency to achieve fault detection of the hydraulic pump. Deng et al. [46] proposed a fault diagnosis method based on EMMD and Teager energy operator demodulation to solve the problem of weak early fault vibration signals of the hydraulic piston pump.

In the process of feature extraction of vibration signal, the original primary method has some limitations, so it needs to be improved. Zheng et al. [47] proposed an IEWT-based signal processing method for hydraulic pump fault diagnosis in view of the serious over-decomposition problem of EWT. Jiang et al. [48] proposed a method of hydraulic pump fault signal demodulation based on LMD and IAMMA. Li et al. [49] proposed a hydraulic pump fault feature extraction method based on MCS and RE. According to the maximum relational entropy criterion and the progressive fusion strategy, a relative entropy algorithm was established to fuse the initial features into new degraded features.

Some comparison methods and processing of vibration signals from different angles can still play a role in fault diagnosis of hydraulic pumps. Gao et al. [50] compared and analyzed the two fault diagnosis methods of WT and spectrum analysis, and concluded that when analyzing the same vibration signal dataset, the diagnosis ability of the method based on WT was more accurate. Sun et al. [51] proposed a fault diagnosis method for hydraulic pumps based on a fusion algorithm that processes vibration signals successively through LCD and DCS to improve the characteristic performance of signals. Siyuan et al. [52] proposed a hydraulic pump fault diagnosis method based on PCA of Q statistics, which uses normal vibration signals to establish a principal component model and then compares it with the test samples obtained by Q statistics to diagnose faults. Wang et al. [53] proposed a fault diagnosis method based on WP and MTS. This method performs WPT on the collected vibration signals, removes redundant features by the Taguchi method, extracts principal components, and then uses an MD-based calculation method to diagnose hydraulic pump faults. Chen et al. [54] proposed a hydraulic pump fault diagnosis method based on compression sensing theory, which uses the original vibration signal of the hydraulic pump to construct a compression dictionary matrix, uses the Gaussian random matrix to compress the vibration monitoring data of the hydraulic pump and uses a SOMP algorithm to reconstruct the test data. Tang et al. [55] proposed a fault diagnosis method for hydraulic pump fault under variable load in order to solve the problem of dynamic characteristic analysis of hydraulic pumps, which collects vibration signals and uses the axial RMS trend gradient for fault diagnosis.

The fault diagnosis methods of hydraulic pumps based on signal processing have their own limitations, such as time domain analysis, which is easy to cause misjudgment when the fault is serious, has large randomness, and is not suitable for non-stationary signals; Frequency domain analysis cannot reflect time characteristics and is not sensitive to early faults; The multi-sensor information fusion method has some limitations, such as the difficulty of sensor configuration and management, and the complexity of fault information fusion algorithm design.

(2) Methods based on artificial intelligence

Although the signal processing method of vibration signal can effectively extract and express the fault information of hydraulic pumps, the speed and accuracy of its method to diagnose the fault of hydraulic pumps are not ideal. However, with the rapid development of artificial intelligence, more and more intelligent algorithms and models can quickly diagnose faults, and the self-learning ability of artificial intelligence makes the accuracy of diagnosis algorithms and models a high level. Therefore, the artificial intelligence method combining the signal processing method based on vibration signal feature extraction with the artificial intelligence diagnosis algorithm and model is more effective.

① Artificial intelligence method based on neural network

With the generalization ability of a neural network, more and more neural network models are applied to fault diagnosis of hydraulic pumps. The fully connected neural network has the ability of self-learning and searching for optimal solutions at high speed. It has the advantages of high accuracy and rapidity in the fault diagnosis of hydraulic pumps. Gao et al. [56] proposed a fault diagnosis method based on EMD and NN. Sun et al. [57] proposed a hydraulic pump fault diagnosis method based on ITD and softmax regression, which uses ITD to process the vibration signal of the hydraulic pump and trains the softmax regression model to diagnose possible fault modes. Ding et al. [58] used LMD to process the collected vibration signal data of the hydraulic pump to form a feature vector, trained the Softmax regression model with the reduced features, and obtained the fault diagnosis model of the hydraulic pump. Jikun et al. [59] proposed a fault diagnosis method for hydraulic pumps based on WPT and SOM-NN. This method uses WPT to extract features from vibration signals, and SOM-NN trains through normal samples and fault samples to diagnose faults when they occur.

Although a fully connected neural network has high accuracy, it needs a lot of trainable variables, which is prone to model overfitting, and model convergence speed needs to be improved. The convolutional neural network can further extract the features of the input through the convolution kernel, and the trainable parameters of the model are greatly reduced by sharing the convolution kernel. Tang et al. [60] proposed an intelligent fault diagnosis method for hydraulic pumps based on CNN and CWT, which uses CWT to convert the original vibration signal into image features, and establishes a new deep convolutional neural network framework that combines feature extraction and classification, and can further improve the convergence speed of the model by optimizing the CNN's hyperparameters. Zhu et al. [61] proposed an improved AlexNet intelligent fault diagnosis method based on WPA combined with changing the network structure, reducing the number of parameters and computational complexity. Tang et al. [62] proposed a normalized convolutional neural network (NCNN) framework based on a batch normalization strategy for feature extraction, and then used a Bayesian algorithm to automatically adjust the model hyperparameters. BP neural network was used for fault diagnosis based on synchronous noise wavelet transform of vibration signals. Yan et al. [63] proposed a simple 7-layer CNN network setting method based on a base-period to realize fault diagnosis of hydraulic pumps. Zhu et al. [64] improved the core size and number based on the standard LENet-5 model, added a batch normalization layer to the network architecture, and built a PSO-Improve-CNN fault diagnosis model based on vibration signals by automatically optimizing the model's hyperparameters through PSO. Tang et al. [65] established an adaptive CNN hydraulic pump fault diagnosis model using Bayesian Optimization hyperparameters based on the Gaussian process by taking the time-frequency image of the vibration signal after CWT as input data. Tang et al. [66] converted the vibration signal into an image through CWT, preliminarily extracted effective features from the converted time-frequency image, built a CNN model to achieve fault diagnosis, and realized the visualization of simplified features by using T-DSNE.

In addition, there is also a new neural network model based on the improved functions in the neural network. Luc et al. [67] proposed a CPRBF-NN composed of multiple parallel-connected RBF subnets in combination with chaos theory and applied the proposed method in combination with vibration signals to fault diagnosis of hydraulic pumps. Huijie et al. [68] proposed to integrate the RELU activation function and Dropout strategy into SAE to directly train and identify vibration signals, forming a SAE-based fault diagnosis method for hydraulic pumps. Du et al. [69] proposed a method to extract 17 time-domain features of vibration signals, analyzed the sensitivity of features to the failure to select sensitive feature parameters, built a neural network diagnosis model, and formed a hydraulic pump fault diagnosis method based on sensitivity analysis and PNN. Dongmei et al. [70] took the vibration data as the input and the failure mode matrix as the target output to obtain a PARD-BP-based fault diagnosis method.

② Artificial intelligence method based on a support vector machine

Support vector machine (SVM), which originates from statistical learning theory, can be used for supervised learning, unsupervised learning, and semi-supervised learning, and it has an outstanding ability for both linear and nonlinear signals. Casoli et al. [71] collected vibration signals and used them to extract features for fault diagnosis, reduced the obtained features to reduce the amount of calculation, and used them to train different types of support vector mechanisms to build hydraulic pump fault diagnosis models. Tian et al. [72] proposed a fault diagnosis method based on WPT, SVD, and SVM. Lu et al. [73] proposed a new method for hydraulic pump fault diagnosis that combines EEMD and SVR models. This method uses a combination of GA and grid search to optimize the parameters of SVM. Fei et al. [74] proposed a fault extraction method combining WPA, FE, and LLTSA, and then proposed a hydraulic pump fault diagnosis method combining SVM. Niu et al. [75] proposed a hybrid fault diagnosis method for hydraulic pumps that combines the RNS algorithm and SVM. Zhao et al. [76] proposed that CEEMD is used to decompose the signal, then STFT and TFE are used to extract the fault features, and multi-class SVM is used to diagnose the fault of the hydraulic pump. Hu et al. [77] proposed the SS-SVM fault diagnosis algorithm, which constitutes a multi-fault classifier for hydraulic pump fault diagnosis. This method requires only a few fault data samples for training the classifier and has strong fault diagnosis ability in the case of small samples. Tian et al. [78] proposed a degradation feature extraction method for hydraulic pumps based on ILCD and MF, and input the degradation feature into BT-SVM for fault diagnosis of hydraulic pumps.

③ Artificial intelligence method based on a limit learning machine

In essence, the limit learning machine maps the input feature data to the random space and then uses the least square linear regression. Its advantages are that the hidden layer does not need iteration, the learning speed is fast, and the generalization performance is good. Li et al. [79] proposed a comprehensive fault diagnosis method for hydraulic pumps based on MEEMD, AR spectral energy, and WKELM method. Ding et al. [80] proposed a fault diagnosis method combining EWT, PCA signal processing method, and ELM. Liu et al. [81] proposed a time series dynamic feature extraction method based on CEEMDAN and CMBSE, based on a hydraulic pump fault diagnosis method combining t-SNE and WOA-KELM was proposed. Lan et al. [82] proposed an intelligent fault diagnosis method for hydraulic pumps based on WPT, LTSA, EMD, LMD multiple signal processing technology, and ELM identification technology.

④ Artificial intelligence method based on fuzzy theory

The structure of the hydraulic pump is complex, and the causes of the failure of the hydraulic pump cannot be completely divided, which has certain fuzziness. Therefore, the fuzzy set and membership function of the hydraulic pump can be constructed, and the fault of the hydraulic pump can be diagnosed using the method of fuzzy theory. Wang et al. [83] proposed a method to capture the degraded characteristic signal of SIE and then used the vibration signal combined with the FCM algorithm to build a hydraulic pump fault diagnosis method. Wang et al. [84] proposed a rough set method for mechanical fault diagnosis, which extracts the spectral features of vibration signals as the attributes of learning samples, and uses a set of decision rules obtained from the upper and lower approximation of decision classes as a rough classifier. Wang et al. [85] extracted diagnostic features from the spectrum of vibration signals, processed the spectrum representing a variety of different fault states using fuzzy membership function, and made fuzzy comprehensive discrimination according to anti-fuzzy diagnostic rules, thus realizing correct diagnosis of different fault spectra. Mollazade et al. [86] studied a new method of hydraulic pump fault diagnosis based on vibration signal PSD combined with DT and FIS.

The method based on a neural network is to extract fault features by signal processing, then use a neural network as the fault diagnosis model, that is, the fault mode analysis after fault signal processing, so as to realize the nonlinear mapping from fault symptoms to fault causes. The diagnosis reasoning process of this method is not clear and the diagnosis explanation is not intuitive. The fuzzy reasoning method is suitable for dealing with uncertain and incomplete information in pump fault diagnosis. Its disadvantage is that it is difficult to establish complete rules and membership functions, and its learning ability is poor.

3.1.2. Fault Diagnosis Based on Other Signals

In addition to the frequent vibration signals, some other condition monitoring signals also contain fault information about the hydraulic pump, and the new monitoring signals are accompanied by new analysis methods, which makes the fault diagnosis methods of the hydraulic pump more diversified. Shengqiang et al. [87] proposed a KPCA fault diagnosis method based on the sound signal, described the feature extraction of the acoustic signal, and used the KPCA method to diagnose the hydraulic pump fault in view of the unsuitable use of the hydraulic pump vibration sensor and the limitations of the fault diagnosis method based on vibration signal processing. Jiang et al. [88] proposed a fault diagnosis method for an axial piston hydraulic pump based on the combination of the MFCC feature extraction method and ELM. The MFCC voiceprint feature of the processed sound signal is extracted from the acoustic signal, and the ELM model is established for fault diagnosis. Based on the standard LeNet, Zhu et al. [89] used PSO to automatically select the hyperparameters of the diagnosis model and built a PSO-CNN hydraulic pump fault diagnosis model with acoustic signals as input.

Tang et al. [90] used CWT to obtain the time-frequency characteristics of the pressure signal, set the initial hyperparameters to establish a deep CNN, and then used the Bayesian optimization method to realize automatic learning of the main important hyperparameters to build an adaptive CNN-based hydraulic pump fault diagnosis method. Wang et al. [91] used FEMD to decompose the pressure signal and then extracted useful fault information from the signal through RE. This method also has a good ability to suppress noise. Liu et al. [92] proposed to use the instantaneous angular speed (IAS) signal obtained by the equal angle method to diagnose the hydraulic pump fault under non-stationary conditions.

The four major wear faults of hydraulic pumps summarized in the literature research are classified as Fault I: friction wear faults; Fault II: abrasive wear fault; Fault III: pit wear fault; Fault IV: corrosive wear fault. In addition, it further evaluates the paper from the following points:

Index I: enhance fault characteristics;
Index II: optimization of fault diagnosis algorithm;
Index III: adapt to strong noise environment;
Index IV: high diagnostic accuracy.

The above four types of faults and four types of evaluation indicators are applicable to this chapter. The application of fault diagnosis based on a single signal is shown in Table 2.

3.2. *Fault Diagnosis Based on Multiple Signals*

The fault information contained in the current single signal processing is limited. In order to increase the collection of fault information, the characteristic signals of multiple signals can contain more and higher dimensional fault information, which is conducive to improving the accuracy of fault diagnosis of hydraulic pumps and introducing more innovative ways for fault diagnosis of hydraulic pumps.

Table 2. Fault diagnosis method based on a single signal.

Year	Faults Studied				Signal Used	Method Used	Index Evaluation				Reference
	Fault I	Fault II	Fault III	Fault IV			Index I	Index II	Index III	Index IV	
2005			√		Vibration	WT+MRA (multi-resolution analysis)	√			√	[50]
2006	√		√		Vibration	fuzzy logic principle+Spectrum analysis			√		[84]
2008	√				Vibration	RNS+SVM		√			[75]
2008		√	√		Vibration	WA+PCA	√		√		[69]
2008	√			√	Vibration	PSD+DT+FIS	√	√			[86]
2009	√		√		Vibration	WPD			√		[41]
2011	√		√		Vibration	PCA				√	[52]
2011	√		√		Vibration	CPRBF		√		√	[67]
2011		√	√		Sound	KPCA			√		[87]
2012	√		√		Vibration	WP+MTS		√			[53]
2012	√		√		Vibration	SSSVN	√	√			[77]
2013		√		√	Vibration	EMD+NN	√		√		[56]
2013	√			√	Vibration	Spectrum analysis + rough set theory			√		[85]
2014	√	√			Vibration	PARD-BP		√			[70]
2014		√	√		Vibration	WPT+SOM		√		√	[59]
2015	√	√			Vibration	WPT+SVD+SVM	√		√		[72]
2015		√	√		Vibration	RELU-Dropout+SAE		√			[68]
2015	√		√		Vibration	LMD+Softmax	√			√	[58]
2015			√		Vibration	SIE+FCM		√			[83]
2015	√		√		Vibration	SOMP+compressive sensing theory		√			[54]
2015	√				Vibration	LMD+IAMMA	√	√			[48]
2015	√		√		Vibration	EMD+CEEMD+STFT+TFE+SVM	√		√		[76]
2015		√	√		Vibration	DCT+CNC+HHT	√		√		[38]
2016	√		√		Vibration	ITD+Softmax				√	[57]
2016	√	√		√	Vibration	7-layer CNN		√			[63]
2016		√	√		Vibration	HFHLCSD+BSS+DCTS+DCTHSE	√		√		[39]
2016		√	√		Vibration	WNC+CNN+HHT	√		√		[42]
2016	√				Vibration	ILCD+MF+BT-SVM	√		√		[78]

Table 2. *Cont.*

Year	Faults Studied				Signal Used	Method Used	Index Evaluation				Reference
	Fault I	Fault II	Fault III	Fault IV			Index I	Index II	Index III	Index IV	
2017	√		√		Vibration	sensitivity analysis+PNN	√			√	[69]
2017			√		Vibration	EEMD+GA+SVR		√		√	[73]
2018		√			Vibration	LCD+DCS	√				[51]
2018			√		Vibration	SPIP+HMM		√			[43]
2018	√		√		Vibration	WPA+FE+LLTSA+SVM	√	√			[74]
2018	√	√			Vibration	WPT+LTSA+EMD+LMD+ELM				√	[82]
2019	√				Vibration	EWT+VCR	√		√		[35]
2019	√				Vibration	EMMD+Teager	√				[46]
2019			√		Vibration	FFT			√		[71]
2019	√		√		Sound	MFCC+ELM	√	√			[88]
2019	√		√		Vibration	IEWT		√			[47]
2019	√		√		Vibration	MCS+RE	√				[49]
2020	√		√		Vibration	EWT+PCA+ELM	√		√		[80]
2020	√	√			Vibration	CWT+CNN	√	√			[60]
2020	√				Vibration	EWT+VCR+HT	√				[45]
2020		√			Vibration	PSE	√				[37]
2020		√	√		Pressure	FEMD+RE	√		√		[91]
2020		√	√		Vibration	CWT+CNN+T-DSNE	√	√			[66]
2021	√				Vibration	MEEMD+AR+WKELM	√	√			[79]
2021	√		√		Vibration	CEEMDAN+CMBSE+t-SNE+WOA-KELM		√	√	√	[81]
2021	√		√		Vibration	WPA+AlexNet-CNN		√			[61]
2021	√		√		Vibration	PSO-Improve-CNN	√	√			[64]
2021	√				Angular velocity	IAS+NST			√		[92]
2021	√		√		Vibration	EEMD+Pearson	√		√		[36]
2021	√		√		Vibration	RMS	√				[55]
2022	√		√		Vibration	NCNN+Bayes+BP		√			[62]
2022	√		√		Vibration	WT+Bayes+CNN		√			[65]
2022		√	√		Pressure	CWT+Bayes+CNN		√			[90]
2022		√	√		Sound	CNN+PSO		√			[89]

(1) Method based on signal processing

The essence of the multi-signal hydraulic pump fault diagnosis method is to process each input signal separately, and then use a certain fusion method to fuse the feature information contained in the multi signals, so that the extracted fault information is enough to diagnose the fault state. Liu et al. [93] proposed a fault diagnosis method for hydraulic gear pumps based on EEMD and the Bayesian network. This scheme is a method based on multi-source information fusion. Compared with the traditional fault diagnosis method using only EEMD, this method can comprehensively utilize all useful information other than sensor signals. Lu et al. [94] proposed a multi-source information fusion fault diagnosis method based on D-S evidence theory, which uses a fuzzy membership function to construct the basic probability assignment of three evidence bodies. Based on the acceleration, power consumption, flow, and pressure signals under different states, Buiges et al. [95] used the collected signals to compare with the normal state signals for fault diagnosis. Przystupa et al. [96] considered displaying the changes of pressure and flow on FFT and STFT spectrum to realize the application of short-time Fourier transform to fault diagnosis of hydraulic pumps under different operating conditions. Ma Z. et al. [97] established a variable rate inverse gaussian process model to describe the deterioration behavior of the pump, and proposed a Bayesian statistical fault diagnosis method for pressure and flow degradation data analysis. Ruixiang et al. [98] used pressure spectrum signal, temperature signal, and motion signal as diagnostic features, and then used information fusion technology to diagnose hydraulic pump faults. Du et al. [99] proposed a hierarchical clustering fault diagnosis scheme that distinguishes obvious faults through single signal processing of vibration and flow and uses data fusion technology to find fuzzy information. Zengshou et al. [100] proposed an information fusion diagnosis method based on improved D-S evidence theory and space-time domain. Du et al. [101] proposed a clustering diagnosis algorithm based on statistical ARPD in the diagnosis method based on vibration, flow, and pressure signals. Fu et al. [102] studied the relationship between the Bayesian network algorithm and the fault components of the hydraulic pump and then used the Bayesian network algorithm to diagnose the fault when the simulation data of vibration, pressure, temperature, and flow are incomplete.

(2) Methods based on artificial intelligence

Similar to intelligent methods in Section 3.1, the multi-signal hydraulic pump fault diagnosis method is divided into neural network-based method, classifier-based method, and migration learning-based method.

① Artificial intelligence method based on neural network

In the structure of neural networks, the number of neurons in the input layer often exceeds one, so the multi-signal input is compatible with the multi-input characteristics of the input layer of the neural network structure.

The convolutional neural network has exceeded the discrimination ability of human eyes in the accuracy of image recognition, so the digital signal of the hydraulic pump can be converted into an image signal for the convolutional neural network to diagnose the fault of the hydraulic pump. Tang et al. [103] proposed an intelligent fault diagnosis method based on the adaptive learning rate of a neural network to diagnose different fault types by using CWT to convert the three original signals of vibration signal, pressure signal, and sound signal into two-dimensional time-frequency images, and using adaptive learning rate strategy to establish an improved deep CNN model. Taking the vibration signals and pressure signals of hydraulic pumps as the analysis objects. Jiang et al. [104] proposed a fault diagnosis algorithm for hydraulic pumps based on EWT and one-dimensional CNN and deployed the one-dimensional CNN model to the cloud platform to achieve real-time fault diagnosis based on the cloud platform. When based on one-dimensional input signals, there is also a high-precision neural network structure to improve the accuracy of hydraulic pump fault diagnosis. An RBF neural network adopts a linear optimization strategy and has fast learning speed and can approach any nonlinear function with arbitrary accuracy. Zuo et al. [105] built a hydraulic pump fault diagnosis method based on RBF neural network, which takes the pump shell vibration signal and pumps outlet pressure pulse signal as input characteristics.

There is also PNN with RBF neural network function, which is a neural network based on Bayesian decision rules. Zuo et al. [106] built a hydraulic pump fault diagnosis method based on PNN, which takes the pump casing vibration signal and pump outlet pressure pulse signal as input characteristics. Dong et al. [107] used WPT to extract the main fault information contained in the power signal in the historical data, combined with the parameters such as force, oil pressure, casing pressure, and dynamic liquid level to build the fault feature vector, established the PNN model, obtained the mapping relationship between the fault feature vector and the fault form through training the model, and diagnosed the fault form to be entered according to the fault feature vector to be entered. Jiao et al. [108] collected vibration signals and pressure signals to establish a fault diagnosis model based on EMD and PNN. Li et al. [109] proposed a hydraulic pump fault diagnosis method based on the combination of kernel principal components and PNN. This method uses KPCA to reduce the dimension of multi-source data and then diagnoses the fault mode through the PNN network.

② Classifier based approach

The function of a classifier is to classify chaotic targets into different categories according to different input signals. In the fault diagnosis of hydraulic pumps, the input signal mapped faults can be classified by the classifier to diagnose the faults. Lakshmanan et al. [110] proposed a hydraulic pump fault diagnosis method that takes the pressure signal, flow signal, and torque signal of the pump as original real-time data for feature extraction, and inputs them into SVM after CWT. Jiang et al. [111] used the decision tree to build a random forest model, trained six continuous variables of the hydraulic screw pump system as input characteristics, and built a hydraulic pump fault diagnosis method based on the random forest model. Hu et al. [112] built a multi-fault diagnosis system based on data fusion according to the D-S evidence theory and used DMM to build a fault diagnosis feature with a basic probability assignment function, ensuring the objectivity of reliability distribution evaluation.

③ Methods based on Transfer Learning

In order to generalize the ability of the model, the trained model parameters can be migrated to the new model to help train, which can make the initialization performance of the model higher, the promotion rate faster, and the convergence better. Miao et al. [113] used CEEMD and SVD to decompose pressure signal, vibration signal, and flow signal to construct feature vectors and built a hydraulic pump fault diagnosis method through a TrAdaBoost migration learning algorithm. He et al. [114] proposed a migration learning algorithm based on deep MFAM and designed a multi-signal fusion module that assigns weights to vibration signals and acoustic signals, improving the dynamic adjustment ability of the method.

The application of multi-signal-based fault diagnosis is shown in Table 3.

3.3. Other Fault Diagnosis Methods

Whether it is based on signal processing or artificial intelligence, it is based on the data-driven fault diagnosis method of hydraulic pumps. This method realizes fault diagnosis of a hydraulic pump by using the mapping relationship between digital signal and fault and does not describe the mechanism function of fault in detail. Some studies have proposed new knowledge or concepts based on the relationship between non digital signal information and hydraulic pump fault mapping [115–119].

Table 3. Fault diagnosis method based on multiple signals.

Year	Faults Studied				Signal Used	Index Evaluation				Reference
	Fault I	Fault II	Fault III	Fault IV		Index I	Index II	Index III	Index IV	
2002			√		Information fusion technology	√				[98]
2010		√	√		Hierarchical clustering analysis	√		√		[99]
2011	√		√		Improved DS evidence theory and spatiotemporal information fusion		√	√		[100]
2012	√		√		D-S+DMM	√		√		[111]
2013		√	√		Clustering diagnosis algorithm based on ARPD	√	√			[101]
2013	√	√			MFAM+Transfer learning		√			[113]
2014		√	√		PNN				√	[105]
2014	√				RBF-NN	√			√	[104]
2015	√		√		EEMD+Bayes+NN	√	√			[93]
2017	√		√		DS evidence theory	√				[94]
2017		√	√		EMD+PNN	√			√	[107]
2019	√		√		Inverse gaussian model + Bayes optimization		√		√	[97]
2020	√			√	PCA	√				[95]
2020		√		√	STFT+FFT	√				[96]
2020	√		√		SVM+Multilayer Perceptron(MLP)	√				[109]
2020	√				Stochastic forest neural network				√	[110]
2020	√		√		Singular value decomposition + transfer learning	√	√			[112]
2020	√				Reliability analysis + Bayesian network		√			[102]
2021	√				CNN based on improved adaptive learning rate		√		√	[103]
2021		√	√		KPCA+PNN		√			[108]
2021		√	√		CNN+EWT+WISE-PaaS	√	√			[114]
2022	√		√		Wavelet packet analysis+PNN	√			√	[106]

On the basis of an accelerated life test, Guo et al. [120] proposed a dynamic grid technology to simulate the internal flow field of hydraulic pumps in detail. On the basis of film thickness analysis, Ma et al. [121] put forward a hydraulic pump diagnosis method based on elastohydrodynamic lubrication model analysis by comprehensively considering structural parameters, working condition parameters, and material performance parameters. In view of the multi-crack fault of the hydraulic gear pump gear, Zhao et al. [122] established the vibration wavelet finite element calculation formula of complete gear and cracked gear, studied the fault diagnosis of blind source separation and particle swarm optimization algorithm, and correctly diagnosed the location of multiple cracks of the gear.

3.4. Centrifugal Pump Fault Diagnosis Method

The above content is mainly a detailed analysis of the fault diagnosis method of the hydraulic pump, and as a centrifugal pump that also transports liquid, it is also of comparative significance to analyze it. In centrifugal pumps, it is necessary not only to identify the fault but also to discover the severity of the failure and classify it.

Muralidharan et al. [123] used the DWT to calculate the wavelet characteristics of the vibration signal, used rough sets to generate rules, and used fuzzy logic to classify. Sakthivel et al. [124] used the C4.5 decision tree algorithm to extract statistical features from vibration signals in good and fault states for fault diagnosis. Muralidharan et al. [125] studied the vibration-based fault diagnosis method of a monoblock centrifugal pump and found the best wavelet suitable for single-block centrifugal pump fault diagnosis by calculating and comparing. Nagendra et al. [126] used two different machine learning techniques, SVM and ANN, for centrifugal pump fault diagnosis. It was found that the machine learning method based on ANN combined with chi-square and XGBoost feature ranking techniques is superior to the SVM. Wang et al. [127] proposed a centrifugal pump fault diagnosis method based on CEEMD-sample entropy (SampEn) combined with RF. Based on the characteristic evaluation of the information ratio combined with principal component analysis, Ahmad et al. [128] proposed a new Ir-PCA method. The comparison results found the method was superior to existing advanced methods in terms of fault classification accuracy. ALTobi et al. [129] used MLP and SVM to classify the six fault states and normal states of the centrifugal pump. Therefore, an MLP hybrid training method based on the combination of Back Propagation (BP) and Genetic Algorithm (GA) was proposed.

3.5. Fault Diagnosis Block Diagram

Based on the fault diagnosis methods proposed in the above literature, I have summarized the following fault diagnosis block diagram, as shown in Figure 2. Since there are many types of diagnosis methods and many expand on the basic methods, I just list the basic methods for reference.

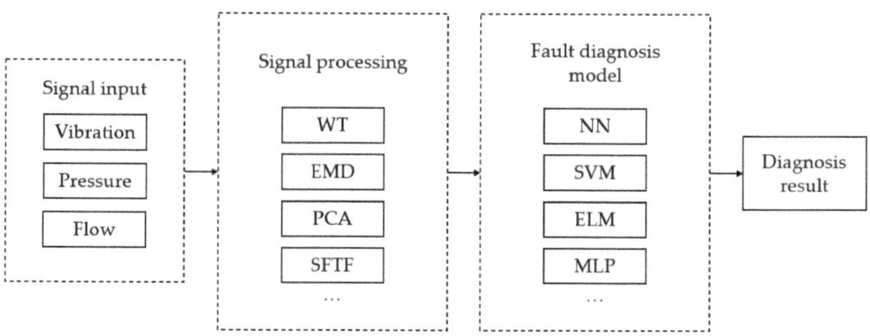

Figure 2. Fault diagnosis block diagram.

4. Fault Prediction and Health Management

On the basis of fault diagnosis, appropriate prediction and analysis methods can be used to achieve fault prediction. Furthermore, for the health management of the whole life cycle of the hydraulic pump, the remaining service life of the hydraulic pump can be predicted and the whole process of health status monitoring of the hydraulic pump can be studied.

4.1. Fault Prediction

To maintain the stable operation of the hydraulic pump in its whole life cycle, the failure prediction of the hydraulic pump can predict the failure that will occur in the early stage of the failure, so as to timely repair the failure in the early stage of low cost and reduce the expansion of loss. The methods of hydraulic pump fault prediction can be roughly divided into two parts, intelligent prediction, and non-intelligent prediction.

The non-intelligent prediction method refers to that the prediction method has no self-learning ability. In short, the non-intelligent prediction method does not use mechanical learning or neural network, which makes the usability of this method relatively weak. Gomes et al. [130] used the empirical model of degradation evolution combined with Kalman filter technology to predict the failure of hydraulic pumps, and successfully predicted two-time series from actual operation to failure data. Amin et al. [131] developed an online health monitoring system for hydraulic pumps by using feature extraction, a fuzzy reasoning system, and knowledge fusion technology. Bykov et al. [132] described the analysis of the state data set of the hydraulic system and tried to diagnose the failure in the valve switching mode, so as to further study the possibility of predicting the failure. Ma et al. [133] analyzed the key failure modes of aircraft hydraulic pumps based on operation and maintenance statistics and proposed a failure prediction method based on multi-source information fusion. Lisowski et al. [134] constructed a function-component matrix (EC) and a component-failure matrix (CF) by using the quality method and then multiplied the two matrices to obtain a function-failure EF matrix containing potential failure information, thus realizing the failure prediction of hydraulic pumps.

Intelligent prediction methods mainly include prediction methods with self-learning ability using neural networks or machine learning. To improve the accuracy of fault prediction, Li et al. [135] proposed a hydraulic pump fault prediction method based on BE and DBN, which is based on the DBN model of constraint limit RBM as a prediction model and introduces QPSO to search the optimal value of the initial parameters of the network. Xu et al. [136] analyzed the cause and mechanism of hydraulic pump degradation due to wear, established a degradation model through joint simulation of Simulink and AMESim, and predicted the failure of the hydraulic pump using a multi-step SVM algorithm. Ding et al. [137] proposed a fault prediction method based on logistic regression that obtains a hydraulic pump fault prediction model by LMD processing of the pump vibration signal, feature reduction using PCA, and training the LR model with the reduced features. Tian [138] used the method of combining EEMD and SEOS to envelope demodulate the vibration signal of the hydraulic pump, and then used WPA to extract the fault features, to establish a hydraulic pump fault prediction model combining WPA and SVM. Sun et al. [139] proposed a multi-channel vibration signal fusion method based on DCS. This method takes the synthetic spectral entropy as the feature and uses the extracted feature to establish an ESN model for prediction, which can be used for fault prediction of hydraulic pumps.

4.2. Prediction of Remaining Useful Life

During the normal use of the hydraulic pump, the remaining useful life of the current hydraulic pump can be predicted in time, and the working condition of the hydraulic pump can be adjusted in time through the working time, which is conducive to extending the normal useful life of the hydraulic pump. The remaining useful life prediction methods of hydraulic pumps can be roughly divided into two categories, data-driven methods and model-driven methods.

① Data-driven approach

The data-driven methods can be divided into neural network methods and non-neural network methods. Lee et al. [140] constructed HI through vibration signal and pressure signal, and trained a Bi-LSTM neural network using different performance indicators for RUL prediction of hydraulic pumps. Wang et al. [141] used DCAE to characterize the vibration data of hydraulic pumps, constructed HI to determine the degradation state, and input the health index as a tag into the RUL prediction model based on the Bi-LSTM network. Guo et al. [142] used VMD, Hilbert, and FA to process the vibration data of the hydraulic pump, established the degradation evaluation index, trained the Trainbr-RBFNN model with the degradation evaluation index, and obtained the RUL prediction model for the hydraulic pump.

The non-neural network method can still achieve the RUL prediction of hydraulic pumps. Yu et al. [143] proposed a MAAKR method for information fusion, using 3B-Spline with monotonic constraints to build Hi, and using the MCPF method to monotonically update the random coefficients of the model to achieve RUL prediction of hydraulic pumps. Tongyang et al. [144] proposed an AOPF prediction method to improve the long-term prediction accuracy of RUL and used the MCS method to estimate the posterior probability density function of the future state of the hydraulic pump. Li et al. [145] proposed a new method for RUL prediction of hydraulic pumps based on KPCA and JITL. This method uses WT to extract features, KPCA to fuse features, and constructs an RUL prediction method based on k-VNN and JITL methods.

② Model-driven methods

The data-driven method is to use the data information to map the tag of the target fault of the hydraulic pump through the processing and analysis of the data. This method completely bypasses the professional knowledge of the hydraulic pump and only has the mapping relationship from input to output. Based on the model-driven approach, starting from the expertise of hydraulic pumps, mathematical explicit relationships are constructed. Geng et al. [146] proposed a life assessment method that combines SMOTE algorithm, KS test, and cumulative damage theory. The SMOTE algorithm is used to solve the imbalance problem between sample groups, and KS is the classic method for evaluating the goodness of fit. Zhonghaim et al. [147] obtained the fatigue life of the piston by using DLDR through the analysis of the actual load spectrum of the hydraulic piston pump and simulated the fatigue life of the piston by using the finite element analysis software. Wang et al. [148] described the performance degradation model with the Wiener process, predicted the remaining useful life (RUL) of the pump, estimated the initial parameters of the wiener process by MLE using the EM algorithm, estimated the drift coefficient of the wiener process by recursive estimation using Kalman filter method and calculated the RUL of the pump according to the performance degradation model based on wiener process. Wang et al. [149] used the contaminant sensitivity theory of the hydraulic system to derive the mathematical explicit relationship between oil pollution and the useful life of the piston pump and predicted the useful life of the piston pump under certain pollution conditions using a group of experimental data. Sun et al. [150] proposed an improved IG process model to describe the wear degradation of hydraulic pumps and used Monte Carlo integration and EM algorithm to estimate the model parameters.

4.3. Health Status Detection

The real-time health monitoring of the hydraulic pump can diagnose whether the operating state of the hydraulic pump is healthy at each time, which is conducive to the timely adjustment of the hydraulic pump in response to emergencies and the management and use of the hydraulic pump.

The detection of the health state of the hydraulic pump is not limited to the detection of the fault state, so the amount of data required is very large. A neural network can achieve considerable effect in processing large sample data. According to different health states of hydraulic pumps, Shaowu et al. [151] proposed that after collecting vibration signal data of hydraulic pumps, STFT, WT, and Wigner-Will distributions are used to form time-frequency maps, and then CNN is used to classify and identify time-frequency images of different volumetric efficiencies of hydraulic pumps, so as to monitor the health status of hydraulic pumps. Lin et al. [152] proposed that according to the distribution of the information entropy of the characteristic parameters of the hydraulic pump, various state characteristic parameters can be obtained to characterize the contribution of the hydraulic pump in health, so as to realize the fusion of various characteristic parameters, and then use the grey theory to detect the health state of the hydraulic pump. Hancock et al. [153] researched and developed a method to decompose the vibration signal of vertical hydraulic pumps using WPA, and input the characteristic signal into the adaptive neuro-fuzzy inference fault detection system for pump health state detection. Succi et al. [154] take the fundamental pumping frequency and its harmonics as the input features of the neural network model and use the multilayer neural network model of back propagation and Kohonen feature map to detect the health state of the hydraulic pump.

There are also some studies that use non-neural network methods, which can also achieve the purpose of detecting the health state of hydraulic pumps. Zhouf et al. [155] proposed a WOA-based RSDD method to extract feature parameters, which combined with the modified hierarchical amplitude aware displacement entropy MHAPE to form a health state detection method for hydraulic pumps. Gao et al. [156] proposed a health diagnosis method for hydraulic pumps based on WPD and WCRA and developed a health detection system based on WPD residual analysis. Shapping et al. [157] used the method of combining WPD and Hilbert envelope demodulation to eliminate the interference effect of radial and axial acceleration signals, replaced Shannon entropy with NE for state identification, and proposed a WPNE-based method for identifying the health state of hydraulic pumps.

5. Analysis of the Summary Paper

5.1. Statistical Analysis

Figure 3 shows the statistics of different research directions of hydraulic pump faults in recent years in the literature listed in this paper, and it can be seen that the mainstream research direction is still a fault diagnosis. Equipment fault diagnosis technology has developed to today and has become an independent interdisciplinary comprehensive information processing technology, it is based on reliability theory, cybernetics, information theory, and system theory as the theoretical basis, modern test instruments and computers as a means, combined with the special laws of various diagnostic objects and gradually formed a new discipline, so it is loved by many scholars for research.

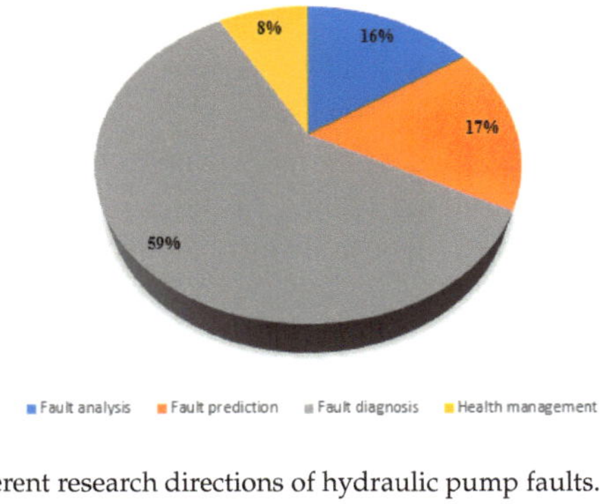

Figure 3. Different research directions of hydraulic pump faults.

Figure 4 shows that among the fault diagnosis methods of hydraulic pumps based on single signals, the fault diagnosis method uses vibration signals to diagnose the faults of hydraulic pumps, which is the first choice for most studies at present. More than 90% of scholars in the selected articles use vibration signals.

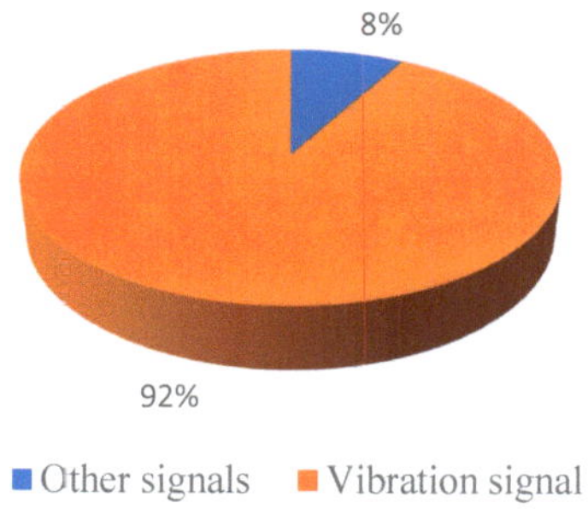

Figure 4. Single signal scale.

With the development of fault diagnosis algorithms in recent years, more and more research on hydraulic pump fault diagnosis has been carried out, which is almost a straight-line trend. As shown in Figures 5 and 6, it can be concluded from the analysis of the two figures that the research on fault diagnosis of hydraulic pumps will continue to increase in the future. With the development of detection signals from simplicity to complexity, it can be seen that the research of single signal fault diagnosis is more than that of multi-signal methods. However, with the development of signal fusion technology, the research of multi-signal fault diagnosis is also increasing year by year.

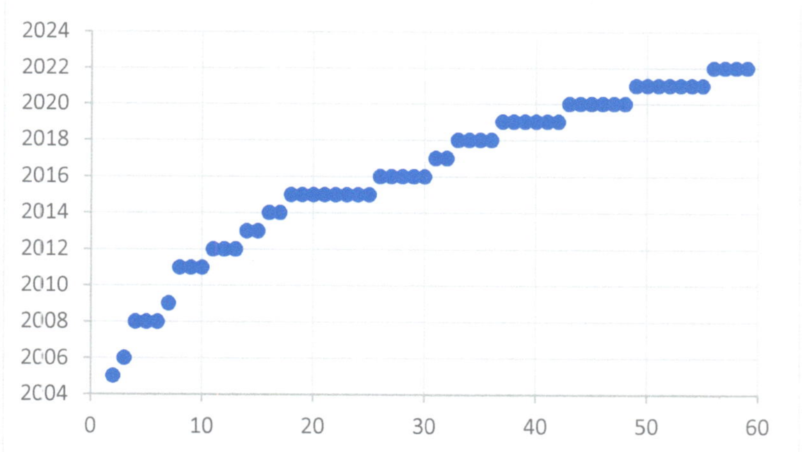

Figure 5. Development trend of single signal articles.

Figure 7 shows the proportion of signal processing and artificial intelligence, which shows that diagnosis methods based on artificial intelligence are more and more popular. Although the signal processing methods are developing year by year, most of the research focuses on the composite method of signal processing methods to deal with fault characteristics and human intelligent algorithms to build diagnosis models.

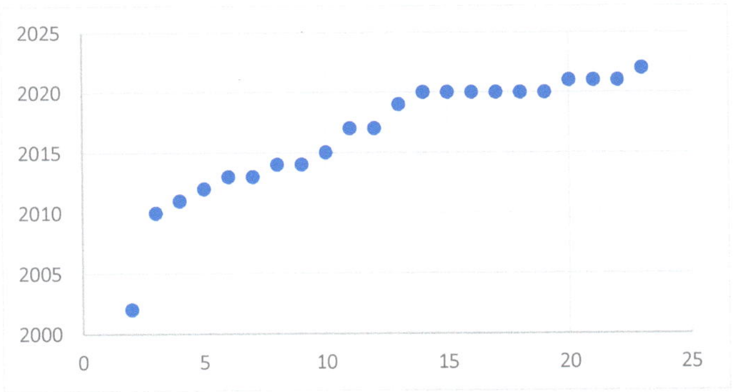

Figure 6. Development trend of multi-signal articles.

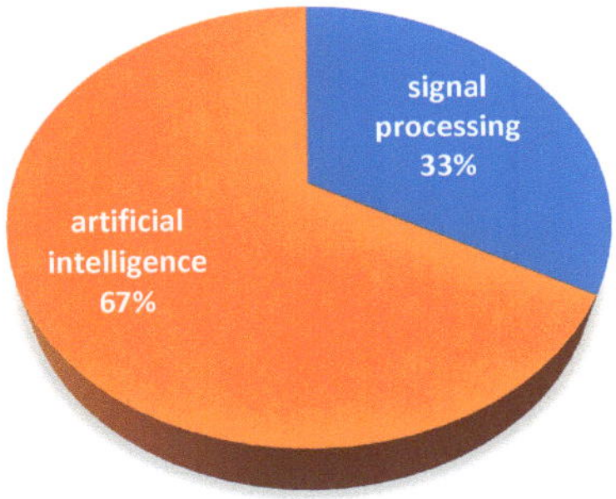

Figure 7. The ratio of signal processing to artificial intelligence.

5.2. Discussion on Future Development

This paper summarizes the application of hydraulic pump fault research, but there are some inevitable omissions. To sum up, through the statistical analysis of the selected documents, it can be concluded that in the actual environment, it is difficult to obtain high-quality fault data from a single signal and extract the fault information contained. On the contrary, the multi-signal method is useful because it contains more information. The artificial intelligence method is useful because it has high feasibility in dealing with complex situations (such as compound faults). In order to better promote the development of hydraulic pump fault diagnosis, the following aspects can be carried out in the future:

(1) Because of the weak signal features in the early stage of fault, it is difficult to extract fault features, so fault feature extraction is still a direction that needs further exploration. Because of the powerful function of the deep learning method, fault feature extraction based on the deep learning method will be an important research direction.
(2) Although multi-data signals contain more information, the efficient information fusion methods for multi-data signals are still insufficient, so more efficient information fusion methods are also the direction to be further explored.
(3) From the statistical analysis of the review papers, it can be concluded that the diagnosis method of artificial intelligence will become mainstream. However, each intelligent method also has defects, and the combination of multiple intelligent methods can be used to fill the defects, such as reverse neural networks combined with multilayer perceptrons.

6. Conclusions

Fault diagnosis is the key to the health management of hydraulic pumps. It can improve the reliability of the hydraulic pump from the aspect of the data signal, and significantly reduce the risk of operation collapse and catastrophic failure. In recent years, the research on hydraulic pump fault diagnosis has been very active, but there is a lack of systematic analysis and summary of the developed methods. In order to make up for this gap, this paper systematically summarizes the relevant methods from the two aspects of fault diagnosis and health management. Finally, through the statistical analysis of the literature, some development prospects in this field are pointed out, which provides reference and guidance for researchers and practitioners to further carry out and apply relevant research. Nowadays, with the rapid development of machine learning algorithms and deep learning, data and signal-based methods are becoming the main direction in the future. The same trend applies to feature extraction methods. Therefore, the powerful ability of machine learning algorithms, especially deep learning algorithms, obviously has great potential in the future.

Author Contributions: Conceptualization, J.X. and Y.Y.; methodology, J.X. and Y.Y.; software, L.D.; validation, L.D., Y.Y. and J.X.; formal analysis, J.X. and L.D.; investigation, G.F.; resources, Y.Y. and J.L.; data curation, L.D.; writing—original draft preparation, Y.Y. and L.D.; writing—review and editing, J.X.; visualization, L.D.; supervision, J.X.; project administration, J.X.; funding acquisition, J.X. and Y.Y. All authors have read and agreed to the published version of the manuscript.

Funding: This research received no external funding.

Institutional Review Board Statement: Not applicable.

Informed Consent Statement: Not applicable.

Data Availability Statement: There is no any data involved.

Conflicts of Interest: The authors declare no conflict of interest.

Nomenclature

AOPF	Adaptive-Order Particle Filter
AR	Autoregressive
BE	Bispectrum Entropy
BI-LSTM	Bi-Directional Long-Short Term Memory
BT-SVM	Binary Tree Support Vector Machine
CEEMD	Complementary Ensemble Empirical Mode Decomposition
CEEMDAN	Complete Ensemble Empirical Mode Decomposition with Adaptive Noise
CMBSE	Composite Multi-Scale Basic Scale Entropy
CNC	Cosine Neighboring Coefficients
CNN	Convolutional Neural Network
CPRBF-NN	Radial Basis Function Network In Conjunction With Chaos Theory
CWT	Continuous Wavelet Transform
DBN	Deep Belief Network
DCAE	Deep Convolutional Autoencoder
DCS	Discrete Cosine Transform–Composite Spectrum
DCT	Discrete Cosine Transform
DLDR	Double Linear Damage Rule
DT	Decision Trees
EEMD	Ensemble Empirical Mode Decomposition
ELM	Extreme Learning Machine
EM	Expectation Maximization
EMD	Empirical Mode Decomposition
EMMD	Extremum Field Mean Mode Decomposition
ESN	Modified Echo State Networks
EWT	Empirical Wavelet Transform
FA	Factor Analysis
FCM	Fuzzy C-Means
FE	Fuzzy Entropy

FEMD	Fast Empirical Mode Decomposition
FFT	Fast Fourier Transform
FIS	Fuzzy Inference System
FMEA	Failure Mode And Effects Analysis
FMECA	Modes, Effects, And Criticality Analysis
GA	Genetic Algorithm
HHT	Hilbert–Huang Transform
HMM	Hidden Markov Model
HT	Hilbert Transform
IAMMA	Improved Adaptive Multiscale Morphology Analysis
Ir-PCA	Informative ratio-Principal component analysis
IEWT	Improved Empirical Wavelet Transform
IG	Inverse Gaussian
ILCD	Improved Local Characteristic-Scale Decomposition
ITD	Intrinsic Time-Scale Decomposition
JITL	Just In Time Learning
KPCA	Kernel Principal Component Analysis
KS	Kolmogorov-Smirnov
K-VNN	K-Vector Nearest Neighbor
LCD	Local Characteristic-Scale Decomposition
LLTSA	Liner Local Tangent Space Alignment
LMD	Local Mean Decomposition
LR	Logistic Regression
LTSA	Local Tangent Space Alignment
MAAKR	Modified Auto-Associative Kernel Regression
MCPF	Monotonicity-Constrained Particle Filtering
MCS	Monte Carlo Simulation
MD	Mahalanobis Distance
MEEMD	Modified Ensemble Empirical Mode Decomposition
MF	Multi-Fractal Spectrum
MFAM	Multi-Signal Fusion Adversarial Model
MFCC	Mel-Frequency Cepstral Coefficient
MHAPE	Modified Hierarchical Amplitude-Aware Permutation Entropy
MLE	Maximum Likelihood Estimation
MLP	Multilayer Perceptron
MMPE	Mean Of Multi-Scale Permutation Entropy
MPE	Multi-Scale Permutation Entropy
MTS	Mahalanobis–Taguchi System
NCNN	Normalized Convolutional Neural Network
NE	Norm Entropy
NN	Neural Network
PARD	Pruning Algorithm Based Random Degree
PCA	Principal Component Analysis
PCC	Pearson Correlation Coefficient
PHM	Prognostics And Health Management
PNN	Probabilistic Neural Network
PSD	Power Spectral Density
PSE	Power Spectral Entropy
PSO	Particle Swarm Optimization
QPSO	Quantum Particle Swarm Optimization
RBF	Radial Basis Function
RBM	Boltzmann Machine
RE	Relative Entropy
RFC	The Random Forest Classifier
RMS	Root Mean Square

RNS	Real-Valued Negative Selection
RSDD	Resonance-Based Sparse Signal Decomposition
RUL	Remaining Useful Life
SAE	Stacked Autoencoders
SEOS	Smoothed Energy Operation Separation
SIE	Spatial Information Entropy
SMOTE	Synthetic Minority Over-Sampling Technique
SOM-NN	Self-Organizing Mapping Neural Network
SOMP	Stagewise Orthogonal Matching Pursuit
SPIP	Symbolic Perceptually Important Point
SS-SVM	Sphere-Structured Support Vector Machines
STFT	Short Time Fourier Transform
SVD	Singular Value Decomposition
SVM	Support Vector Machine
SVR	Support Vector Regression
T-DSNE	T-Distributed Stochastic Neighbor Embedding
TFE	Time-Frequency Entropy
T-SNE	T-Distributed Stochastic Neighbor Embedding
VCR	Variance Contribution Rate
VMD	Variation Mode Decomposition
WA	Wavelet Analysis
WCRA	Wavelet Coefficient Residual Analysis
WKELM	Wavelet Kernel Extreme Learning Machine
WNC	Wavelet Neighboring Coefficients
WOA	Whale Optimization Algorithm
WOA-KELM	Whale Optimization Algorithm Kernel Extreme Learning Machine
WP	Wavelet Packet
WPA	Wavelet Packet Analysis
WPD	Wavelet Packet Decomposition
WPNE	Wavelet Packet Norm Entropy
WPT	Wavelet Packet Transform
WT	Wavelet Transform

References

1. Jani, D.B.; Ashish, S.; Aditya, S.; Yash, S.; Bishambhar, S.; Nikhil, S.; Manmohan, S. An overview on aircraft hydraulic system. *Renew. Sustain. Energy Rev.* **2019**, *6*, 29–35.
2. Huang, K.; Wu, S.; Li, F.; Yang, C.; Gui, W. Fault Diagnosis of Hydraulic Systems Based on Deep Learning Model With Multirate Data Samples. *IEEE Trans. Neural Networks Learn. Syst.* **2021**, *33*, 6789–6801. [CrossRef] [PubMed]
3. Khan, K.; Sohaib, M.; Rashid, A.; Ali, S.; Akbar, H.; Basit, A.; Ahmad, T. Recent trends and challenges in predictive maintenance of aircraft's engine and hydraulic system. *J. Braz. Soc. Mech. Sci. Eng.* **2021**, *43*, 403. [CrossRef]
4. Kim, S.; Choi, J.-H. Convolutional neural network for gear fault diagnosis based on signal segmentation approach. *Struct. Health Monit.* **2018**, *18*, 1401–1415. [CrossRef]
5. Peng, Z.K.; Tse, P.W.; Chu, F.L. An improved Hilbert-Huang transform and its application in vibration signal analysis. *J. Sound Vib.* **2005**, *286*, 187–205. [CrossRef]
6. Ortega, A.; Marco, S.; Perera, A.; Šundic, T.; Pardo, A.; Samitier, J. An intelligent detector based on temperature modulation of a gas sensor with a digital signal processor. *Sens. Actuators B Chem.* **2001**, *78*, 32–39. [CrossRef]
7. van Esch, B.P.M. Performance and Radial Loading of a Mixed-Flow Pump Under Non-Uniform Suction Flow. *J. Fluids Eng.* **2009**, *131*, 051101. [CrossRef]
8. Helgestad, B.O.; Foster, K.; Bannister, F.K. Pressure Transients in an Axial Piston Hydraulic Pump. *Proc. Inst. Mech. Eng.* **1974**, *188*, 189–199. [CrossRef]
9. Al-Hashmi, S. Statistical Analysis of Acoustic Signal for Cavitation Detection. *Int. J. Emerg. Technol. Adv. Eng.* **2013**, *3*, 55–66.
10. Zheng, H.; Li, Z.; Chen, X. Gear fault diagnosis based on continuous wavelet transform. *Mech. Syst. Signal Process.* **2002**, *16*, 447–457. [CrossRef]
11. Chen, Z.; Mauricio, A.; Li, W.; Gryllias, K. A deep learning method for bearing fault diagnosis based on Cyclic Spectral Coherence and Convolutional Neural Networks. *Mech. Syst. Signal Process.* **2020**, *140*, 106683. [CrossRef]
12. Liu, J.; Shi, D.; Li, G.; Xie, Y.; Li, K.; Liu, B.; Ru, Z. Data-driven and association rule mining-based fault diagnosis and action mechanism analysis for building chillers. *Energy Build.* **2020**, *216*, 109957. [CrossRef]
13. Xu, L.; Yu, X.; Hou, Z. Analysis of the Causes of Driving Gear Shaft Fractures in Gear Pumps. *J. Fail. Anal. Prev.* **2020**, *20*, 242–248. [CrossRef]

14. Qin, X.; Liu, J.; Zhao, X.; Feng, L.; Pang, R. Fracture failure analysis of transmission gear shaft in a bidirectional gear pump. *Eng. Fail. Anal.* **2020**, *118*, 104886. [CrossRef]
15. Shawki, S. Early failure of high pressure screw pumps: Shaft fracture. *J. Fail. Anal. Prev.* **2013**, *13*, 595–600. [CrossRef]
16. Weber, M. Some Safety Aspects on the Design of Sparger Systems for the. *Process Saf. Prog.* **2006**, *25*, 326–330. [CrossRef]
17. Yordanov, B.; Krastev, D.; Mitov, I. Exploitation Fatigue Fraction of Hydraulic Pump Shaft. *Int. J. NDT Days* **2019**, *II*, 554–561.
18. Li, P.; Zhao, Y.; Ma, C. Fracture analysis of an aluminum alloy gear pump housing. *Key Eng. Mater.* **2017**, *723*, 288–293. [CrossRef]
19. Sekercioglu, T. Fracture analysis of gear pump used for polymer production. *Eng. Fail. Anal.* **2006**, *13*, 835–842. [CrossRef]
20. Scott Pflumm, J.; Banks, J.C. Utilizing dynamic fuel pressure sensor for detecting bearing Spalling and gear pump failure modes in Cummins pressure time (PT) Pumps. *Annu. Conf. Progn. Health Manag. Soc.* **2011**, 3, 490–501.
21. Hemati, A.; Shooshtari, A. Gear Pump Root Cause Failure Analysis Using Vibrations Analysis and Signal Processing. *J. Fail. Anal. Prev.* **2020**, *20*, 1815–1818. [CrossRef]
22. Lee, Y.B.; Lee, G.C.; Yang, J.D.; Park, J.W.; Baek, D. cheon Failure analysis of a hydraulic power system in the wind turbine. *Eng. Fail. Anal.* **2020**, *107*, 104218. [CrossRef]
23. Wang, Y.; Wang, Y. Fracture Failure Analysis of Flameproof Enclosure of Hydraulic Pump Regulator. *IOP Conf. Ser. Mater. Sci. Eng.* **2020**, *768*, 022064. [CrossRef]
24. Zhang, W.G.; Lin, G.M.; Shang, M.; Chen, Q. Analysis of external gear pump failure. *Adv. Mater. Res.* **2013**, *791*, 623–626. [CrossRef]
25. Das, S.K.; Munda, P.; Chowdhury, S.G.; Das, G.; Singh, R. Effect of microstructures on corrosion and erosion of an alloy steel gear pump. *Eng. Fail. Anal.* **2014**, *40*, 89–96. [CrossRef]
26. Jiang, J.S.; Zhao, Z.Y. Analysis on shaft failure of the twin screw pump. *Appl. Mech. Mater.* **2014**, *526*, 253–256. [CrossRef]
27. Milović, L.; Zrilić, M. Failure analysis of rotary screw pumps. *Integritet Vek Konstr.* **2009**, *9*, 57–62.
28. Shang, L.; Wang, N. Cause and analysis of hydraulic pump failure. *J. Phys. Conf. Ser.* **2021**, *1885*, 022028. [CrossRef]
29. Hidayat, H.; Aviva, D.; Muis, A.; Halik, A.; Sudarsono, S.; Pranoto, S.; Cahyadi, D. Failure analysis of excavator hydraulic pump. *IOP Conf. Ser. Mater. Sci. Eng.* **2022**, *1212*, 012052. [CrossRef]
30. Ułanowicz, L.; Jastrzębski, G.; Szczepaniak, P.; Sabak, R.; Rykaczewski, D. Malfunctions of Aviation Hydraulic Pumps. *J. KONBiN* **2020**, *50*, 257–276. [CrossRef]
31. Fabiś-Domagała, J. Analysis of Defects and Failu Res of Hydraulic Gear Pumps With the Use of Selected Qualitative Methods. *Tribologia* **2017**, *272*, 33–38. [CrossRef]
32. Major, S.; Cyrus, P.; Dostal, R. Numerical Analysis of Fatigue Degradation of Screw Pump 4 Finite Element Model 3 Material Characterization. *Wseas Trans. Appl. Theor. Mech.* **2016**, *11*, 142–147.
33. Ma, Q.; Song, D.; Liu, B. Fault mode analysis and simulation verification of hydraulic system based on AMEsim. *J. Phys. Conf. Ser.* **2021**, *2006*, 012013. [CrossRef]
34. Lee, G.H.; Akpudo, U.E.; Hur, J.W. FMECA and MFCC-based early wear detection in gear pumps in cost-aware monitoring systems. *Electronics* **2021**, *10*, 2939. [CrossRef]
35. Yu, H.; Li, H.; Li, Y.; Li, Y. A novel improved full vector spectrum algorithm and its application in multi-sensor data fusion for hydraulic pumps. *Meas. J. Int. Meas. Confed.* **2019**, *133*, 145–161. [CrossRef]
36. Jiang, W.L.; Zhang, P.Y.; Li, M.; Zhang, S.Q. Axial Piston Pump Fault Diagnosis Method Based on Symmetrical Polar Coordinate Image and Fuzzy C-Means Clustering Algorithm. *Shock Vib.* **2021**, *2021*, 6681751. [CrossRef]
37. Zheng, Z.; Li, X.; Zhu, Y. Feature extraction of the hydraulic pump fault based on improved Autogram. *Meas. J. Int. Meas. Confed.* **2020**, *163*, 107908. [CrossRef]
38. Wang, Y.; Li, H.; Ye, P. Fault Feature Extraction of Hydraulic Pump Based on CNC De-noising and HHT. *J. Fail. Anal. Prev.* **2015**, *15*, 139–151. [CrossRef]
39. Sun, J.; Li, H.; Xu, B. Degradation feature extraction of the hydraulic pump based on high-frequency harmonic local characteristic-scale decomposition sub-signal separation and discrete cosine transform high-order singular entropy. *Adv. Mech. Eng.* **2016**, *8*, 1687814016659601. [CrossRef]
40. Liu, S.; Jiang, W.; Niu, H. Fault diagnosis of hydraulic pump based on rough set and PCA algorithm. In Proceedings of the 2008 Fifth International Conference on Fuzzy Systems and Knowledge Discovery, Jinan, China, 18–20 October 2008; Volume 5, pp. 256–260. [CrossRef]
41. Hou, W.; Lu, C.; Liu, H.; Lu, C. Fault diagnosis based on wavelet package for hydraulic pump. In Proceedings of the 2009 8th International Conference on Reliability, Maintainability and Safety, Chengdu, China, 20–24 July 2009; pp. 831–835. [CrossRef]
42. Wang, Y.; Huang, Z.; Zhao, X.; Zhu, Y.; Wei, D. A novel de-noising method based on discrete cosine transform and its application in the fault feature extraction of hydraulic pump. *J. Shanghai Jiaotong Univ.* **2016**, *21*, 297–306. [CrossRef]
43. Jia, Y.; Xu, M.; Wang, R. Symbolic important point perceptually and hidden markov model based hydraulic pump fault diagnosis method. *Sensors* **2018**, *18*, 4460. [CrossRef] [PubMed]
44. Wang, L.; Book, W.J.; Huggins, J.D. Application of singular perturbation theory to hydraulic pump controlled systems. *IEEE/ASME Trans. Mechatron.* **2012**, *17*, 251–259. [CrossRef]
45. Yu, H.; Li, H.; Li, Y. Vibration signal fusion using improved empirical wavelet transform and variance contribution rate for weak fault detection of hydraulic pumps. *ISA Trans.* **2020**, *107*, 385–401. [CrossRef] [PubMed]

46. Deng, S.; Tang, L.; Su, X.; Che, J. Fault Diagnosis Technology of Plunger Pump based on EMMD-Teager. *Int. J. Perform. Eng.* **2019**, *15*, 1912–1919. [CrossRef]
47. Zheng, Z.; Wang, Z.; Zhu, Y.; Tang, S.; Wang, B. Feature extraction method for hydraulic pump fault signal based on improved empirical wavelet transform. *Processes* **2019**, *7*, 824. [CrossRef]
48. Jiang, W.; Zheng, Z.; Zhu, Y.; Li, Y. Demodulation for hydraulic pump fault signals based on local mean decomposition and improved adaptive multiscale morphology analysis. *Mech. Syst. Signal Process.* **2015**, *58*, 179–205. [CrossRef]
49. Li, H.; Sun, J.; Ma, H.; Tian, Z.; Li, Y. A novel method based upon modified composite spectrum and relative entropy for degradation feature extraction of hydraulic pump. *Mech. Syst. Signal Process.* **2019**, *114*, 399–412. [CrossRef]
50. Gao, Y. IMECE2005-81711 Comparison of Hydraulic Pump Faults Diagnosis Methods. *ASME Int. Mech. Eng. Congr. Expo.* **2005**, *42207*, 73–78.
51. Sun, J.; Li, H.; Tian, Z. Degradation feature extraction of hydraulic pump based on LCD-DCS fusion algorithm. *Proc. Inst. Mech. Eng. Part B J. Eng. Manuf.* **2018**, *232*, 1460–1470. [CrossRef]
52. Liu, S.; Ding, L.; Jiang, W. Study on application of Principal Component Analysis to fault detection in hydraulic pump. In Proceedings of the 2011 International Conference on Fluid Power and Mechatronics, Beijing, China, 17–20 August 2011; pp. 173–178. [CrossRef]
53. Wang, Z.; Lu, C.; Tao, X.; Fan, H.; Wang, Z. Application of Wavelet Packet and Mahalanobis-Taguchi System for Hydraulic Pump Fault Diagnosis. *Lect. Notes Inf. Technol.* **2012**, *14*, 41–46.
54. Chen, Z.; Lu, C.; Yuan, H. Hydraulic pump fault diagnosis with compressed signals based on stagewise orthogonal matching pursuit. *Vibroeng. Procedia* **2015**, *5*, 223–228.
55. Tang, H.; Fu, Z.; Huang, Y. A fault diagnosis method for loose slipper failure of piston pump in construction machinery under changing load. *Appl. Acoust.* **2021**, *172*, 107634. [CrossRef]
56. Gao, S.X.; Chen, X.H.; Ding, Y. Fault recognition of gear pump based on EMD neural network. *Appl. Mech. Mater.* **2013**, *333–335*, 1635–1639. [CrossRef]
57. Sun, J.; Lu, C.; Ding, Y. Fault diagnosis for hydraulic pump based on intrinsic time-scale decomposition & softmax regression. *Vibroeng. Procedia* **2016**, *10*, 229–234.
58. Ding, Y.; Ma, J.; Tian, Y. Health assessment and fault classification for hydraulic pump based on LR and softmax regression. *J. Vibroeng.* **2015**, *17*, 1805–1816.
59. Jikun, B.; Chen, L.; Zhipeng, W.; Zili, W. An approach to performance assessment and fault diagnosis for hydraulic pumps. *J. Vibroeng.* **2014**, *16*, 1444–1454.
60. Tang, S.; Yuan, S.; Zhu, Y.; Li, G. An integrated deep learning method towards fault diagnosis of hydraulic axial piston pump. *Sensors* **2020**, *20*, 6576. [CrossRef]
61. Zhu, Y.; Li, G.; Wang, R.; Tang, S.; Su, H.; Cao, K. Intelligent fault diagnosis of hydraulic piston pump based on wavelet analysis and improved alexnet. *Sensors* **2021**, *21*, 549. [CrossRef]
62. Tang, S.; Zhu, Y.; Yuan, S. Intelligent fault identification of hydraulic pump using deep adaptive normalized CNN and synchrosqueezed wavelet transform. *Reliab. Eng. Syst. Saf.* **2022**, *224*, 108560. [CrossRef]
63. Yan, J.; Zhu, H.; Yang, X.; Cao, Y.; Shao, L. Research on fault diagnosis of hydraulic pump using convolutional neural network. *J. Vibroeng.* **2016**, *18*, 5141–5152. [CrossRef]
64. Zhu, Y.; Li, G.; Wang, R.; Tang, S.; Su, H.; Cao, K. Intelligent fault diagnosis of hydraulic piston pump combining improved LeNet-5 and PSO hyperparameter optimization. *Appl. Acoust.* **2021**, *183*, 108336. [CrossRef]
65. Tang, S.; Zhu, Y.; Yuan, S. Intelligent fault diagnosis of hydraulic piston pump based on deep learning and Bayesian optimization. *ISA Trans.* **2022**, *129*, 555–563. [CrossRef] [PubMed]
66. Tang, S.; Zhu, Y.; Yuan, S.; Li, G. Intelligent diagnosis towards hydraulic axial piston pump using a novel integrated cnn model. *Sensors* **2020**, *20*, 7152. [CrossRef] [PubMed]
67. Lu, C.; Ma, N.; Wang, Z. Fault detection for hydraulic pump based on chaotic parallel RBF network. *EURASIP J. Adv. Signal Process.* **2011**, *2011*, 49. [CrossRef]
68. Zhu, H.; Ting, R.; Wang, X.; Zhou, Y.; Fang, H. Fault diagnosis of hydraulic pump based on stacked autoencoders. In Proceedings of the 2015 12th IEEE International Conference on Electronic Measurement & Instruments, Qingdao, China, 16–18 July 2015; Volume 1, pp. 58–62. [CrossRef]
69. Du, Z.; Zhao, J.; Zhang, X. A recognition method of plunger wear degree of plunger pump using probability neural network. *Vibroeng. Procedia* **2017**, *14*, 45–50. [CrossRef]
70. Lv, D.; Yang, L. Hydraulic Pump Fault Diagnosis Control Research Based on PARD-BP Algorithm. *Sens. Transducers* **2014**, *183*, 221–226.
71. Casoli, P.; Pastori, M.; Scolari, F.; Rundo, M. A vibration signal-based method for fault identification and classification in hydraulic axial piston pumps. *Energies* **2019**, *12*, 953. [CrossRef]
72. Tian, Y.; Lu, C.; Wang, Z.L. Approach for Hydraulic Pump Fault Diagnosis Based on WPT-SVD and SVM. *Appl. Mech. Mater.* **2015**, *764–765*, 191–197. [CrossRef]
73. Lu, C.; Wang, S.; Makis, V. Fault severity recognition of aviation piston pump based on feature extraction of EEMD paving and optimized support vector regression model. *Aerosp. Sci. Technol.* **2017**, *67*, 105–117. [CrossRef]

74. Fei, W.; Liqing, F.; Ziyuan, Q. Fault diagnosis method for hydraulic pump based on fuzzy entropy of wavelet packet and LLTSA. *Int. J. Online Eng.* **2018**, *14*, 60–75. [CrossRef]
75. Niu, H.; Jiang, W. Application of hybrid approach based on immune algorithm and support vector machine for fault diagnosis of hydraulic pump. *Zhongguo Jixie Gongcheng China Mech. Eng.* **2008**, *19*, 1–6. [CrossRef]
76. Zhao, W.; Wang, Z.; Ma, J.; Li, L. Fault Diagnosis of a Hydraulic Pump Based on the CEEMD-STFT Time-Frequency Entropy Method and Multiclass SVM Classifier. *Shock Vib.* **2016**, *2016*, 2609856. [CrossRef]
77. Hu, X. Study on fault diagnosis of hydraulic pump based on sphere-structured support vector machines. In Proceedings of the 2012 2nd International Conference on Consumer Electronics, Communications and Networks, Yichang, China, 21–23 April 2012; pp. 2894–2896. [CrossRef]
78. Tian, Z.; Li, H.; Sun, J.; Li, Y. Degradation feature extraction of the hydraulic pump based on local characteristic-scale decomposition and multi-fractal spectrum. *Adv. Mech. Eng.* **2016**, *8*, 1687814016676679. [CrossRef]
79. Li, Z.; Jiang, W.; Zhang, S.; Sun, Y.; Zhang, S. A hydraulic pump fault diagnosis method based on the modified ensemble empirical mode decomposition and wavelet kernel extreme learning machine methods. *Sensors* **2021**, *21*, 2599. [CrossRef]
80. Ding, Y.; Ma, L.; Wang, C.; Tao, L. An EWT-PCA and extreme learning machine based diagnosis approach for hydraulic pump. *IFAC Pap.* **2020**, *53*, 43–47. [CrossRef]
81. Liu, X.; Yang, X.; Shao, F.; Liu, W.; Zhou, F.; Hu, C. Composite Multi-Scale Basic Scale Entropy Based on CEEMDAN and Its Application in Hydraulic Pump Fault Diagnosis. *IEEE Access* **2021**, *9*, 60564–60576. [CrossRef]
82. Lan, Y.; Hu, J.; Huang, J.; Niu, L.; Zeng, X.; Xiong, X.; Wu, B. Fault diagnosis on slipper abrasion of axial piston pump based on Extreme Learning Machine. *Meas. J. Int. Meas. Confed.* **2018**, *124*, 378–385. [CrossRef]
83. Wang, Y.K.; Li, H.R.; Wang, B.; Xu, B.H. Spatial Information Entropy and Its Application in the Degradation State Identification of Hydraulic Pump. *Math. Probl. Eng.* **2015**, *2015*, 532684. [CrossRef]
84. Wang, J.; Hu, H. Vibration-based fault diagnosis of pump using fuzzy technique. *Meas. J. Int. Meas. Confed.* **2006**, *39*, 176–185. [CrossRef]
85. Wang, J. A rough set approach of mechanical fault diagnosis for five-plunger pump. *Adv. Mech. Eng.* **2013**, *5*, 174987. [CrossRef]
86. Mollazade, K.; Ahmadi, H.; Omid, M.; Alimardani, R. An Intelligent Combined Method Based on Power Spectral Density, Decision Trees and Fuzzy Logic for Hydraulic Pumps Fault Diagnosis. *Int. J. Intell. Syst. Technol.* **2008**, *2*, 986–998.
87. Wu, S.; Meng, Y.; Jiang, W.; Zhang, S. Kernel principal component analysis fault diagnosis method based on sound signal processing and its application in hydraulic pump. In Proceedings of the 2011 International Conference on Fluid Power and Mechatronics, Beijing, China, 17–20 August 2011; Volume 1, pp. 98–101. [CrossRef]
88. Jiang, W.; Li, Z.; Li, J.; Zhu, Y.; Zhang, P. Study on a fault identification method of the hydraulic pump based on a combination of voiceprint characteristics and extreme learning machine. *Processes* **2019**, *7*, 894. [CrossRef]
89. Zhu, Y.; Li, G.; Tang, S.; Wang, R.; Su, H.; Wang, C. Acoustic signal-based fault detection of hydraulic piston pump using a particle swarm optimization enhancement CNN. *Appl. Acoust.* **2022**, *192*, 108718. [CrossRef]
90. Tang, S.; Zhu, Y.; Yuan, S. An adaptive deep learning model towards fault diagnosis of hydraulic piston pump using pressure signal. *Eng. Fail. Anal.* **2022**, *138*, 106300. [CrossRef]
91. Wang, Y.; Zhu, Y.; Wang, Q.; Yuan, S.; Tang, S.; Zheng, Z. Effective component extraction for hydraulic pump pressure signal based on fast empirical mode decomposition and relative entropy. *AIP Adv.* **2020**, *10*, 075103. [CrossRef]
92. Liu, J.M.; Gu, L.C.; Geng, B.L.; Shi, Y. Hydraulic pump fault diagnosis based on chaotic characteristics of speed signals under non-stationary conditions. *Proc. Inst. Mech. Eng. Part C J. Mech. Eng. Sci.* **2021**, *235*, 3468–3482. [CrossRef]
93. Liu, Z.; Liu, Y.; Shan, H.; Cai, B.; Huang, Q. A fault diagnosis methodology for gear pump based on EEMD and bayesian network. *PLoS ONE* **2015**, *10*, e0125703. [CrossRef]
94. Lu, C.; Wang, S.; Wang, X. A multi-source information fusion fault diagnosis for aviation hydraulic pump based on the new evidence similarity distance. *Aerosp. Sci. Technol.* **2017**, *71*, 392–401. [CrossRef]
95. Buiges, C.G.; König, C. A Sensor Data-Based Approach for the Definition of Condition Taxonomies for a Hydraulic Pump. *Eng. Proc.* **2020**, *2*, 82. [CrossRef]
96. Przystupa, K.; Ambrożkiewicz, B.; Litak, G. Diagnostics of Transient States in Hydraulic Pump System with Short Time Fourier Transform. *Adv. Sci. Technol. Res. J.* **2020**, *14*, 178–183. [CrossRef]
97. Ma, Z.; Wang, S.; Liao, H.; Zhang, C. Engineering-driven performance degradation analysis of hydraulic piston pump based on the inverse Gaussian process. *Qual. Reliab. Eng. Int.* **2019**, *35*, 2278–2296. [CrossRef]
98. Tingqi, L. Fault diagnosis of airplane hydraulic pump. In Proceedings of the 4th World Congress on Intelligent Control and Automation, Shanghai, China, 10–14 June 2002; pp. 3150–3152.
99. Du, J.; Wang, S. Hiberarchy clustering fault diagnosis of hydraulic pump. In Proceedings of the 2010 Prognostics and System Health Management Conference, Macao, China, 12–14 January 2010. [CrossRef]
100. Dong, Z.; Zhang, X. Modified D-S evidential theory in hydraulic system fault diagnosis. *Procedia Environ. Sci.* **2011**, *11*, 98–102. [CrossRef]
101. Du, J.; Wang, S.; Zhang, H. Layered clustering multi-fault diagnosis for hydraulic piston pump. *Mech. Syst. Signal Process.* **2013**, *36*, 487–504. [CrossRef]
102. Fu, X. Bayesian network based fault diagnosis of aero hydraulic pump. In Proceedings of the CSAA/IET International Conference on Aircraft Utility Systems, Online, 18–21 September 2020; pp. 1–5.

103. Tang, S.; Zhu, Y.; Yuan, S. An improved convolutional neural network with an adaptable learning rate towards multi-signal fault diagnosis of hydraulic piston pump. *Adv. Eng. Inform.* **2021**, *50*, 101406. [CrossRef]
104. Jiang, W.; Li, Z.; Zhang, S.; Wang, T.; Zhang, S. Hydraulic Pump Fault Diagnosis Method Based on EWT Decomposition Denoising and Deep Learning on Cloud Platform. *Shock Vib.* **2021**, *2021*, 6674351. [CrossRef]
105. Zuo, G.L.; Niu, F.L.; Cheng, Y.; Zhang, Y.X. Study on the fault diagnosis of gear pump based on RBF neural network. *Appl. Mech. Mater.* **2014**, *556–562*, 2957–2961. [CrossRef]
106. Zuo, G.L.; Lai, S.D.; Cheng, Y. Study on the fault diagnosis of gear pump based on PNN neural network. *Adv. Mater. Res.* **2014**, *1044–1045*, 873–876. [CrossRef]
107. Dong, K.; Li, Q.; Zhang, Z.; Jiang, M.; Xu, S. Submersible Screw Pump Fault Diagnosis Method Based on a Probabilistic Neural Network. *J. Appl. Sci. Eng.* **2022**, *25*, 915–923. [CrossRef]
108. Jiao, X.; Jing, B.; Huang, Y.; Li, J.; Xu, G Research on fault diagnosis of airborne fuel pump based on EMD and probabilistic neural networks. *Microelectron. Reliab.* **2017**, *75*, 296–308. [CrossRef]
109. Li, B. Fault Diagnosis Method of Ship Hydraulic System Based on KPCA-PNN. In Proceedings of the 2021 International Conference on Networking, Communications and Information Technology, Manchester, UK, 26–27 December 2021; pp. 73–77. [CrossRef]
110. Lakshmanan, K.; Gil, A.J.; Auricchio, F.; Tessicini, F. A Fault Diagnosis Methodology for An External Gear Pump with the Use of Machine Learning Classification Algorithms: Support Vector Machine and Multilayer Perceptron. Available online: https://repository.lboro.ac.uk/articles/conference_contribution/A_fault_diagnosis_methodology_for_an_external_gear_pump_with_the_use_of_Machine_Learning_classification_algorithms_Support_Vector_Machine_and_Multilayer_Perceptron/12097668 (accessed on 15 November 2022).
111. Jiang, M.; Cheng, T.; Dong, K.; Xu, S.; Geng, Y. Fault diagnosis method of submersible screw pump based on random forest. *PLoS ONE* **2020**, *15*, e0242458. [CrossRef]
112. Hu, X. The fault diagnosis of hydraulic pump based on the data fusion of D-S evidence theory. In Proceedings of the 2012 2nd International Conference on Consumer Electronics, Communications and Networks, Yichang, China, 21–23 April 2012; pp. 2982–2984. [CrossRef]
113. Miao, Y.; Jiang, Y.; Huang, J.; Zhang, X.; Han, L. Application of Fault Diagnosis of Seawater Hydraulic Pump Based on Transfer Learning. *Shock Vib.* **2020**, *2020*, 9630986. [CrossRef]
114. He, Y.; Tang, H.; Ren, Y.; Kumar, A. A deep multi-signal fusion adversarial model based transfer learning and residual network for axial piston pump fault diagnosis. *Meas. J. Int. Meas. Confed.* **2022**, *192*, 110889. [CrossRef]
115. Sun, J.; Li, H.; Xu, B. A Degradation Feature Extraction Method for Hydraulic Pumps Based Upon MUWDF and MF-DFA. *J. Fail. Anal. Prev.* **2016**, *16*, 583–593. [CrossRef]
116. Dong, H.; Xin, H.B. Application of fuzzy Petri nets in hydraulic pump fault diagnosis. *Adv. Mater. Res.* **2014**, *1008–1009*, 1176–1179. [CrossRef]
117. Ma, Z.; Wang, S.; Zhang, C. Life evaluation based on double linear damage rule for hydraulic pump piston fatigue. In Proceedings of the 2016 IEEE International Conference on Aircraft Utility Systems (AUS), Beijing, China, 10–12 October 2016; pp. 825–830. [CrossRef]
118. Roberto Jose, H.A.; Juan Carlos, C.R.; Adel Alfonso, M.M. Predicting the number of failures of a hydraulic pump with a zero excess model. *Contemp. Eng. Sci.* **2018**, *11*, 4187–4194. [CrossRef]
119. Zhou, J.; Ding, X.; Yang, Y. Plunger pump cavitation fault recognition based on analysis of low frequency energy. In Proceedings of the 2021 Global Reliability and Prognostics and Health Management, Nanjing, China, 15–17 October 2021; pp. 1–5.
120. Guo, R.; Li, Y.; Shi, Y.; Li, H.; Zhao, J.; Gao, D. Research on identification method of wear degradation of external gear pump based on flow field analysis. *Sensors* **2020**, *20*, 4058. [CrossRef]
121. Ma, J.; Chen, J.; Li, J.; Li, Q.; Ren, C. Wear analysis of swash plate/slipper pair of axis piston hydraulic pump. *Tribol. Int.* **2015**, *90*, 467–472. [CrossRef]
122. Zhao, B. The application of wavelet finite element method on multiple cracks identification of gear pump gear. *Eng. Comput.* **2015**, *31*, 281–288. [CrossRef]
123. Muralidharan, V.; Sugumaran, V. Rough set based rule learning and fuzzy classification of wavelet features for fault diagnosis of monoblock centrifugal pump. *Measurement* **2013**, *46*, 3057–3063. [CrossRef]
124. Sakthivel, N.R.; Sugumaran, V.; Babudevasenapati, S. Vibration based fault diagnosis of monoblock centrifugal pump using decision tree. *Expert Syst. Appl.* **2010**, *37*, 4040–4049. [CrossRef]
125. Muralidharan, V.; Sugumaran, V.; Indira, V. Fault diagnosis of monoblock centrifugal pump using SVM. *Eng. Sci. Technol. Int. J.* **2014**, *17*, 152–157. [CrossRef]
126. Ranawat, N.S.; Kankar, P.K.; Miglani, A. Fault diagnosis in centrifugal pump using support vector machine and artificial neural network. *J. Eng. Res. EMSME Spec. Issue* **2020**, *99*, 111. [CrossRef]
127. Wang, Y.; Lu, C.; Liu, H.; Wang, Y. Fault diagnosis for centrifugal pumps based on complementary ensemble empirical mode decomposition, sample entropy and random forest. In Proceedings of the 2016 12th world congress on intelligent control and automation (WCICA), Guilin, China, 12–15 June 2016.
128. Ahmad, Z.; Nguyen, T.K.; Ahmad, S.; Nguyen, C.D.; Kim, J.M. Multistage centrifugal pump fault diagnosis using informative ratio principal component analysis. *Sensors* **2021**, *22*, 179. [CrossRef]

129. Al Tobi, M.; Bevan, G.; Wallace, P.; Harrison, D.; Okedu, K.E. Faults diagnosis of a centrifugal pump using multilayer perceptron genetic algorithm back propagation and support vector machine with discrete wavelet transform-based feature extraction. *Comput. Intell.* **2021**, *37*, 21–46. [CrossRef]
130. Gomes, J.P.P.; Leão, B.P.; Vianna, W.O.L.; Galvão, R.K.H.; Yoneyama, T. Failure prognostics of a hydraulic pump using Kalman Filter. In Proceedings of the Annual Conference of the PHM Society, Minneapolis, MN, USA, 23–27 September 2012; pp. 464–468.
131. Amin, S.; Byington, C.; Watson, M. Fuzzy inference and fusion for health state diagnosis of hydraulic pumps and motors. In Proceedings of the NAFIPS 2005—2005 Annual Meeting of the North American Fuzzy Information Processing Society, Detroit, MI, USA, 26–28 June 2005; pp. 13–18. [CrossRef]
132. Bykov, A.D.; Voronov, V.I.; Voronova, L.I. Machine Learning Methods Applying for Hydraulic System States Classification. In Proceedings of the 2019 Systems of Signals Generating and Processing in the Field of on Board Communications, Moscow, Russia, 20–21 March 2019. [CrossRef]
133. Ma, L.; Yan, X.; Zhao, X. Research on the Failure Prediction of Aircraft Engine-Driven Hydraulic Pump. In Proceedings of the MEMAT 2022; 2nd International Conference on Mechanical Engineering, Intelligent Manufacturing and Automation Technology, Guilin, China, 7–9 January 2022; pp. 7–11.
134. Lisowski, E.; Fabiś, J. Prediction of potential failures in hydraulic gear pumps. *Arch. Foundry Eng.* **2010**, *10*, 73–76.
135. Li, H.; Tian, Z.; Yu, H.; Xu, B. Fault Prognosis of Hydraulic Pump Based on Bispectrum Entropy and Deep Belief Network. *Meas. Sci. Rev.* **2019**, *19*, 195–203. [CrossRef]
136. Xu, G.; Gao, Z.; Hu, X.; Luo, Y. Modeling and simulation of aero-hydraulic pump wear failure. In Proceedings of the 2017 Prognostics and System Health Management Conference (PHM-Harbin), Harbin, China, 9–12 July 2017; pp. 1–7. [CrossRef]
137. Yu, D.; Ma, J.; Tian, Y. Performance assessment and fault classification for hydraulic pump based on LMD and LR. *Vibroeng. Procedia* **2014**, *4*, 194–199.
138. Tian, H.L.; Li, H.R.; Hu, B.H. Fault Prediction for Hydraulic Pump Based on EEMD and SVM. *China Mech. Eng.* **2013**, *24*, 926–931.
139. Sun, J.; Li, H.; Xu, B. Prognostic for hydraulic pump based upon DCT-composite spectrum and the modified echo state network. *Springerplus* **2016**, *5*, 1293. [CrossRef]
140. Lee, M.S.; Shifat, T.A.; Hur, J.W. Kalman Filter Assisted Deep Feature Learning for RUL Prediction of Hydraulic Gear Pump. *IEEE Sens. J.* **2022**, *22*, 11088–11097. [CrossRef]
141. Wang, C.; Jiang, W.; Yue, Y.; Zhang, S. Research on Prediction Method of Gear Pump Remaining Useful Life Based on DCAE and Bi-LSTM. *Symmetry* **2022**, *14*, 1111. [CrossRef]
142. Guo, R.; Li, Y.; Zhao, L.; Zhao, J.; Gao, D. Remaining Useful Life Prediction Based on the Bayesian Regularized Radial Basis Function Neural Network for an External Gear Pump. *IEEE Access* **2020**, *8*, 107498–107509. [CrossRef]
143. Yu, H.; Li, H. Pump remaining useful life prediction based on multi-source fusion and monotonicity-constrained particle filtering. *Mech. Syst. Signal Process.* **2022**, *170*, 108851. [CrossRef]
144. Li, T.; Wang, S.; Shi, J.; Ma, Z. An adaptive-order particle filter for remaining useful life prediction of aviation piston pumps. *Chin. J. Aeronaut.* **2018**, *31*, 941–948. [CrossRef]
145. Li, Z.; Jiang, W.; Zhang, S.; Xue, D.; Zhang, S. Research on prediction method of hydraulic pump remaining useful life based on kpca and jitl. *Appl. Sci.* **2021**, *11*, 9389. [CrossRef]
146. Geng, Y.; Wang, S.; Zhang, C. Life estimation based on unbalanced data for hydraulic pump. In Proceedings of the 2016 IEEE International Conference on Aircraft Utility Systems, Beijing, China, 10–12 October 2016; pp. 796–801. [CrossRef]
147. Ma, Z.; Wang, S.; Shi, J.; Li, T.; Wang, X. Fault diagnosis of an intelligent hydraulic pump based on a nonlinear unknown input observer. *Chin. J. Aeronaut.* **2018**, *31*, 385–394. [CrossRef]
148. Wang, X.; Lin, S.; Wang, S.; He, Z.; Zhang, C. Remaining useful life prediction based on the Wiener process for an aviation axial piston pump. *Chin. J. Aeronaut.* **2016**, *29*, 779–788. [CrossRef]
149. Wang, X.; Lin, S.; Wang, S. Remaining useful life prediction model based on contaminant sensitivity for aviation hydraulic piston pump. In Proceedings of the 2016 IEEE International Conference on Aircraft Utility Systems (AUS), Beijing, China, 10–12 October 2016; pp. 266–272. [CrossRef]
150. Sun, B.; Li, Y.; Wang, Z.; Ren, Y.; Feng, Q.; Yang, D. An improved inverse Gaussian process with random effects and measurement errors for RUL prediction of hydraulic piston pump. *Meas. J. Int. Meas. Confed.* **2021**, *173*, 108604. [CrossRef]
151. Sun, S.; Sheng, Z.; Jiang, W.; Li, Z. Study on the health condition monitoring method of hydraulic pump based on convolutional neural network. In Proceedings of the 2020 12th International Conference on Measuring Technology and Mechatronics Automation (ICMTMA), Phuket, Thailand, 28–29 February 2020; pp. 149–153. [CrossRef]
152. Lin, Z.; Zheng, G.; Wang, J.; Shen, Y.; Chu, B. The method for identifying the health state of aircraft hydraulic pump based on grey prediction. In Proceedings of the 2016 Prognostics and System Health Management Conference, Chengdu, China, 19–21 October 2016; pp. 1–4. [CrossRef]
153. Hancock, K.M.; Zhang, Q. A hybrid approach to hydraulic vane pump condition monitoring and fault detection. *Trans. ASABE* **2006**, *49*, 1203–1211. [CrossRef]
154. Succi, G.P.; Chin, H. Helicopter hydraulic pump condition monitoring using neural net analysis of the vibration signature. *SAE Trans.* **1996**, *105*, 124–131. [CrossRef]

155. Zhou, F.; Liu, W.; Yang, X.; Shen, J.; Gong, P. A new method of health condition detection for hydraulic pump using enhanced whale optimization-resonance-based sparse signal decomposition and modified hierarchical amplitude-aware permutation entropy. *Trans. Inst. Meas. Control.* **2021**, *43*, 3360–3376. [CrossRef]
156. Gao, Y.; Zhang, Q. A wavelet packet and residual analysis based method for hydraulic pump health diagnosis. *Proc. Inst. Mech. Eng. Part D J. Automob. Eng.* **2006**, *220*, 735–745. [CrossRef]
157. Wang, S.P.; Ma, Q.X.; Lu, C.Q. State recognition based on wavelet packet norm entropy for aircraft hydraulic pump. In Proceedings of the 2015 International Conference on Fluid Power and Mechatronics, Harbin, China, 5–8 August 2015; pp. 1318–1323. [CrossRef]

MDPI AG
Grosspeteranlage 5
4052 Basel
Switzerland
Tel.: +41 61 683 77 34

Sensors Editorial Office
E-mail: sensors@mdpi.com
www.mdpi.com/journal/sensors